ECOLOGICAL RISK ASSESSMENT OF CONTAMINATED SEDIMENTS

Edited by
Christopher G. Ingersoll
U.S. Geological Survey

Tom Dillon
EA Engineering

Gregory R. Biddinger
Exxon Company, U.S.A.

Proceedings of the Pellston Workshop on Sediment Ecological Risk Assessment
23–28 April 1995
Pacific Grove, California

SETAC Special Publications Series

Current Series Editor
C. G. Ingersoll, Ph.D.
U.S. Geological Survey, Midwest Science Center

Past Series Editors
T. W. La Point, Ph.D.
The Institute of Wildlife and Environmental Toxicology, Clemson University

B. T. Walton, Ph.D.
U.S. Office of Science and Technology Policy, Executive Office of the President

C. H. Ward, Ph.D.
Department of Environmental Sciences and Engineering, Rice University

Publication sponsored by the Society of Environmental Toxicology
and Chemistry (SETAC) and the SETAC Foundation for Environmental Education

Published by SETAC Press
Pensacola, Florida

Cover by Michael Kenney Graphic Design and Advertising
Copyediting and typesetting by Wordsmiths Unlimited

Library of Congress Cataloging-in-Publication Data

Pellston Workshop on Sediment Ecological Risk Assessment (1995 : Pacific Grove, Calif.)
    Ecological risk assessment of contaminated sediments : proceedings of the Pellston Workshop on Sediment
Ecological Risk Assessment, 23–28 April 1995, Pacific Grove, California / edited by Christopher G. Ingersoll, Tom
Dillon, Gregory R. Biddinger.
        p. cm. -- (SETAC special publications series)
    "Publication sponsored by the Society of Environmental Toxicology and Chemistry (SETAC) and the SETAC
Foundation for Environmental Education."
    Includes bibliographical references and index.
    ISBN 1-880611-09-0 (alk. paper)
    1. Contaminated sediments--Environmental aspects--Congresses. 2. Ecological risk assessment--Congresses.
I. Ingersoll, Christopher G., 1955- . II. Dillon, Tom, 1949- . III. Biddinger, Gregory R. IV. SETAC (Society)
V. SETAC Foundation for Environmental Education. VI. Title. VI. Series.
QH545.C59P45 1995
628.1'68--dc21   97-3808
                                                                                                                      CIP

© 1997 Society of Environmental Toxicology and Chemistry (SETAC)
Published by SETAC Press
SETAC Press is an imprint of the Society of Environmental Toxicology and Chemistry.
No claim is made to original U.S. Government works.

International Standard Book Number 1-880611-09-0
Printed in the United States of America
04 03 02 01 00 99 98 97   10 9 8 7 6 5 4 3 2 1

∞ The paper used in this publication meets the minimum requirements of the American National Standard for
    Information Sciences—Permanence of Paper for Printed Library Materials, ANSI Z39.48-1984.

Reference Listing: Ingersoll CG, Dillon T, Biddinger GR, editors. 1997. Ecological risk assessment of contaminated
sediments. SETAC Pellston Workshop on Sediment Ecological Risk Assessment; 1995 Apr 23–28; Pacific Grove CA.
Pensacola FL: SETAC Pr. 390 p.

The SETAC Special Publications Series

The SETAC Special Publications Series was established by the Society of Environmental Toxicology and Chemistry (SETAC) to provide in-depth reviews and critical appraisals on scientific subjects relevant to understanding the impacts of chemicals and technology on the environment. The series consists of single- and multiple-authored or edited books on topics reviewed and recommended by the SETAC Board of Directors and approved by the Publications Advisory Council for their importance, timeliness, and contribution to multidisciplinary approaches to solving environmental problems. The diversity and breadth of subjects covered in the series reflects the wide range of disciplines encompassed by environmental toxicology, environmental chemistry, and hazard and risk assessment. Despite this diversity, the goals of these volumes are similar; they are to present the reader with authoritative coverage of the literature, as well as paradigms, methodologies and controversies, research needs, and new developments specific to the featured topics. All books in the series are peer reviewed for SETAC by acknowledged experts.

The SETAC Special Publications are useful to environmental scientists in research, research management, chemical manufacturing, regulation, and education, as well as to students considering careers in these areas. The series provides information for keeping abreast of recent developments in familiar subject areas and for rapid introduction to principles and approaches in new subject areas.

Ecological Risk Assessment of Contaminated Sediments presents the collected papers stemming from a SETAC-sponsored Pellston Workshop on Sediment Ecological Risk Assessment, held in Pacific Grove, California, 23–28 April 1995. The workshop focused on discussions of unresolved scientific issues and needed research in the area of sediment ecological risk assessment. Like all previous SETAC workshops, participation was limited to invited experts from government, academia, and industry who were selected because of their experience with the workshop topic. The workshop provided a structured environment for the exchange of ideas and debate such that consensus positions would be derived and documented for some of the issues surrounding the science of sediment ecological risk assessment.

Preface

This book presents the proceedings of the 22nd Pellston Workshop, held 23–28 April 1995 in Pacific Grove, California, where the workshop series began in 1977. Like previous workshops, participation was limited to invited experts from government, academia, and industry who were selected because of their experience with the workshop topic. The workshop provided a structured environment for the exchange of ideas and debate such that consensus positions would be derived and documented for some of the issues surrounding the science of sediment ecological risk assessment. The proceedings reflect the current state-of-the-art of these topics and focus on principles and practices designed to improve the scientific and regulatory communities' ability to assess environmental risks associated with contaminated sediments.

Acknowledgments

The Workshop on Sediment Ecological Risk Assessment and publication of the workshop proceedings were made possible through the financial support of Environment Canada; National Water Research Institute, Environment Canada; ECS Ontario Region; the Hong Kong Environmental Protection Department; the National Oceanic and Atmospheric Administration; the U.S. Army (Grant No. DAMD17-95-1-5039); and the U.S. Army Corps of Engineers. The content of this publication does not necessarily reflect the position or the policy of any of these organizations or the U.S. government, and no official endorsement should be inferred.

The workshop organizers gratefully acknowledge the participants for their enthusiastic commitment to the workshop's objectives and for their cooperative efforts in the preparation of this book. We are also deeply indebted to the SETAC / SETAC Foundation Office staff and volunteers who contributed to the success of the workshop. Special thanks go to Greg Schiefer, Rod Parrish, Cheri Mertins, Linda Longsworth, and Tanya Shaffer for their hard work in planning and implementing the workshop. The workshop participants are listed in Table 1 of the Executive Summary.

Greg Schiefer supervised the peer review of this volume for consideration as a Special Publication of the Society of Environmental Toxicology and Chemistry. Gary Pascoe served as the peer reviewer for the entire volume and at least two individuals reviewed each individual chapter. Mimi Meredith reviewed the volume for editorial style.

Foreword

This workshop was a continuation of a series of successful workshops called the "Pellston Workshop Series." Since 1977, twenty-six workshops have been held at Pellston and several other locations to evaluate current and prospective environmental issues. Each has focused on a relevant environmental topic, and the proceedings of each have been published as a peer-reviewed or informal report. These documents have been widely distributed and are valued by environmental scientists, engineers, regulators, and managers because of their technical basis and their comprehensive, state-of-the-science reviews. The workshops in the Pellston Series are as follows:

- *Estimating the Hazard of Chemical Substances to Aquatic Life*. Held in Pellston, Michigan, June 13–17, 1977. Proceedings published by the American Society for Testing and Materials, STP 657, in 1978.

- *Analyzing the Hazard Evaluation Process*. Held in Waterville Valley, New Hampshire, August 14–18, 1978. Proceedings published by The American Fisheries Society in 1979.

- *Biotransformation and Fate of Chemicals in the Aquatic Environment*. Held in Pellston, Michigan, August 14–18, 1979. Proceedings published by The American Society of Microbiology in 1980.

- *Modeling the Fate of Chemicals in the Aquatic Environment*. Held in Pellston, Michigan, August 16–21, 1981. Proceedings published by Ann Arbor Science in 1982.

- *Environmental Hazard Assessment of Effluents*. Held in Cody, Wyoming, August 23–27, 1982. Proceedings published in a SETAC Special Publication by Pergamon Press in 1986.

- *Fate and Effects of Sediment-Bound Chemicals in Aquatic Systems*. Held in Florissant, Colorado, August 11–18, 1984. Proceedings published in a SETAC Special Publication by Pergamon Press in 1987.

- *Research Priorities in Environmental Risk Assessment*. Held in Breckenridge, Colorado, August 16-21, 1987. Proceedings published by SETAC in 1987.

- *Biomarkers: Biochemical, Physiological, and Histological Markers of Anthropogenic Stress*. Held in Keystone, Colorado, July 23–28, 1989. Proceedings published in a SETAC Special Publication by Lewis Publishers in 1992.

- *Population Ecology and Wildlife Toxicology of Agricultural Pesticide Use: A Modeling Initiative for Avian Species*. Held in Kiawah Island, South Carolina, July 22–27, 1990. Proceedings published in a SETAC Special Publication by Lewis Publishers in 1993.

- *A Technical Framework for [Product] Life-Cycle Assessments*. Held in Smuggler's Notch, Vermont, August 18-23, 1990. Proceedings published by SETAC in January 1991, with second printing in September 1991 and third printing in March 1994.

- *Aquatic Microcosms for Ecological Assessment of Pesticides*. Held in Wintergreen, Virginia, October 7–11, 1991. Interim Report published February, 1992. Proceedings to be published by SETAC in 1997.
- *A Conceptual Framework for Life-Cycle Assessment Impact Assessment*. Held in Sandestin, Florida, February 1–6, 1992. Proceedings published by SETAC in 1993.
- *A Mechanistic Understanding of Bioavailability: Physical-Chemical Interactions*. Held in Pellston, Michigan, August 17–22, 1992. Proceedings published in a SETAC Special Publication by Lewis Publishers in 1994.
- *Life-Cycle Assessment Data Quality Workshop*. Held in Wintergreen, Virginia, October 4–9, 1992. Proceedings published by SETAC in 1994.
- *Avian Radio Telemetry in Support of Pesticide Field Studies*. Held in Pacific Grove, California, January 5-8, 1993. Proceedings to be published by SETAC in 1997.
- *Sustainability-Based Environmental Management*. Held in Pellston, Michigan, August 25–31, 1993. Co-sponsored by the Ecological Society of America. Proceedings to be published by SETAC in 1997.
- *Environmental Risk Assessment for Organochlorine Compounds*. Held in Alliston, Ontario, Canada, July 25–29, 1994. Proceedings to be published by SETAC in 1997.
- *Application of Life-Cycle Assessment to Public Policy*. Held in Wintergreen, Virginia, August 14–19, 1994. Proceedings to be published by SETAC in 1997.
- *Ecological Risk Assessment Modeling Systems*. Held in Pellston, Michigan, August 23–28, 1994. Proceedings to be published by SETAC in 1997.
- *Avian Toxicity Testing*. Held in Pensacola, Florida, December 4–7, 1994. Co-sponsored by OECD. Proceedings published by OECD in 1997.
- *Chemical Ranking and Scoring*. Held in Sandestin, Florida, February 12–16, 1995. Proceedings to be published by SETAC in 1997.
- *Sediments Risk Assessment*. Held in Pacific Grove, California, April 23–28, 1995. Proceedings published by SETAC in 1997.
- *Ecotoxicology and Risk Assessment for Wetlands*. Held in Gregson, Montana, July 30–August 3, 1995. Proceedings to be published by SETAC in 1997.
- *Uncertainty in Ecological Risk Assessment*. Held in Pellston, Michigan, August 23–28, 1995. Proceedings to be published by SETAC in 1997.
- *Whole Effluent Toxicity Testing: An Evaluation of Methods and Prediction of Receiving System Impacts*. Held in Pellston, Michigan, September 16–21, 1995. Proceedings published by SETAC in 1996.

SETAC Press

- *Reassessment of Metals Criteria for Aquatic Life Protection: Priorities for Research and Implementation.* Held in Pensacola, Florida, February 10–14, 1996. Proceedings published by SETAC in 1997.

Information about the availability of workshop reports can be obtained by contacting:

Society of Environmental Toxicology and Chemistry (SETAC)
1010 North 12th Avenue
Pensacola, FL 32501-3370
U.S.A.
T 904 469 1500   F 904 469 9778
E setac@setac.org
http://www.setac.org

About the Editors

Christopher G. Ingersoll, Ph.D., is presently an aquatic toxicologist with the U.S. Geological Survey at the Midwest Science Center in Columbia, Missouri. He received his bachelors' (1978) and masters' (1982) degrees from Miami University in Oxford, Ohio, and his doctorate (1986) from the University of Wyoming in Laramie. Since 1986, his research has focused on developing and applying methods for investigating the bioavailability of contaminants in sediment. He has coordinated the development of chronic toxicity tests that have been used to evaluate contaminated sediments in several locations including the Great Lakes, the upper Mississippi River, and the Clark Fork River in Montana. He is also working with several other investigators to develop and apply sediment quality guides for use in assessing contaminated sediments. Chris is involved in developing standard methods for conducting toxicity tests through the American Society for Testing and Materials (ASTM) and the U.S. Environmental Protection Agency (USEPA). He is the chair of ASTM Committee E47 on Biological Effects and Environmental Fate and is the editor of the SETAC Special Publication Series.

Tom M. Dillon is a Project Director with EA Engineering, Science and Technology, Inc. in Hunt Valley, Maryland. His major technical and research interests include sediment and aquatic toxicology, fate of chemical contaminants in the environment, ecological risk assessment, and risk management. He has been a strong advocate for applying technically sound science in the regulatory environment and in public policy formulation. He has authored over 70 technical publications in these areas. Dr. Dillon was a founding member of the U.S. Army and Department of Defense Biological Technical Assistance Groups (BTAG). He is a longtime member of SETAC and serves on its Long-Range Planning and Technical Committees. Prior to joining the private sector, Dr. Dillon was a research team leader with the U.S. Army Corps of Engineers, investigating contaminated sediment effects on aquatic biota.

Gregory R. Biddinger, Ph.D., is presently an Advisor for Exxon Company, USA, in their Environmental and Safety Department. He obtained his doctoral training in Aquatic Ecology and Physiology at Indiana State University and subsequently trained as an environmental toxicologist while a post-doctoral associate at Cornell University. Since 1983, Dr. Biddinger has worked as an environmental toxicologist for the Illinois EPA and Exxon developing and directing programs to manage the risks of chemicals in the marketplace and the safe disposal of wastes from manufacturing processes. During the time of this workshop, Dr. Biddinger was the Environmental Compliance Manager for the Exxon refinery in Benicia, California.

During his career, Dr. Biddinger has been actively involved in the advancement and standardization of testing methods in environmental toxicology and risk assessment. Since 1992, he has held the position of Chair for SETAC's Ecological Risk Assessment Advisory Group, which was the principal sponsor for this workshop.

Executive Summary

Tom Dillon, Gregory R. Biddinger, Christopher G. Ingersoll

Overview

Sediments are both a source and a sink for persistent contaminants entering the aquatic environment. Sediment quality assessment methods involving sediment toxicity and bioaccumulation testing began in the early 1980s at about the same time as did environmental risk assessment (ERA) procedures, and the two processes have evolved in parallel but separately. To date, sediment assessment procedures have not been formally integrated into a single process. The purposes of the SETAC Sediment Ecological Risk Assessment Workshop, held 23–28 April 1995 in Pacific Grove, California, were to provide

1) a framework for integrating sediment quality assessment with ERA,
2) guidance for three specific applications (product assessment, dredging, and site cleanups), and
3) a list of actions that can be taken to advance the application of ERA to contaminated sediments.

Steering Committee and Workgroup Organization

The Steering Committee (Table 1) comprised nine individuals and a representative from the SETAC Foundation. Membership on the Steering Committee was based on

1) knowledge of and experience with sediments and risk assessment procedures,
2) experience with as broad a range as possible of sediment applications,
3) knowledge of individuals working in this and related fields (the Steering Committee chose the remaining participants), and
4) willingness and proven ability to perform organizational tasks before, during, and after the workshop.

Participants (Table 1) were chosen by the Steering Committee based on requirements for stakeholder representation and international expertise in all areas of ERA and sediment quality assessment. The workshop was organized into applications and issues workgroups. The applications workgroups were Product Safety Assessment (Chapter 4), Navigational Dredging (Chapter 5), and Contaminated Site Cleanup (Chapter 7). The issues workgroups were Ecological Relevance (Chapter 12) and Methodological Uncertainty (Chapters 17 and 18). The first full day of the workshop comprised plenary presentations, beginning with an overview of ERA (Chapter 1) and proceeding through the applications and issues papers. Three subsequent days were spent primarily in workgroups. At the end of the second day, workgroups provided draft outlines of their prospective chapters, which subsequently were discussed in a plenary session. The third and fourth days were spent in workgroups, with brief morning plenaries during which the Steering Committee provided direction and clarification. The plenaries also helped

Table 1 Steering committee and workshop participants

Steering Committee	
Gregory R. Biddinger (chair)	Exxon Company USA, Benicia, California
Tom Dillon (chair)	EA Engineering, Hunt Valley, Maryland
William J. Adams	Kennecott Corporation, Salt Lake City, Utah
G. Allen Burton	Wright State University, Dayton, Ohio
Peter M. Chapman	EVS Environmental Consultants, Vancouver, British Columbia
Kristin E. Day	Environment Canada, Burlington, Ontario
Alyce T. Fritz	National Oceanic and Atmospheric Administration, Seattle, Washington
Christopher G. Ingersoll	U.S. Geological Survey, Columbia, Missouri
William van der Schalie	U.S. Environmental Protection Agency, Washington, DC
Gregory Schiefer	SETAC Foundation, Pensacola, Florida

Product Safety Assessment Workgroup	
Charles A. Pittinger (chair)	Procter and Gamble Company, Cincinnati, Ohio
William J. Adams	Kennecott Corporation, Salt Lake City, Utah
Joseph J. Dulka	Du Pont Agricultural Products, Wilmington, Delaware
Rachel Fleming	Water Research Center, United Kingdom
Rich Kimerle	Monsanto, St. Louis, Missouri
Patricia E. King	Sierra Club, Madison, Wisconsin
Thomas W. La Point	Clemson University, Pendleton, South Carolina
Anthony F. Maciorowski	U.S. Environmental Protection Agency, Washington, DC
Gregory Schiefer	SETAC, Pensacola, Florida

Navigational Dredging Workgroup	
Richard Peddicord (chair)	EA Engineering, Hunt Valley, Maryland
Tom Chase	U.S. Environmental Protection Agency, Washington, DC
Tom Dillon	EA Engineering, Hunt Valley, Maryland
Jim McGrath	Port of Oakland, Oakland, California
Wayne R. Munns	U.S. Environmental Protection Agency, Narragansett, Rhode Island
Kees van de Gucthe	Ministry of Transport, Public Works and Water, The Netherlands
William van der Schalie	U.S. Environmental Protection Agency, Washington, DC

Table 1 continued

Contaminated Site Cleanup Workgroup	
Peter M. Chapman (chair)	EVS Environmental Consultants, Vancouver, British, Columbia
Manuel Cano	Shell Development Company, Houston, Texas
Alyce T. Fritz	National Oceanic and Atmospheric Administration, Seattle, Washington
Connie Gaudet	Environment Canada, Hull, Quebec
Charles A. Menzie	Menzie-Cura and Associates, Chelmsford, Massachusetts
Mark Sprenger	U.S. Environmental Protection Agency, Edison, New Jersey
William A. Stubblefield	ENSR Consulting and Engineering, Ft. Collins, Colorado

Ecological Relevance Workgroup	
Ted DeWitt (chair)	Battelle Pacific Northwest Laboratory, Sequim, Washington
Gregory R. Biddinger	Exxon Company USA, Benicia, California
William H. Clements	Colorado State University, Ft. Collins, Colorado
Kristin E. Day	Environment Canada, Burlington, Ontario
Roger Green	University of Western Ontario, London, Ontario
Wayne G. Landis	Western Washington University, Bellingham, Washington
Peter Landrum	National Oceanic and Atmospheric Administration, Ann Arbor, Michigan
Donald J. Morrisey	National Institute for Water and Atmospheric Research, Hamilton, New Zealand
Mary Reiley	U.S. Environmental Protection Agency, Washington DC
David M. Rosenburg	Department of Fisheries and Oceans, Winnipeg, Manitoba
Glenn W. Suter II	Oak Ridge National Laboratory, Oak Ridge, Tennessee

Methodological Uncertainty Workgroup	
Keith R. Solomon (chair)	University of Guelph, Guelph, Ontario
Gerald T. Ankley	U.S. Environmental Protection Agency, Duluth, Minnesota
Renato Baudo	Institute Italiano Idrobioligia, Pallanza, Italy
G. Allen Burton	Wright State University, Dayton, Ohio
Christopher G. Ingersoll	U.S. Geological Survey, Columbia, Missouri
Wibert Lick	University of California-Santa Barbara, Santa Barbara, California
Samuel N. Luoma	U.S. Geological Survey, Menlo Park, California
Donald D. MacDonald	MacDonald Environmental Sciences, Ladysmith, British Columbia
Trefor B. Reynoldson	Environment Canada, Burlington, Ontario
Richard C. Swartz	U.S. Environmental Protection Agency, Newport, Oregon

identify and promote discussion of crosscutting issues and provided a venue for progress reports. The final day was spent in plenary, with each workgroup presenting and discussing the results of their deliberations. Before participants left the workshop, draft products were provided on diskette and hard copy and were subsequently finalized by mail, e-mail, and fax. Editing by the Steering Committee, peer review of the invited papers, and writing of the Executive Summary followed the completion of the workshop.

Workshop Objective and Charges to Participants

The Pellston Sediment Ecological Risk Assessment Workshop brought together two disciplines that have evolved more or less separately: sediment quality assessment and ecological risk assessment. The intent was to provide a forum for synthesis and synergy among workshop participants from the two disciplines. The ultimate goal was to advance the state-of-the-art and the state-of-the-practice of sediment assessment by incorporating appropriate risk-based principles and practices and by attempting to balance environmental risks with the cost of environmental protection. Although concerns regarding human health were discussed at the workshop, the primary focus was on procedures for conducting sediment ecological risk assessments (SERAs).

All participants were given the charge to address the following issue:

- Critically analyze existing approaches for evaluating sediment quality, and identify principles and practices of SERA that will improve the scientific and regulatory communities' ability to assess environmental risks associated with contaminated sediments.

Participants also were asked to address the following subsidiary charges as appropriate:

- Identify appropriate and ecologically relevant assessment and measurement endpoints.
- Identify technically sound models and appropriate extrapolations.
- Identify significant areas of uncertainty.
- Recommend research and development needs, especially those reducing significant uncertainties.
- Recommend strategic modifications to improve sediment quality assessments through the use of risk-based approaches.

The issues workgroups were specifically charged to critically examine and identify where, in the risk assessment process, methodological uncertainty and ecological relevance most influence the outcome.

The applications workgroups were specifically charged to develop a generic risk assessment process appropriate to a specific application area for evaluating environmental impacts.

Workshop summary

The workshop was held in Pacific Grove, California, 23–28 April 1995. Participants were assigned to one of five working groups, three devoted to specific applications and two to special issues. The applications workgroups were Navigational Dredging, Product Safety Assessment, and Contaminated Site Cleanup. Special issue workgroups were Ecological Relevance and Methodological Uncertainty. Workshop sessions alternated between plenary and individual workgroup meetings.

Findings of the individual workgroups are reported in the chapters that follow. This summary describes major crosscutting issues as well as important findings from individual workgroups. The first issue to emerge was a clear consensus that the current ERA paradigm (*e.g.*, U.S. Environmental Protection Agency [USEPA] 1992) is appropriate for assessing sediment quality (Chapter 1). It is a very useful guide to organizing issues and identifying site-specific data gaps. This is an important and highly desirable characteristic because SERAs vary so greatly in scope, content, and purpose.

The second major issue to emerge was the interaction between risk assessor and risk manager. There is longstanding precedent in human health risk assessments (HHRAs) and ERAs to separate risk assessment from risk management. This is done for the correct and commendable goal of ensuring scientific integrity in the risk assessment process. However, this separation has, in practice, too often meant a lack of communication between the risk assessor and risk manager. As a result, risk assessments too often do not meet the needs of the risk manager (Chapter 2). Most workshop participants agreed that an active risk assessor–risk manager dialogue is important but were unsure as to the structure of such a dialogue or where respective roles and responsibilities overlapped. In the opinion of many participants, this topic has sufficient merit to justify its own workshop.

The third major crosscutting issue was the development of and relationship between assessment and measurement endpoints for SERAs. *Assessment endpoints* are highly valued characteristics of the site or system that should be protected, restored, or remediated. These valued characteristics reflect ecological concerns as well as social and political issues. *Measurement endpoints* are specific observations reflecting change in the valued characteristic of the assessment endpoint. During SERAs, data from multiple measurement endpoints are gathered to assess change in the assessment endpoint. The assessment endpoint is related to the measurement endpoints through exposure (pathways and routes) as well as through chemical-specific mechanisms of effect.

Another major theme to emerge from the workshop was the recommendation to use a weight-of-evidence approach in SERAs. In a weight-of-evidence approach, multiple lines of evidence are generated to support decision-making. The implicit corollary is that no single line of evidence should drive decision-making in the weight-of-evidence approach. Some lines of evidence (measurement endpoints) can be "weighted," that is, valued more highly than others. For example, data from American Society for Testing and Materials (ASTM)-standardized sediment bioassays may receive greater emphasis than would results from bioassays still under development. Likewise, an exposure model validated in the field would be more highly valued than one that was not. While workshop participants expressed a clear preference for the weight-of-evidence approach, there was no consensus as to how the approach would incorporate uncertainties or relate assessment and measurement endpoints. This is an important gap because the multiple lines of evidence (measurement endpoints) generated during the SERA must all be relational to the assessment endpoint.

Using a tiered approach in SERAs was also a strong consensus theme at the workshop. This recommendation has consistently emerged at previous Pellston Workshops (see the list of publications from Pellston Workshops in the Foreword). In a tiered approach, increasingly complex and usually more costly, time-consuming evaluations are undertaken only as required to quantify and reduce uncertainties associated with risk estimates. One proceeds through the tiers until sufficient information with an acceptable level of uncertainty exists to make risk management decisions. For sediments, one of the best working examples of a tiered assessment is the U.S. Environmental Protection Agency/U.S. Army Corps of Engineers (USACE) four-tiered system for evaluating dredged material disposal in the aquatic environment (USEPA and USACE 1991, 1994). Although not currently risk-based, the USEPA/USACE tiered assessment has been used for many years to successfully evaluate diverse sediments under a range of disposal conditions (Chapter 5).

Discussions at the workshop suggested that the most pressing technical advancements are needed in the area of exposure assessment. All workgroups cited significant data gaps and modeling deficiencies for predicting spatial–temporal distributions of sediments and sediment-associated contaminants. Sediment transport models have been developed and field validated only for coarse-grain sandy material. Similar models for fine-grain sediments (where most contaminants tend to reside) lag far behind. Models predicting the fate of sediment-associated chemicals, for the most part, are not field validated and do not consider kinetics (*i.e.*, they assume chemical equilibrium). Having access to predictive exposure models is especially critical to risk managers because sediment risks are managed by reducing or eliminating exposure, not by altering toxicity. Thus, the paucity of well-developed, predictive exposure models hampers our ability to manage sediment risks.

What constitutes a suitable "reference" for SERAs was an important issue for several workgroups. Unlike HHRA, where numerical frames of reference exist (*i.e.*, 10^{-4} to 10^{-6} excess cancer risks), ERA lacks corresponding guidance. For sediment assessments, comparison to a reference provides the primary basis for data interpretation (Chapter 5). The "reference" can take the form of a reference sediment, a reference benthic community, or a toxicity reference value. Several workgroups concluded that the ecological risk of sediments should be evaluated not in the absolute sense but in the context of the receiving environment. This strongly suggests that the sediment reference should be based on characteristics of the receiving ecosystem.

The relationship between goals of the sediment risk assessment and statistical design was discussed in the context of uncertainty analysis. Risk assessments are typically structured to be environmentally conservative. That is, highly conservative screening values, calculations, and model assumptions are used throughout. This minimizes the chance of false negative errors (concluding there is no problem, when in fact one exists). Statistical designs, however, usually focus on the probability of making a false positive error (concluding there is a problem, when in fact one does not exist) by establishing a low α level. Future statistical designs should also focus on addressing the goal of the risk assessment to minimize false negative errors (Chapter 18).

The remainder of this summary provides a historical perspective *vis a vis* another Pellston Workshop on sediments ("Historical perspective: comparison to Pellston VI [1984]") and major SERA research and development needs emerging from this workshop ("Priority research and development needs").

Historical perspective: comparison to Pellston VI (1984)

About ten years ago, another SETAC workshop focused on sediment-associated chemicals. It was the sixth Pellston Workshop held in Florissant, Colorado, in 1984 (Dickson *et al*. 1987). The five preceding Pellston Workshops had dealt with the hazard assessment process and how chemicals and effluents might impact the aquatic environment. The emphasis was generally on dissolved chemicals and water column impacts. Pellston VI represented the increasing realization that 1) many chemicals released to the aquatic environment ultimately become associated with bottom sediments, and 2) the assumption of irreversibility of chemicals sorbed to sediments was wrong. The scientific and regulatory communities began to redirect their attention away from the water column toward sediment-associated chemicals and benthic community impacts. The terms *bioavailability*, *partitioning*, and *bioaccumulation* came into more common and widespread use.

Because the 1995 Pellston SERA Workshop in Pacific Grove dealt with the ecological risk of sediments, we thought it would be insightful to draw a historical perspective with Pellston VI by examining its major technical conclusions of ten years ago. Those conclusions from Pellston VI are restated below as they originally appeared.

a) Determining exposure concentrations of sediment-associated chemicals is essential for assessing the impact of contaminated sediments in the aquatic environment.

b) Equilibrium partitioning models provide an estimate of the maximum amount of sorbed material that is bioavailable.

c) A hydrophobic sorption model: Bioavailability of hydrophobic solutes is dependent on organic carbon content.

d) A means of modeling metal sorption to sediment has been proposed.

e) Sorption models must be fully assessed in the laboratory before field testing.

f) The relationship between fate, distribution, and bioavailability of sediment-associated chemicals, and the oxidized state of sediment should be thoroughly investigated.

g) Test methods and appropriate organisms should be recommended for assessing the toxicity of chemicals that sorb to sediments.

h) Bioaccumulation of sediment-associated contaminants by aquatic organisms is not presently an effective measure of adverse ecological impact because of the paucity of residue-effects correlations.

i) Toxic chemicals that sorb to sediment are a potential hazard to aquatic systems.

j) At this time, it is not feasible to develop numerical sediment quality criteria.

k) An approach to assess the hazard of sediment-bound chemicals is needed.

The first six conclusions (a to f) involve exposure assessment issues. This demonstrates that ten years ago, like today, exposure was a central issue in assessing the potential impacts of sediment-associated chemicals. A major contribution that emerged from Pellston VI was the concept of equilibrium partitioning (EqP). The workshop participants developed a consensus partitioning model for nonpolar organic chemicals using EqP. Since that time, numerous laboratory and field investigations have examined the validity of EqP for nonpolar organics under equilibrium conditions. A sorption EqP-based model for metals was also proposed at Pellston VI. However, since that time, the importance of acid volatile sulfides (AVS) on metal bioavailability has been more widely recognized and intensely studied (*e.g.*, Di Toro *et al.* 1990).

Two conclusions of Pellston VI (g, h) dealt with effects assessment. "Test methods and appropriate organisms" have been the focus of considerable research and development since that time, and as a result, sediment bioassays for both freshwater and saltwater sediments are now available. Most tests measure survival following short-term exposures to bedded sediment. Fewer address sublethal impacts (*e.g.*, growth, reproduction) following chronic exposures. Continued development of more sensitive and ecologically relevant endpoints (*e.g.*, chronic effects on growth, reproduction, and population endpoints) has the potential to produce better, more ecologically relevant measurement endpoints for sediment risk assessments. The "paucity of residue-effects correlations" noted at the Pellston VI Workshop still exists today.

The last three conclusions (i, j, k) focus on the question of how to assess sediments containing anthropogenic chemicals. At the time of Pellston VI, workshop participants felt it would not be feasible to develop numerical sediment quality criteria (SQC). They cited too many uncertainties in methodology and the incomplete validation of supporting theories. They concluded that sediment assessments should be made on a case-by-case basis because the physical and chemical characteristics of each sediment are practically unique. Development of numerical SQC, however, proceeded after Pellston VI, and today a number of chemical-specific sediment quality values exist to evaluate sediments (Chapter 18). These chemical values should be used as one of the several measures in the weight-of-evidence approach to sediment risk assessments. Finally, the Pellston VI participants cited the need for an "approach to assess the hazard of sediment-bound chemicals." Until the 1995 Pellston SERA Workshop, developing such an approach received little formal attention. Results of the 1995 workshop suggest that a generic approach now exists, that it can be adapted for diverse applications, but that much research and development remains to be done.

Priority research and development needs

Research and development (R&D) needs generated by individual workgroups at the workshop are reported in the chapters that follow. Important crosscutting R&D issues are discussed in the paragraph below; specific R&D needs are listed at the end of this section. Both are organized per the major elements of the SERA paradigm: problem formulation, exposure assessment, effects assessment, and risk characterization.

Resolving exposure assessment issues appears to be a problematic, recurring theme and therefore should receive priority in any R&D effort. Not only are exposure models technically challenging but also, when properly developed and validated, they permit risk managers to assess alternative solutions. This is because it is exposure, not toxicity, that is managed to reduce risk. Problem formulation issues (assessment/measurement endpoints, risk assessor–risk manager interface) were viewed as critical but easier to address from an R&D perspective. The problem formulation issues are important to address because they define the scope and technical direction for SERAs. Further development of chronic sublethal sediment bioassays with ecologically relevant endpoints (survival, growth, reproduction, population-level endpoints) and residue-effects relationships for persistent, bioaccumulative chemicals were cited as top priorities for effects assessment research. Finally, major R&D needs for risk characterization were techniques to 1) quantitatively integrate effects and exposure data, 2) combine multiple lines of evidence in a weight-of-evidence approach, and 3) assess and communicate uncertainties associated with estimates of ecological risk.

Problem formulation
The major R&D needs associated with problem formulation are these:
- Develop guidance for selecting ecologically relevant assessment and measurement endpoints.
- Develop a weight-of-evidence approach for linking assessment and measurement endpoints. The approach should be consistent yet flexible enough for a wide range of SERAs.
- Develop guidance for selecting what is an appropriate reference (*i.e.*, reference sediments, reference areas, reference toxicity values, reference benthic communities).
- More fully develop the concept of the risk assessor–risk manager interface. Help define roles and responsibilities for scientists, managers, stakeholders, and the public.

Exposure assessment
The major R&D needs associated with exposure assessment are these:
- Develop and field-validate exposure models for predicting space–time distributions of a) fine-grain sediments and b) sediment-associated chemicals in the food web.

- Develop techniques to ensure exposure model outputs are in units consistent with effects assessments data (*e.g.*, mg contaminant/kg sediment, mg suspended sediment/liter, proportion of contaminated bedded sediment).
- Develop predictive exposure models for metals and ionic chemicals sorbed to sediments. Field-validate existing models for neutral, hydrophobic organic chemicals.
- Develop exposure models/techniques to address complex mixtures of chemicals embedded in the sediment matrix.
- Develop quantitative tissue residue–biological effects relationships for persistent bioaccumulative chemicals.

Effects assessment

The major R&D needs associated with effects assessment are these:
- Pursue further development of chronic sediment toxicity tests that measure survival, growth, reproduction, and population-level endpoints.
- Develop technically sound interpretive guidance for effects assessment tools (*e.g.*, individual-to-population and lab-to-field extrapolations).
- Develop techniques for simulating field exposures in laboratory sediment toxicity tests (*e.g.*, time-variant suspended sediment exposures, multiple disposal events, field gradients representing dilution with other sediments).
- Develop quantitative tissue residue–biological effects relationships for persistent bioaccumulative chemicals.
- Refine approaches for selecting what constitutes an appropriate reference or range of reference values.

Risk characterization

The major R&D needs associated with risk characterization are these:
- Develop quantitative techniques to integrate effects assessment (*e.g.*, exposure-response relationships) and exposure assessment data (*e.g.*, field gradients).
- Establish qualitative and quantitative uncertainty analysis procedures, including the appropriate use of uncertainty factors, the integration of uncertainty from multiple lines of evidence (*e.g.*, toxicity, chemistry, benthic analysis), and the identification/quantification of false negative and false positive errors.
- Evaluate the impact of extrapolations (*e.g.*, lab-to-field, species-to-species, response-to-response) on estimates of ecological risk.
- Adapt toxicity identification evaluation (TIE) procedures currently used to evaluate effluents and sediment pore water for whole sediment toxicity tests.
- Develop methods to assess recovery potential following exposure to sediment-associated chemicals.
- Use carefully designed field studies to evaluate the predictive ability of sediment quality guidelines (SQG), including the potential for generating false negative and false positive errors.

References

Dickson KL, Maki AW, Brungs WA. 1987. Fate and effects of sediment-bound chemicals in aquatic systems. New York: Pergamon Pr.

Di Toro DM, Mahony JH, Hansen DJ, Scott KJ, Hicks MB, Mayr SM, Redmond M. 1990. Toxicity of cadmium in sediments: the role of acid volatile sulfides. *Environ Toxicol Chem* 9:1487–1502.

[USEPA] U.S. Environmental Protection Agency. 1992. Framework for ecological risk assessment. Washington DC: Risk Assessment Forum. EPA/630/R-92/001.

[USEPA and USACE] U.S. Environmental Protection Agency and U.S. Army Corps of Engineers. 1991. Evaluation of dredge material proposed for ocean disposal. Washington DC. EPA-503/8-91/001.

[USEPA and USACE] U.S. Environmental Protection Agency and U.S. Army Corps of Engineers. 1994. Evaluation of dredged material proposed for discharge in inland and near coastal waters (draft). Washington DC. EPA-823-B-94-002.

Contents

List of Tables

List of Figures

Acronyms and Initialisms

ACPA	American Crop Protection Association
ADDAMS	Automated Dredging and Disposal Alternatives Management System
AE	assimilation efficiency
AET	apparent effects threshold
AF	accumulation factor
AGDISP	aerial spray drift model, derivative of U.S. Forest Service Spray Drift Model
AI	artificial intelligence
ANOVA	analysis of variance
API	American Petroleum Institute
ARCS	assessment and remediation of contaminated sediment
ASTM	American Society for Testing and Materials
AVS	acid volatile sulfide
BACI	before-after/control-impact
BBA/IVA	Biologische Bundesanstalt Für Land-Und Forstwirtschaft
BCF	bioconcentration factor
BEAST	BEnthic Assessment of SedimenT
BMWP	Biological Monitoring Working Party (Great Britain)
BOD	biological oxygen demand
BSAF	biota-to-sediment accumulation factor
CBR	critical body residue
CCME	Canadian Council of the Ministers of the Environment
CEPA	Australian Commonwealth Environmental Protection Agency
CFR	U.S. Code of Federal Regulations
CHN	carbon–hydrogen–nitrogen
COD	chemical oxygen demand
CWA	U.S. Clean Water Act
CWQC	Canadian water quality criteria
D-NAPL	denser-than-water non-aqueous phase liquids
DDT	dichlorodiphenyltrichloroethane
DDTR	DDT residue
DESDM	DowElanco Spray Drift Model
DOC	dissolved organic carbon
DoENI	Department of the Environment in Northern Ireland

EC	European Commission
EC	expected concentration
ECETOC	European Centre for Ecotoxicology and Toxicology of Chemicals
ECPA	European Crop Protection Association
EDTA	ethylenediamine tetraacetic acid
EEC	estimated environmental concentration
ELA	Experimental Lakes Area (northern Ontario)
EMAP	USEPA's Environmental Monitoring and Assessment Program
EPIC	Erosion Productivity Impact Calculator
EQC	environmental quality criteria
EqP	equilibrium partitioning
ER	effects range
ERA	ecological risk assessment
ERL	effects range low
ERM	effects range median
EXAMS	Exposure Analysis Modeling System
FDA	U.S. Food and Drug Administration
FIFRA	U.S. Federal Insecticide, Fungicide and Rodenticide Act
GCSDM	Group Co-ordinating Sea Disposal Monitoring
GLEAMS	Groundwater Loading Effects of Agricultural Management Systems
HC	hazard concentration
HHRA	human health risk assessment
HPAH	high molecular weight PAH
IJC	International Joint Commission
IODS	International Ocean Disposal Symposium
L-NAPL	lighter-than-water non-aqueous phase liquids
LC	lethal concentration
LD	lethal dose
LOEC	lowest-observed-effect concentration
LPAH	low molecular weight PAH
LTER	National Science Foundation's Long-Term Ecological Research

MA DEP	Massachusetts Department of Environmental Protection
MAFF	Ministry of Agriculture, Fisheries and Food (England)
MATC	maximum acceptable toxicant concentration
MDD	minimum detectable difference
MFC	mixed flask culture
MPRSA	U.S. Marine Protection, Research, and Sanctuaries Act
NAPL	non-aqueous phase liquids
NEPA	U.S. National Environmental Policy Act
NOAA	U.S. National Oceanic and Atmospheric Administration
NOEC	no-observed-effects concentration
NPDES	U.S. National Pollutant Discharge Elimination System
NRC	National Research Council
NRDA	natural resources damage assessment
NS&T	U.S. NOAA's National Status and Trends Program
NTIS	U.S. National Technical Information Service
OC	organic carbon
OECD	Organization of Economic Cooperation and Development
ORNL	Oak Ridge National Laboratory
PAH	polycyclic aromatic hydrocarbon
PARCOM	Paris Commission
PCB	polychlorinated biphenyl
PEC	predicted environmental concentration
PEL	probable effect level
PMN	premanufacture notification
PNJ	Pierson, Newmann, and James (method)
POTW	publicly owned treatment works
PRP	potentially responsible party
PRZM 2	Plant Root Zone Model
QA/QC	quality assurance/quality control
QSAR	quantitative structure-activity relationship
RI/FS	remedial investigation and feasibility studies
RIVPACS	River InVertebrate Prediction And Classification System

SAM	standardized aquatic microcosm
SEM	simultaneously extracted metal
SERA	sediment ecological risk assessment
SLC	screening level concentration
SMB	Sverdrup, Munk, and Bretschneider (method)
SOAEFD	Scottish Office Agriculture and Fisheries Department
SQC	sediment quality criteria
SQG	sediment quality guideline
SSSA	Soil Science Society of America
SWRRB	Simulator for Water Resources in Rural Basins
TBP	Theoretical Bioaccumulation Potential model
TDDT	total DDT
TEL	threshold effect level
TIE	toxicity identification evaluation
TMDL	total maximum daily load
TOC	total organic carbon
TP	total phosphorus
TPAH	total PAH
TPCB	total PCB
TSCA	Toxic Substances Control Act
TU	toxic unit
U.S.	United States
USACE	U.S. Army Corps of Engineers (formerly "USCOE")
USC	United States Code
USEPA	U.S. Environmental Protection Agency
USGS	U.S. Geological Survey
UV	ultraviolet
VOC	volatile organic compound
VROM	Dutch Ministry of Housing, Physical Planning & Environment
WACSL	Washington state cleanup screening level
WASQS	Washington state sediment quality standards
WASP5	Water Quality Analysis Simulation Program
WQC	water quality criteria

SESSION 1
PLENARY OVERVIEW OF
SEDIMENT ECOLOGICAL RISK ASSESSMENT

Chapter 1
Overview of the ecological risk assessment framework

Glenn W. Suter II

1.1 Introduction

Ecological risk assessment (ERA) is the estimation of the likelihood of undesired effects of human actions or natural events and the accompanying risks to nonhuman organisms, populations, and ecosystems. Ecological risk assessment began with efforts to apply the concepts and rigor of human health, engineering, and financial risk assessment to ecological hazards. Those efforts led to the development of a consensus standard framework for ERA by the National Research Council (NRC), U.S. Environmental Protection Agency (USEPA), and others (Barnthouse and Suter 1986; USEPA 1992; Suter 1993). Ecological risk assessment differs from impact assessment, hazard assessment, and other environmental assessment techniques in the following points:

- It has a standard logical structure.
- It separates assessment from management.
- It has clearly defined endpoints.
- It explicitly recognizes the role of uncertainty in decision-making.

1.2 Standard logical structure

Ecological risk assessment is characterized by a standard logical structure or paradigm (Barnthouse and Suter 1986; USEPA 1992; Suter 1993). This structure is derived from a paradigm for human health risk assessment (HHRA) but has been modified to accommodate differences between ecological systems and humans (Figure 1-1). The principal one is that, unlike HHRA, which begins by identifying the hazard (_e.g._, the chemical is a carcinogen), ERA begins by dealing with the diversity of entities and responses that may be affected, of interactions and secondary effects that may occur, of scales at which effects may be considered, and of modes of exposure. These issues are dealt with by combining them into a conceptual model of the relationships among sources, agents, transport processes, and receptors.

The second component of the ecorisk paradigm involves parallel characterizations of ecological effects and exposure. The need to consider both the magnitude of exposure and the effects associated with varying levels of exposure may seem self-evident, but it precludes some commonly employed regulatory approaches. The requirement that effects be considered precludes using exceedance of background or detection limits as a criterion for action. Similarly, the requirement that exposure be considered precludes

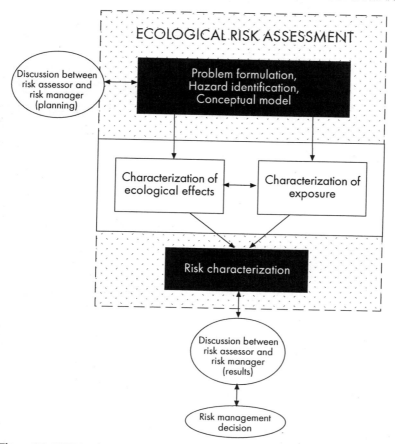

Figure 1-1   USEPA risk assessment procedure (redrawn after USEPA 1992)

banning chemicals simply because they have certain toxicological properties such as car-cinogenicity or teratogenicity.

The third component of an ERA is risk characterization. It includes combining exposure and effects information to estimate the magnitude of realized effects, to estimate the un-certainties, and to interpret the risks.

1.3   Risk assessors and risk managers

Probably the most contentious issue in risk assessment is the proper relationship of risk assessors to risk managers, who decide what actions to take in response to risks. The conflict arises from two contravening issues. First, decision-makers should not have the opportunity to manipulate the data so that they support a desired decision. This concern is an argument for keeping the risk manager out of the process until the assessors hand him or her the results. Second, if risk assessors are given free rein, they may provide an unbiased answer to the wrong question or may introduce their own biases. This concern is an argument for making the risk manager a participant in the risk assessment to ensure that it is relevant and that any value judgments are those of the responsible party. The currently favored solution to this problem is to involve the risk manager in the problem formulation and then keep him or her out of the process until the results are presented.

Many scientists involved in ecotoxicology and applied ecology do not appreciate the role of the risk manager. They view the ERA process as one in which scientists decide what is important to assess and whether significant effects are occurring, leaving the risk manager simply to decide what to do about them. However, risk managers, not scientists, are designated representatives of the public. Scientists advise and educate risk managers, but they are not the responsible parties.

1.4 Clearly defined endpoints

One of the characteristics of risk assessment is that endpoints are clearly and operationally defined, *e.g.*, the likelihood of cancer or the frequency of crashes. Ecotoxicologists have difficulty defining equivalent endpoints. However, if ERA is to be rigorous and persuasive, it is necessary to define exactly what it is that the assessment is attempting to estimate (Suter 1989, 1993; USEPA 1992). Vague phrases like "ecosystem health" will not do; neither will clearly defined but arcane properties like levels of heat shock proteins. The selection of assessment endpoints is probably the most important and difficult aspect of the scientist's interaction with the risk manager during problem formulation.

Another aspect of endpoint definition that is often neglected is the relationship of the effects measures to the assessment endpoints. In some cases, the relationship is one of correspondence. If the assessment endpoint is the percent reduction in species richness or abundance of benthic macroinvertebrates, then those properties are measurable in many contaminated sites. However, if the assessment is based on toxicity testing, the relationship of measurement endpoints to assessment endpoints is more problematical. What does it mean to benthic macroinvertebrate species richness or abundance that a given percentage of *Hyalella azteca* die in a sediment toxicity test?

1.5 Uncertainty

The concept of risk implies some degree of uncertainty concerning actual effects of an action. The frank acceptance of uncertainty is different from conventional science and prior ecotoxicological assessment paradigms. Conventional science requires that the investigators continue to perform studies until they can demonstrate with very high confidence (*i.e.*, 95%) that the hypothesized phenomenon is real. The hazard assessment paradigm similarly requires that one continue to do more and more complex toxicity tests and fate studies until one is confident of the acceptability of a chemical release (Cairns *et al.* 1979). However, risk assessment accepts the common sense proposition that decisions must be made under conditions of significant uncertainty. Therefore, there is often a nontrivial risk that an undesired outcome will occur.

Estimating uncertainties is a difficult problem, but the real trick is determining which uncertainties are relevant to the decision and presenting those uncertainties in a comprehensible and useful manner. Dealing with uncertainty by employing conservative assumptions is no longer acceptable (NRC 1994). It is incumbent upon us to estimate effects and uncertainties separately. That is, we must estimate the most likely outcome and the likelihood of more or less severe outcomes.

1.6 Variety in assessment practices

Within the ecorisk formalism, there is considerable variety in actual practice relative to other types of risk assessment. This is in part because of the relative novelty of ERA and the lack of guidance from regulatory agencies. More fundamentally, this novelty is due to the variety of ecological receptors, hazardous agents to be assessed, and mandates for assessment.

The most important distinction is the one between entirely predictive assessments, such as new product registration, and retrospective assessments that address existing contamination. Predictive assessments rely on laboratory testing and modeling. Retrospective assessments not only use conventional laboratory testing and modeling but also can use biological field surveys and toxicity tests of contaminated media.

1.7 Inference in risk assessment

Conclusions concerning ecological risks are based on weight of evidence. The need for weighing of evidence is obvious in ERAs for contaminated sites where information may be available concerning the concentrations of contaminants in various media, the toxicity of those media, and the biotic communities inhabiting those media. However, weighing of evidence is needed even when predictive assessments of new chemicals are performed. For new chemicals, evidence to be weighed includes results of various toxicity tests, physical-chemical properties of the chemical, statistical and mathematical models, and effects of analogous chemicals that have been released in the past.

As far as possible, the weighing of evidence should be performed by an *a priori* logic, not an *ad hoc* one. The best known example of such a logical structure is Chapman's (1990) "sediment quality triad" (Table 1-1). The principal limitation of that method is that it does not explicitly incorporate variance in the quality of the lines of evidence. For example, if chemicals are not measured at toxic concentrations and toxicity tests are negative but the community is altered (Table 1-1, line 5), it may be that the alteration is not due to toxic chemicals or it may be that both the analytical methods and the toxicity tests are not sufficiently sensitive.

1.8 Phasing of the assessment

Phasing is a desirable feature of ERAs, but unlike hazard assessments, ERAs do not require it. ERAs use whatever data are available and present the risk manager with the choice of making a decision with the current level of uncertainty or of reducing uncertainty by performing more studies. In general, it is advisable to gather some data and

Table 1-1   Inference based on the "sediment quality triad"

Situation	Chemicals present	Toxicity	Community alteration	Possible conclusions
1	+	+	+	Strong evidence for pollution-induced degradation
2	–	–	–	Strong evidence that there is no pollution-induced degradation
3	+	–	–	Contaminants are not bioavailable, or are present at nontoxic levels
4	–	+	–	Unmeasured chemicals or conditions exist, with potential to cause degradation
5	–	–	+	Alteration is not due to toxic chemicals
6	+	+	–	Toxic chemicals are stressing the system but are not sufficient to significantly modify the community
7	–	+	+	Unmeasured toxic chemicals are causing degradation
8	+	–	+	Chemicals are not bioavailable, or alteration is not due to toxic chemicals

Source: Chapman 1990.

Responses are shown as either positive (+) or negative (–), indicating whether or not measurable (*e.g.,* statistically significant) differences from control/reference conditions are determined.

perform a preliminary assessment in order to properly perform the problem formulation phase of the definitive ERA. That assessment is commonly termed a screening assessment because it screens out certain chemicals, media, receptors, and areas from further consideration.

Care must be taken in the design of phased studies. The logic that has been used in phasing assessments of new chemicals (start with quick but insensitive tests) does not work for other sorts of assessments. If we collect sediments from a contaminated site and perform acute lethality tests, we have no appropriate response to negative results. Unless we are unconcerned about sublethal effects or effects of extended exposures, negative results do not allow us to either declare the sediment acceptable or identify an effect.

1.9 References

Barnthouse LW, Suter II GW. 1986. User's manual for ecological risk assessment. Oak Ridge TN: Oak Ridge National Laboratory.

Cairns Jr J, Dickson KL, Maki AW. 1979. Estimating the hazard of chemical substances to aquatic life. *Hydrobiologia* 64:157–166.

Chapman PM. 1990. The sediment quality triad approach to determining pollution-induced degradation. *Sci Total Environ* 97/98:815–825.

[NRC] National Research Council. 1994. Science and judgement in risk assessment. Washington DC: National Academy Pr.

Suter II GW. 1989. Ecological endpoints. Ecological assessment of hazardous waste sites: a field and laboratory reference document. Corvallis OR: Corvallis Environmental Research Laboratory. EPA 600/3-89/013.

Suter II GW. 1993. Ecological risk assessment. Boca Raton FL: Lewis.

[USEPA] U.S. Environmental Protection Agency. 1992. Framework for ecological risk assessment. Washington DC: Risk Assessment Forum. EPA/630/R-92/001.

SESSION 1
PLENARY OVERVIEW OF
SEDIMENT ECOLOGICAL RISK ASSESSMENT

Chapter 2

Integration of risk assessment and risk management

Anthony F. Maciorowski

Scientists and risk managers must interact and communicate in the development and interpretation of risk assessments and their final applications. However, they must also learn to respect each other's spheres of expertise, influence, and control in the overall decision-making process. Integrated decision-making is a recent and rapidly evolving process. As such, the respective roles, responsibilities, information needs, and process boundary points are neither well defined nor well understood. As an example, the Framework for Ecological Risk Assessment (USEPA 1992) clearly emphasizes the need for discussions between risk managers and risk assessors at the planning stage, at key decision points, and on completion of the risk characterization. However, the framework provides little specific information concerning communication and process issues at the risk assessor–risk manager interface or concerning how to incorporate risk mitigation interactions into the original risk assessment.

Improved understanding of the different perspectives of risk assessors and risk managers is crucial to the ultimate success of integrated decision-making processes. Risk assessors are often concerned with performing risk assessments in the most scientifically credible manner and identifying additional data or research to better characterize risk. Risk managers, on the other hand, may be more interested in integrating ecological risk conclusions in a broader risk or risk–benefit framework to finalize decisions with the information in hand. This may include opting to impose risk reduction or risk mitigation control practices rather than undergoing successive iterations of the original risk assessment.

Risk reduction or _risk mitigation_ activities are defined as actions to reduce or eliminate adverse human health or environmental effects and are becoming increasingly important risk management tools. Underlying policy implications behind risk reduction and integrated decision-making are detailed in the strategic initiatives and guiding principles recently released by USEPA (1994), including ecosystem protection, pollution prevention, strong science and data, partnerships, and environmental accountability. In essence, the emerging policies are directed toward greater participation in environmental problem-solving. This process is intended to engage greater participation in decision-making, including parties affected by the decision (the regulated community, user groups, environmental interest groups, and the general public, as well as scientists.)

The point to underscore is that while ERA is advancing as a science, so too are risk management environmental policy goals and initiatives that include greater public participation. Under these circumstances, it is imperative that risk assessors and risk managers better understand each other and their respective processes. Fortunately, there are some precedents for this understanding. Risk assessment is largely a scientific process that must retain elements of scientific independence and rigor to ensure that risk characterizations provide an objective evaluation of the available data and information. There is longstanding precedent for separating risk assessment and risk management to ensure that scientific integrity is maintained in decision-making (NRC 1983, 1993; Thomas 1987; USEPA 1992). However, separation does not imply an absence of communication or a failure to understand each other's processes and needs.

Risk managers are the ultimate users of risk assessments, which are often subject to constraints within existing legal, policy, and economic realities. As such, risk assessors must be aware of risk management needs in the problem formulation stage, to ensure that assessment endpoints and resolving power that the decision-maker requires are understood. Once agreement on assessment endpoints necessary to decision-making is reached, it is within the realm of the risk assessor to select the measurement endpoints necessary to achieve the goals and objectives dictated by the assessment endpoints. Ideally, this would be agreed to during a formal *a priori* problem-formulation step in the risk assessment process. Unfortunately, routine problem formulation that engages both risk assessors and risk managers has not been a common practice. Indeed, much conflict and controversy surrounding the application of risk assessments can often be traced to the lack of a formal problem-formulation step in the decision process. As such, the importance of promoting formal problem formulation cannot be overstated.

Once a risk characterization is passed to a risk manager, additional risk assessment–risk management discussion is necessary. Presented with a scientific evaluation of risk, the risk manager may want additional information or study, or may need to act on the information in hand regardless of its scientific strengths or shortcomings. Rather than refine the risk assessment, the risk manager may opt to impose mitigation to reduce the risk, even in the face of uncertainty that the mitigation will be effective. When such situations occur, risk assessors must clearly and succinctly summarize both risk and mitigation options for the benefit of risk managers, stakeholders in the decision, and the public at large. Further, risk assessors must be willing to discuss the relative merits of risk mitigation even in the absence of data.

Although there is general agreement that risk assessors need to be involved in risk management decisions, it is also important to ensure the scientific integrity of the risk assessment process. Once a risk characterization is used to reach a decision, the risk assessor rarely has an opportunity to request more data or information on which to base opinions or recommendations. More importantly, the risk assessor has now moved into the risk management arena. In the risk management decision process, the risk assessor may be asked to analyze or judge the effect of the proposed risk mitigation on the original risk assessment. This analysis does not change the original risk assessment but begins to analyze whether management actions such as mitigation will reduce risk to acceptable levels.

Until the overall integrated decision-making process is better defined and understood by both risk assessors and risk managers, there will undoubtedly be some conflict and misunderstanding. However, recognizing and understanding that risk assessors and risk managers have different roles and responsibilities should go a long way toward improving the decision process.

This paper was originally a contribution to the Product Safety Assessment Workgroup (Chapter 4); the editors judged the issue of risk assessment–risk management integration to be a good introduction to general sediment risk assessment issues.

References

[NRC] National Research Council. 1983. Risk assessment in the federal government. Washington DC: National Academy Pr. 191 p.

[NRC] National Research Council. 1993. Issues in risk assessment. Washington DC: National Academy Pr. 356 p.

Thomas LM. 1987. Environmental decision-making today. *Environmental Protection Agency Journal* 13:2-5.

[USEPA] U.S. Environmental Protection Agency. 1992. Framework for ecological risk assessment. Washington DC: Risk Assessment Forum. EPA/630/R-92/001.

[USEPA] U.S. Environmental Protection Agency. 1994. The new generation of environmental protection. Washington DC: USEPA. EPA/200/B-94-002.

SESSION 2
PRODUCT SAFETY ASSESSMENT

_____*Chapter 3*

Assessing ecological risks to benthic species in product and technology development

Charles A. Pittinger, William J. Adams

3.1 Introduction

Those involved in technology development in both the private and public sectors share responsibility for ensuring its safety in all relevant environmental compartments. Technological development of new products, ingredients, industrial processes, and emissions requires a systematic evaluation of potential adverse effects to freshwater and marine, pelagic, and benthic communities. Ecological risk assessment as originally described in the first SETAC Pellston Workshop (Cairns *et al.* 1978) and refined by the USEPA (1992) provides a sound and quantitative framework for these evaluations. Consideration of risks pertaining to benthic communities and the sediments they occupy is an integral element in the broader process of ERA.

Recent scientific, engineering, and marketing developments have greatly improved technical capabilities and available resources for testing and research and for performing ERA. These developments have enabled companies and governmental institutions to better evaluate and manage sediment-related risks of new technologies.

Major developments have occurred in the areas of environmental fate and ecotoxicological evaluation:

- Increased understanding of natural sorption processes and major environmental parameters that affect sorption and chemical bioavailability
- Development of sediment sorption theories for nonionic organics, metals, and some polar organics
- Development of sophisticated models for estimating sediment partitioning, quantitative structure-activity relationships (QSARs), and the expansion of personal computing technology enabling easier access and broader use of these models
- Increased understanding of the routes of exposure to benthic organisms through pore water and sediment ingestion, and better extrapolations among sensitivities of pelagic and benthic organisms to chemicals
- Improved techniques for estimating toxicity to benthic organisms, including the development and standardization of sensitive, acute, and chronic sediment toxicity tests for a broad range of freshwater and marine organisms
- Increased availability of competent contract laboratories capable of conducting sediment toxicity tests

- Development of more reliable procedures for sampling, handling, and storing sediment samples

While considerable progress has been made in understanding sediment sorption processes as they relate to bioavailability, more work remains to be done if practical, routine, and protective test methods and criteria for screening technological impacts to benthic communities are to be developed. This chapter provides an overview of how sediment evaluations, as an element of the ERA process, can successfully be integrated into technology development. A fundamental approach for screening new chemical technologies is presented with a description of typical sediment-related testing methodologies. Some policy considerations pertaining to the management of sediment-related risks are also presented.

3.2 Integrating risk assessment with technology development needs

Ultimate responsibility for ensuring the ecological safety of a new commercial technology, such as a new product or product ingredient, lies with the producer as well as with the users or consumers of that technology. The distinction between ERA and risk management (USEPA 1992) is entirely appropriate for sediment-related assessments of new technologies. As with all ERAs, early communication by the risk managers to those involved in assessing potential technological risks to benthic communities is essential (see Chapter 2). The risk manager should convey the intended use of the assessment (including potential regulatory submissions), the resources available and deemed appropriate, and the critical timelines, depending upon other steps involved in technology development (Moore and Biddinger 1995). These interactions help to ensure that the ultimate risk characterizations are relevant, timely, and cost-effective. These properties are critical to institutions that must apply risk assessments to support environmental management and decision-making as well as to comply with regulatory requirements (White *et al.* 1995).

Though ERA is sometimes viewed as predominantly a function of regulatory agencies, the same assessment approach is often applied by industry as an internal technology development tool. Innovative companies rely upon ERA to identify real or potential problems associated with new products and new technologies, recognizing that environmental factors must be considered in conjunction with conventional marketing and manufacturing factors. For this reason, many companies have developed and implemented environmental policies to address ecological risks throughout the development, manufacturing, distribution, and marketing process.

Effective implementation of these environmental policies requires the following:
- A high-level commitment by the company or institution to environmental quality, involving full ownership and "buy-in" at the company's highest management level

- Access to advanced testing capabilities in environmental toxicology, microbiology, and chemistry. Screening-level tests that can reliably predict vulnerabilities in a compound's safety profile are essential to keep pace with myriad other variables and demands in bringing a new technology to market. Prediction of biodegradability, sorptivity, and bioaccumulation potential are additional important screening-level testing areas that can identify potential sediment issues.

- An effective technology development process that incorporates early environmental screening and assessment. Companies must ensure an effective management system that incorporates environmental and human risk assessment into technology development, process development, product development, and manufacturing processes. Integration at an early stage is preferred, as competitive market forces dictate ambitious and precisely controlled timelines.

- Effective communications and information feedback systems across product development, process development, manufacturing, and marketing operations. Communications between or among companies engaged in customer–supplier relationships are necessary to ensure comprehensive understanding and management of environmental safety across the overall life cycle of the technology. Corporations involved in customer–supplier relationships frequently exchange safety information and cooperate in the conduct of ERAs. Companies supplying ingredients to consumer product manufacturers often work in tandem to generate ecological fate-and-effects data and to conduct joint risk assessments.

3.3 Predicting sediment risk through early screening and chemical property evaluation

Predictive and efficient screening of chemicals for sediment-related fate and effects is indispensable in technology development because early product development efforts often generate large numbers of candidate substances. Early recognition of sediment partitioning potential is necessary to plan an appropriate risk assessment strategy before major commercial decisions and investments are made. Indiscriminate benthic testing of all candidate substances early in technology development is neither economically feasible nor necessary to ensure safety. Screening-level assessments using physical-chemical parameters, fundamental fate and toxicity trends extrapolated from pelagic testing, and the use of conservative assumptions and assessment factors can effectively be used to evaluate safety to benthic species and sediment processes early in the technology development process.

Initial predictions of sediment and porewater concentrations can often be obtained from octanol-water partition coefficients and structure-activity relationships (Cowan *et al.* 1995). The prediction is typically conservative and assumes no burial or transformation (*e.g.*, biodegradation, hydrolysis) to reduce concentrations and no complexation to reduce bioavailability. For example, screening-level sediment risk assessments may be conducted by comparing porewater concentration estimates with toxicity data for pelagic test organisms, typically fish or daphnids, by assuming similar sensitivities and routes of exposure as benthic organisms. Screening-level assessments based on comparisons with pelagic species, however, cannot simulate biotic and abiotic sediment transformation and redistribution processes that may alter exposure concentrations.

Two chemical properties useful in screening potential exposures to sediment-associated contaminants are persistence and sorptivity. In general, chemicals most likely to occur in appreciable concentrations in sediments exhibit relatively long persistence and high sorptivity. Highly sorptive but readily degradable compounds, under typical conditions, are usually degraded before reaching toxic thresholds in sediments or fauna (Versteeg and Shorter 1992). Recalcitrant but nonsorptive compounds are typically evaluated in ERAs aimed at pelagic organisms in surface waters. The potential for a substance to accumulate in a particular environmental compartment can be estimated by comparisons of the substance's biological half-life with its residence time in that compartment (Larson and Cowan 1995).

While chemicals that readily mineralize have less potential to accumulate in sediments (Fendinger *et al.* 1994), persistent chemicals require broader and more intensive evaluation in all environmental compartments, including sediments. Sorptive chemicals typically accumulate in environmental compartments with longer residence times, *e.g.*, sediments. Given the longer residence in these compartments, biodegradation proceeding at rates slower than in surface waters can still be a practical removal mechanism for commercial chemicals (Ventullo and Larson 1994).

Screening-level tests that measure the biological half-life of a compound under aerobic conditions can be cost-effective and are often conducted at an early stage in technology development. Results of biodegradability tests simulating aerobic conditions in surface waters (*e.g.*, American Society for Testing and Materials [ASTM] 1995a) frequently can be used to help predict chemical fate in sediments. Anaerobic biodegradability testing requires specialized skills and laboratory facilities and is not usually conducted in the course of initial screening.

However, screening-level tests for anaerobic biodegradability are available (European Centre for Ecotoxicology and Toxicology of Chemicals [ECETOC] 1988). Field investigations of biodegradability (Shimp and Schwab 1991; Federle and Schwab 1992) are usually conducted to confirm results of laboratory-based predictions for commercially important or large-volume product ingredients.

Potential risks of food-chain transfer and ecological (or human health) effects from chemicals that bioaccumulate in benthic organisms can also be evaluated early in screening new chemical technologies. While bioaccumulation concerns are not always specific to chemicals that partition in sediment, the chemicals that partition are often persistent and may bioaccumulate. Structure-activity relationships for predicting bioaccumulation potential have been described for a large number of chemicals (Lyman *et al.* 1982). For this reason and for cost reasons, laboratory bioaccumulation tests are typically conducted in the course of higher-level ERAs and are reserved for the most promising chemical technologies. Statutory requirements (Table 3-1) may also trigger empirical bioaccumulation testing, particularly for certain technology sectors (*e.g.*, agricultural chemical production).

3.4 Empirical approaches for chemical technology evaluation

More definitive exposure estimation techniques and direct testing of benthic organism sensitivity may be necessary when safety questions arise upon initial screening or when regulatory requirements dictate empirical testing. To conserve costs and to expedite technology development, toxicity testing of benthic organisms is usually reserved for materials that have a demonstrated commercial value and a potential to accumulate in sediments (Pittinger *et al.* 1989).

Definitive exposure testing, unless required by law, is similarly reserved for "mature" or highly commercialized technology markets in order to verify or validate model predictions. These often employ sediment-water systems that vary in size from small aquaria to experimental streams, in which material concentrations can be directly measured in pore water, sediment solids, whole sediment, and overlying water (Pittinger *et al.* 1988). With this information, better mechanistic understanding of sediment and porewater exposures and first-order biodegradation properties can be gained.

A number of sediment toxicity tests have been developed and standardized (Table 3-2) and are available through contract laboratories. Toxicity testing of new materials with benthic organisms is conceptually similar to testing pelagic organisms; however, it is often considerably more involved (and expensive) due to 1) the potential for multiple routes of exposure to the organism (*i.e.*, respiration of pore water, ingestion of sediment particles, and direct epidermal contact) and 2) complex partitioning and accelerated transformation processes that may greatly alter bioavailability and introduce a plethora of secondary compounds. For this reason, direct benthic toxicity tests ideally should be designed to account for transformation processes and ecological and life history characteristics of the organisms most likely to be exposed.

Field monitoring of concentrations of commercial chemicals in sediments can provide the most realistic estimation of direct exposure to benthic organisms but typically is cost-intensive, time-intensive, and retroactive in nature. As such, monitoring is normally conducted for higher-tiered risk assessments. Monitoring is often constrained by the availability of sensitive and specific analytical methods. The reliability of monitoring data is usually determined by selection of representative sites, by sampling and storage procedures, and by the use of accurate and precise analytical methods (ASTM 1995b).

Table 3-1   Summary of U.S. and European toxicity test requirements by regulatory requirement

Regulation or statute	Type of testing required
• Clean Water Act (CWA)	Aquatic tests for the protection of surface waters
Water Quality Standards	Aquatic tests for the development of water quality criteria (WQC)
Sediment Quality Criteria (SQC)	No aquatic testing required (use WQC and K_{ow})
USEPA National Pollutant Discharge Elimination System (NPDES) Regulations	No tests required at this time
• Toxic Substances Control Act (TSCA)	Industrial and specialty chemicals: aquatic assessments
Premanufacture Notification (PMN)	Sediment testing usually not required. Tests can be required for high K_{ow} chemicals
Section Four Test Rule	Sediment tests with *Chironomus tentans* have been used
Aquatic Test Guideline Number:	
795.12	*Hyalella azteca* flow-through acute test
797.131	Gammarid acute test (*Gammarus* sp.)
797.193	Mysid shrimp acute test
797.195	Mysid shrimp chronic test
797.197	Penaeid shrimp acute test
Adams *et al.* (1985)	Midge partial life-cycle test
• Federal Insecticide, Rodenticide and Fungicide Act (FIFRA)	Pesticide registration: aquatic assessments
Subdivision E - Wildlife and Aquatic Organisms	Simulated or actual field tests for aquatic organisms (mesocosms or outdoor microcosms)
• Food, Drug and Cosmetics Act	Environmental assessments for new food, drug, and cosmetic products
Section 4.10	*Hyalella azteca* acute toxicity test has been required
• Organization of Economic Cooperation and Development (OECD) and European Economic Community	European Community aquatic testing requirements Sediment testing protocols under development
• Paris Commission (PARCOM)	European Community: Paris Commission
Offshore chemical notification/evaluation	Sediment reworker test (*Corophium volutator, Nereis virens,* and *Abra alba*)

Table 3-2 Summary of ASTM standard procedures for conducting aquatic sediment toxicity tests

Test description	Reference [1]
Guide for conducting static acute toxicity tests starting with embryos of four species of saltwater bivalve molluscs [2]	ASTM E 724–94
Practice for conducting bioconcentration tests with fishes and saltwater bivalve molluscs [2]	ASTM E 1022–94
Guide for conducting life-cycle toxicity tests with saltwater mysids [2]	ASTM E 1191–90
Guide for conducting renewal life-cycle toxicity tests with *Daphnia magna* [2]	ASTM E 1193–94
Guide for conducting three-brood, renewal toxicity tests with *Ceriodaphnia dubia* [2]	ASTM E 1295–89 (Reapproved 1995)
Guide for conducting 10-day static sediment toxicity tests with marine and estuarine amphipods	ASTM E 1367–92
Guide for collection, storage, characterization, and manipulation of sediments for toxicological testing	ASTM E 1391–94
Guide for conducting the frog embryo teratogenesis assay-xenopus (fetax)[2]	ASTM E 1439–91
Guide for conducting static and flow-through acute toxicity tests with Mysids from the west coast of the United States[2]	ASTM E 1463–92
Designing biological tests with sediments	ASTM E 1525–94a
Conducting sediment toxicity tests with marine and estuarine polychaetous annelids	ASTM E 1611–94
Guide for determining bioaccumulation of sediment-associated contaminants by benthic invertebrates	ASTM E 1688–95
Test method for measuring toxicity of sediment-associated contaminants with freshwater invertebrates	ASTM E 1706–95

[1] Source: ASTM 1995c.

[2] The aforementioned test is not specific to sediments, but the methodology has frequently been modified to allow the test species to be tested with whole sediments or with sediment pore water.

3.5 Molecular design to alleviate potential sediment-related concerns

Early technology development efforts in designing new chemicals and commercial products are increasingly incorporating environmental considerations to minimize the potential for ecological risk. These environmental criteria must be balanced with important commercial criteria such as efficacy and stability of the ingredient in the product matrix, manufacturing cost and logistical feasibility in manufacturing or formulating, human safety concerns, *etc.* Balancing these and other considerations in technology development is extremely challenging, requiring close coordination among technical professionals involved in product and process development, manufacturing, and marketing, as well as in the life sciences (White *et al.* 1995).

The development of readily biodegradable substances for commercial applications is perhaps the surest approach to minimizing risks to benthic communities. Recent advances in understanding mechanisms of degradation have given rise to "weak-link chemistry." The incorporation of readily hydrolyzable bonds at key structural locations within a molecule may ensure complete and ready degradation through pathways involving labile intermediates. For example, inclusion of two weak ester linkages in the structure of a cationic surfactant enabled the compound to be rapidly biodegraded in standard laboratory tests and a range of environmental compartments (Giolando *et al.* 1995). Developing organic chemicals with less highly substituted (*e.g.*, tertiary carbon atoms) moieties can also facilitate enzyme-mediated decomposition reactions.

To a lesser extent, sorptivity or hydrophobicity can be manipulated for certain chemical classes and technological applications. Sorptivity, or the tendency for a substance to "stick" to surfaces, may however be integral to its function in a commercial product matrix. One example is detergent softening agents, which by nature must cling to fabric surfaces in order to deliver the desired softening benefit (Giolando *et al.* 1995). For product applications requiring the use of sorptive materials, practical biodegradability is the key to satisfying product development needs and ensuring environmental compatibility.

3.6 Science policy considerations in assessing and managing sediment contaminants

Important science policy questions pertaining to the need for and development of sediment quality criteria (SQC) and regulatory standards in the U.S. for registering new chemical technologies and for managing contaminated sediments have been (Adams *et al.* 1990) and are being broadly debated, even as SQC are being prepared for publication by the USEPA. Some of these criteria prompt the following questions:

- Are current water quality criteria (WQC) adequate to prevent sediment contamination? Are sediments allowed to be legally contaminated to toxic levels today? Is another set of national criteria and standards needed?

- Does the severity of contaminated sediments in the U.S. justify a comprehensive national sediment management strategy?

- Do the USEPA's Sediment Quality Criteria and the National Sediment Management Strategy focus on historical problems that may not be readily amenable to resolution through future management policies?

- Are costs and benefits of site remediation appropriately factored into risk management decisions for contaminated sediments, or is "cleanup at any cost" the *de facto* federal and state remediation standard?

Environmental scientists, particularly those in an applied field such as ecotoxicology, must recognize that the scientific methods and questions they pursue have very real economic and social implications. Sound application of scientific methods and results to regulatory decision-making requires a number of elements, including well-validated methods for sediment assessment; continuing and open dialogue among stakeholders involved in and impacted by the process; identification of decision points to reduce unnecessary testing and data generation, as well as criteria that clearly determine when the assessment process is complete; consideration of costs and benefits of instituting further regulatory requirements; and an unbiased peer review process for draft regulations.

3.7 Conclusions

A review of current assessment techniques to evaluate the safety of chemicals in the environment indicates that scientific understanding and tools to perform aquatic risk assessments have significantly improved over the last decade. The need for considering sediment exposures to benthic organisms as part of an overall aquatic risk assessment for new technologies and products is becoming widely understood across both the private and public sectors. Techniques exist and are being developed which can be used for screening and evaluating technologies and products during their development. This allows for early identification of potential environmental safety concerns and informed decision-making in the commercialization process. It is recognized that commercially valuable chemicals can often be designed to be more biodegradable and less sorptive to sediments, thereby improving their environmental compatibility. In looking to the future, it will be important to

- validate the methodologies used to perform sediment risk assessments,
- better evaluate uncertainties of the risk estimates,
- critically evaluate the science policies underlying the regulatory process and their cost/benefits to society, and
- ensure that risk management decisions based upon these assessments reflect sound science.

3.8 References

Adams WJ, Kimerle RA, Barnett Jr JW. 1990. Sediment assessment for the 21st century: an integrated biological and chemical approach. In: Proceedings of the U.S. EPA Workshop on Water Quality Standards for the 21st Century. Washington DC: U.S. Environmental Protection Agency (USEPA) Office of Water.

Adams WJ, Kimerle RA, Mosher RG. 1985. Aquatic safety assessment of chemicals sorbed to sediments. In: Cardwell RD, Purdy R, Bahner RC, editors. Aquatic Toxicology and Hazard Assessment: 7th Symposium. Philadelphia: American Soc of Testing and Materials (ASTM). STP 854. p 429–453.

[ASTM] American Society for Testing and Materials. 1995a. Standard test method for biodegradation by a shake-flask die-away method. 1995 Annual book of standards, Section 11, Water and environmental technology. Volume 11.05, Biological effects and environmental fate; biotechnology; pesticides. Philadelphia: ASTM.

[ASTM] American Society for Testing and Materials. 1995b. Standard guide for collection, storage, characterization, and manipulation of sediments for toxicological testing. 1995 Annual book of standards, Section 11, Water and environmental technology. Volume

11.05, Biological effects and environmental fate; biotechnology; pesticides. Philadelphia: ASTM.

[ASTM] American Society for Testing and Materials. 1995c. 1995 Annual book of standards. Philadelphia: ASTM.

Cairns Jr J, Dickson KL, Maki AW, editors. 1978. Estimating the hazard of chemical substances to aquatic life. Philadelphia: American Soc for Testing and Materials (ASTM). STP 657.

Cowan CE, Versteeg DJ, Larson RJ, Kloepper-Sams PJ. 1995. Integrated approach for environmental assessment of new and existing substances. *Regulatory Toxicol Pharmacol* 21:3-31.

[ECETOC] European Centre for Ecotoxicology and Toxicology of Chemicals. 1988. Evaluation of anaerobic biodegradation. Brussels: ECETOC, 250 Avenue Louise (Bte 63). Technical Report No. 28.

Federle TW, Schwab BS. 1992. Mineralization of surfactants in anaerobic sediments of a laundromat wastewater pond. *Water Res* 26:123–127.

Fendinger NJ, Versteeg DJ, Weeg E, Dyer S, Rapaport RA. 1994. Environmental behavior and fate of anionic surfactants, In: Baker LA, editor. Environmental chemistry of lakes and reservoirs. Washington DC: American Chemical Society (ACS). Advances in Chemistry Series No. 237.

Giolando ST, Rapaport RA, Larson RJ, Federle TW. 1995. Environmental fate and effects of DEEDMAC: a rapidly biodegradable cationic surfactant for use in fabric softeners. *Chemosphere* 30:1067–1083.

Larson RJ, Cowan CE. 1995. Quantitative application of biodegradation data to environmental risk and exposure assessments. *Environ Toxicol Chem* 14:1433–1442.

Lyman WJ, Reehl WF, Rosenblatt DH, editors. 1982. Handbook of chemical property estimation methods. New York: McGraw-Hill. 960 p.

Moore DRJ, Biddinger GR. 1995. The interaction between risk assessors and risk managers during the problem formulation phase. *Environ Toxicol Chem* 14:2013–2014.

Pittinger CA, Hand VC, Masters JA, Davidson LF. 1988. Interstitial water sampling in ecotoxicological testing: partitioning of a cationic surfactant. In: Adams WJ, Chapman GA, Landis WG, editors. Volume 10, Aquatic toxicology and hazard assessment. Philadelphia: American Soc for Testing and Materials (ASTM). STP 971. p 138–148.

Pittinger CA, Woltering DM, Masters JA. 1989. Bioavailability of sediment-sorbed and aqueous surfactants to *Chironomus riparius* (Midge). *Environ Toxicol Chem* 8:1023-1033.

Shimp RJ, Schwab BS. 1991. Use of a flow-through *in situ* environmental chamber to study microbial adaptation processes in riverine sediments and periphyton. *Environ Toxicol Chem* 10:159–167.

[USEPA] U.S. Environmental Protection Agency. 1992. Framework for ecological risk assessment. Washington DC: Risk Assessment Forum. EPA/630/R-92/001. 41 p.

Ventullo RM, Larson RJ. 1994. Biodegradation processes in groundwater and related subsurface systems, In: Zoller U, editor. Groundwater contamination and control. New York: Marcel Dekker. p 57–70.

Versteeg DJ, Shorter SJ. 1992. Effect of organic carbon on the uptake and toxicity of quaternary ammonium compounds to the fathead minnow, *Pimephales promelas*. *Environ Toxicol Chem* 11:571–580.

White PR, DeSmet B, Owens JW, Hindle P. 1995. Environmental management in an international consumer goods company. *Resourc Conserv Recycl* 14:171–184.

SESSION 2
PRODUCT SAFETY ASSESSMENT

_____Chapter 4
Workgroup summary report on sediment risk assessments of commercial products

Charles A. Pittinger, William J. Adams, Joseph J. Dulka, Rachel Fleming,
Rich Kimerle, Patricia E. King, Thomas W. La Point, Tony Maciorowski, Gregory Schiefer

4.1 Introduction

This chapter presents an approach for conducting sediment ecological risk assessments (SERAs) of new and existing products. Sediment ERAs are usually performed as part of an overall risk assessment for a given product, as opposed to an independent assessment of the sediment compartment. *Products* are defined as chemicals. These chemicals form the ingredients for consumer product formulations and for industrial and agricultural products. In this sense, single chemicals, chemical mixtures, impurities, and intermediate compounds formed during their manufacture or formulation are relevant. Excluded from consideration in this context are genetically engineered and nonengineered microorganisms or nonchemical commercialized products (*e.g.*, devices, fabrications, debris, dredged sediments).

Sediment-specific ERAs of products can be either prospective or retrospective in nature. *Prospective assessments* pertain to new chemical products requiring initial safety evaluation prior to their use or commercialization. Often these assessments involve tests with sediments artificially treated with the product or its ingredients in the laboratory. *Retrospective assessments* focus on evaluations of existing chemicals following some period of use and discharge to the environment and typically involve the testing of field-collected samples. The latter type of assessment may be prompted by the need for additional safety testing in light of new information or the development of new methods, or for monitoring to periodically confirm the accuracy of exposure estimates. Given the logistical similarities of retrospective product assessments to site evaluations (Chapter 7), the primary focus of this chapter will be prospective assessments of new chemical products.

Prospective risk assessments of new chemical products are conducted by both the private and public sectors in order to satisfy corporate safety standards or to comply with existing environmental regulations. To date, the scientific and regulatory processes to assess the potential risks of chemical products associated with sediments, both prospectively and retrospectively, have been limited. Aquatic ERAs have historically focused on the protection of water column organisms (fish and invertebrates). Until recently, considerably less attention has been paid to the ecologically relevant benthic community occupying the sediment compartment. The key assessment endpoint of sediment product risk assessments is the protection of benthic species, ecological communities, and associated food webs leading to humans.

The nature and complexity of the sediment environment offers unique challenges in assessing product risk. Chief among these are the difficulty in measuring or estimating the distribution and bioavailability of products among whole sediment and pore waters which determine exposure of benthic organisms. Effects assessments of products to benthic-dwelling organisms may be further complicated by unique habitat requirements which are difficult to simulate under laboratory conditions and by an absence of life history and physiological information for many benthic species. These difficulties have limited the amount of data collected for sediment assessments. Therefore, probabilistic estimations of risk are rare and are largely limited to products in long-term use (*e.g.*, detergent ingredients) or products receiving a particularly high degree of scrutiny for initial registration (*e.g.*, pesticides).

This section proposes a sediment-specific approach for product risk assessment in order to address these limitations and to fulfill a need for clear criteria, or "triggers," for determining when and to what degree formal sediment assessments may be required. A rational approach to problem formulation is presented, which enables public- and private-sector risk assessors and risk managers to determine needs for explicit and tiered sediment risk assessments. Such an approach requires the definition of clear assessment endpoints and reproducible measurement endpoints that meaningfully relate to the assessment endpoints.

The incorporation of evaluation points during problem formulation provides the flexibility needed to ensure the integrity of the process and the protection of the environment, while simultaneously ensuring that this evaluation is done in the most cost-effective and efficient manner. These evaluation steps also determine the sufficiency of the information assessed at each stage, thereby ensuring that the investigative effort and resources are proportional to the product's environmental risk. Additionally, a straightforward evaluation approach for exposure assessment of products to benthic communities is presented. Finally, key research questions and limitations in sediment-related product risk assessments are identified.

4.2 Conceptual framework for prospective product

Several ERA frameworks were reviewed and discussed by workshop participants. These included the Environment Canada (1996), USEPA (1992), European Commission (EC 1994), and Oak Ridge National Laboratory (Barnthouse and Suter 1986) approaches. It was concluded that these approaches were conceptually similar and provided a useful model to develop the product sediment risk assessment approach.

The basic components of an approach for product SERA are depicted in Figure 4-1. The risk assessment approach is patterned after the USEPA Ecological Risk Assessment Framework (USEPA 1992) and is intended to provide guidance for assessing the risk of chemicals sorbed to sediments. The approach assumes that risk assessment is conducted in a risk management framework that emphasizes communication and interaction between risk assessors and decision-makers (risk managers). The approach includes several discrete phases including problem formulation, exposure and effects analysis, risk char-

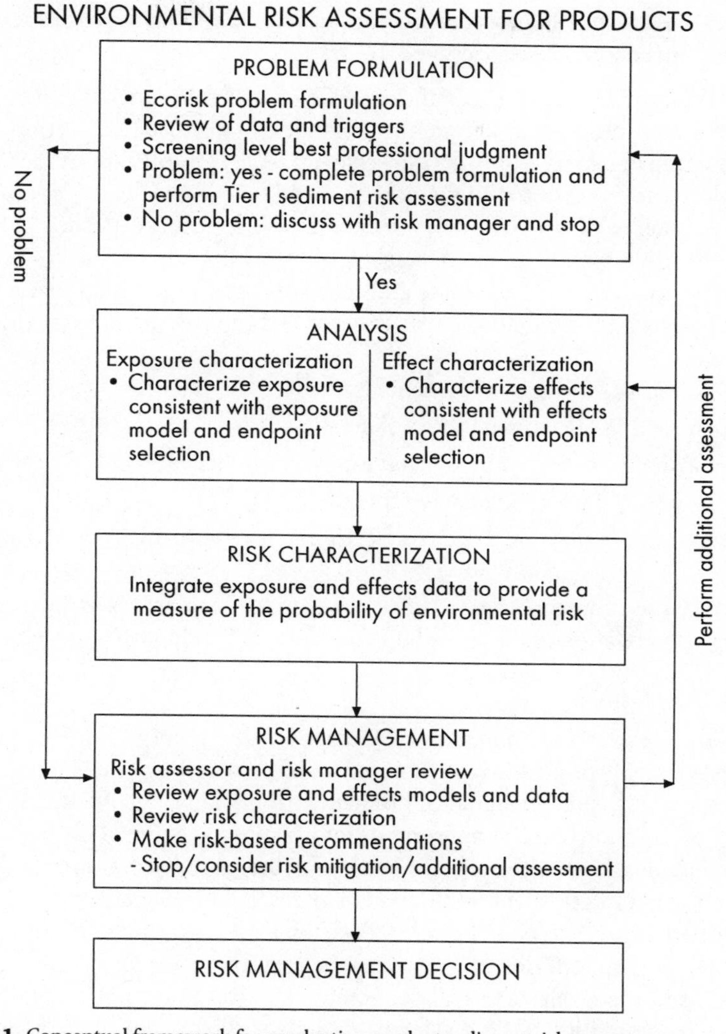

ENVIRONMENTAL RISK ASSESSMENT FOR PRODUCTS

```
                    ┌─────────────────────────────────────────┐
                    │         PROBLEM FORMULATION               │
                    │  • Ecorisk problem formulation            │
                    │  • Review of data and triggers            │
                    │  • Screening level best professional      │
                    │    judgment                               │
                    │  • Problem: yes - complete problem        │
                    │    formulation and perform Tier I         │
                    │    sediment risk assessment               │
                    │  • No problem: discuss with risk          │
                    │    manager and stop                       │
                    └─────────────────────────────────────────┘
                              │ Yes
                    ┌─────────────────────────────────────────┐
                    │               ANALYSIS                    │
                    │ Exposure characterization │ Effect        │
                    │ • Characterize exposure   │ characterization│
                    │   consistent with exposure│ • Characterize │
                    │   model and endpoint      │   effects      │
                    │   selection               │   consistent   │
                    │                           │   with effects │
                    │                           │   model and    │
                    │                           │   endpoint     │
                    │                           │   selection    │
                    └─────────────────────────────────────────┘
                    ┌─────────────────────────────────────────┐
                    │         RISK CHARACTERIZATION             │
                    │  Integrate exposure and effects data to   │
                    │  provide a measure of the probability of  │
                    │  environmental risk                       │
                    └─────────────────────────────────────────┘
                    ┌─────────────────────────────────────────┐
                    │          RISK MANAGEMENT                  │
                    │ Risk assessor and risk manager review     │
                    │ • Review exposure and effects models and  │
                    │   data                                    │
                    │ • Review risk characterization            │
                    │ • Make risk-based recommendations         │
                    │   - Stop/consider risk mitigation/        │
                    │     additional assessment                 │
                    └─────────────────────────────────────────┘
                    ┌─────────────────────────────────────────┐
                    │       RISK MANAGEMENT DECISION            │
                    └─────────────────────────────────────────┘
```

No problem (left side label)

Perform additional assessment (right side label)

Figure 4-1 Conceptual framework for conducting product sediment risk assessment

acterization, and risk assessment–risk management discussions leading to a risk management decision.

As a brief overview, problem formulation is the starting point for the product sediment risk assessment and begins with initial planning discussions between the risk assessor and the risk manager to determine whether or not there is a potential sediment risk and a reason for conducting a formalized assessment. This determination is often based on scientific judgment with a minimum of data. If no further assessment of the sediment environmental compartment is deemed necessary because there is judged to be negligible risk to sediment biota, the process proceeds to a risk assessment–risk management discussion and final decision (see Section 4.3).

If a potential risk is identified, or if insufficient evidence exists to conclude there is a low probability of risk, a formal sediment risk assessment is performed (see Section 4.4). In the analysis step, all necessary exposure and ecological effects studies are conducted and made available to the risk assessor. In the risk characterization step, the risk assessor uses the appropriate tools to perform the risk assessment by integrating exposure and effects data. Finally, the risk assessor and risk manager again enter into discussions to determine whether the risk assessment is adequate and relevant to the impending management decision. Questions addressed by the risk assessor and the risk manager at this point include these:

- Were the goals of the risk assessment attained?
- Is there a need to reformulate the problem?
- Is there a need to collect more data on exposure or effects?
- Were the uncertainties adequately expressed?
- Was the choice of how the risk characterization was performed correct and useful?
- What is the recommendation to the risk manager on the nature and magnitude of the risk in relation to the agreed-upon assessment endpoint?

4.3 Problem formulation

The evaluation of products manufactured and disposed or discharged into the environment contains elements of both science and policy and requires close collaboration between the risk assessor and risk manager. Problem formulation is intended to frame the risk management decision in terms appropriate for hypothesis testing. It involves consideration of potential risks with the product's manufacture, use, distribution, or disposal and of what amount of data might be required to assess that problem (Figure 4-1). There are certain situations and certain chemicals (see Table 4-1) for which no complex risk assessment may be required. However, all products require some consideration of their potential to cause adverse effects to sediment biota as well as other environmental compartments and communities. It is critical that, at this first stage of the ERA, the problem formulation include a discussion between the risk assessor and risk manager of available data related to the physical-chemical properties of the product as well as its likely use and disposal. Prior to the collection of any further exposure or effects characterization, certain triggers based upon the available data (Table 4-1) may be used to determine if further assessment is required. These triggers, generally based upon expert opinion, indicate the likelihood of no environmental concerns or, conversely, the need for additional testing and evaluation. They primarily relate to organic chemicals used as product ingredients or commodity chemicals. Such triggers may include, but are not limited to, a low bioconcentration factor (BCF), short-term persistence (days) in the sediments or water, low sorbtivity to sediments, or other data/considerations to indicate a large margin of safety.

Table 4-1 Screening-level triggers to determine need for additional sediment risk assessments

Screening triggers for further sediment risk assessment	
Risk to benthic organisms	More assessment needed
No significant release to the environment	Significant release to the environment
Low persistence [1] or readily degradable (for organic products or ingredients)	Persistent; not readily degradable
Low sorption to sediments [2]	Sorptive to sediments
High volatility [3]	Nonvolatile
Nontoxic to aquatic organisms [4]	Known toxicity or mechanism of action
Similarity to known chemicals of low environmental concern [5]	Similarity to known chemicals of higher environmental concern

[1] The ratio of a compound's biological half-life to its residence time in an environmental compartment has been used (Larson and Cowan 1995) to assess persistence of continuous releases of consumer products.

[2] $\log K_{ow}$ or $\log K_{oc} > 3.0$ has been used by the USEPA (TSCA) to indicate a need for further assessment.

[3] Henry's Law Constant $\leq 10^{-6}$.

[4] As inferred from other compartments or test species, *i.e.*, water column tests.

[5] Use of structure-activity relationships and best professional judgment may be appropriate for some products.

In addition to physical-chemical characteristics of the product, there are a number of key exposure considerations requiring discussion by the risk assessor and manager, including the following:

- Is there a potential for the material to enter the environment?
- During what stage of the product's life cycle and at what levels does environmental release occur?
- In what other products or processes may the product or its ingredients be discharged?
- If the product is emitted into the environment, does it have the potential to reach an aquatic environment directly or indirectly?
- What is the frequency of occurrence (*e.g.*, continuous outfall versus pulsed events)?
- What is the level of exposure per event?
- Is the sediment exposure likely to be an acute or a chronic exposure issue?
- Upon entering an aquatic environment, does the product have the potential to partition into sediment?
- Do organic constituents of the product readily biodegrade, either aerobically or anaerobically?

- Does the product leave the system through volatilization, hydrolysis, photooxidation, or some other physical degradation mechanism?
- Is the receiving sediment disposed toward biodegradation?
- Is it likely to bind in the receiving sediment due to high organic carbon (OC) content of the sediment or high (high carbon normalized) sediment-water partition coefficient (K_{oc})?
- Once in the sediment, is the product bioavailable?
- Is the product lipophilic, *i.e.*, is the log octanol-water partition coefficient (log P) > 3? Can it bioaccumulate and at what rate?
- Is the product toxic to other aquatic species (*e.g.*, fish, *Daphnia*)?

If it is determined that there are potential risks with the product meriting formal risk assessment, or if it is deemed that insufficient data exist with which to make a definitive decision, the risk assessment proceeds to the exposure and effects characterizations and follows the "typical" risk assessment paradigm.

4.4 Tiered testing and analysis

Tiered approaches in sediment risk assessment of new products and chemicals embody the concept of an iterative process with decision points between iterations, allowing evaluation of the adequacy of the assessment and a decision either to continue with data collection or to finalize the process. Initial, screening-level assessments may lead to more sophisticated iterations involving higher level (*e.g.*, chronic testing, field monitoring) datasets. Needs for higher level testing for effects or exposure assessments may be considered independently; for example, a decision could be made that effects data are sufficient while higher level exposure information is required. Results from each tier or iteration are collated, analyzed, and synthesized. Results from higher level testing or monitoring, if needed, are used to refine the risk characterization. The iterative approach is essential to efficient product testing because it maximizes the use of personnel and resources and minimizes costs to both the manufacturer and the regulatory body responsible for product registration or approval.

Tiered evaluation typically begins with screening-level assessment, employing conservative assumptions and limited datasets to broadly estimate the potential for risk. When there are sufficient data to reasonably conclude that there is a low probability for a chemical to be discharged, reach the sediments, or exert adverse ecological effects, no further or formal assessment may be required. This does not preclude risk considerations for wildlife, aquatic life, or human health. Considerations or triggers for determining the need for additional data are listed in Table 4-1. Any one of these triggers may support a decision to either terminate the assessment or proceed with additional testing.

When review of the available data indicates a formalized sediment risk assessment is justified, the following considerations in problem formulation are incorporated: goal setting; exposure and effects models that include the appropriate assessment and measurement endpoints; criteria to decide whether enough data exist to make a defensible scientific decision on risk; levels of uncertainty; and regulatory, societal, or corporate/policy issues. The approach to conducting the definitive tiered sediment risk assessment is outlined in Sections 4.5 and 4.6.

4.5 Exposure analyses

4.5.1 Estimation of product exposure via modeling

The goals of exposure analysis in performing a product evaluation are to identify and evaluate various sediment exposure scenarios to estimate or predict the levels of exposure associated with each. This results in what is classically termed the estimated or predicted environmental concentration (EEC or PEC). In detailed retrospective risk assessments, the exposure scenarios of greatest interest are often investigated by means of environmental monitoring. In prospective assessments, exposure models are applied with increasing sophistication in a tiered approach.

Most products are complex mixtures of organic ingredients. During consumer usage or disposal, products typically lose their integrity as ingredients, or components follow unique pathways into and through the environment. For example, ingredients of household cleaning products discharged to municipal wastewater treatment are rapidly sequestered into aqueous, solid, or vapor phases where they are individually transported, transformed, or assimilated into various environmental media and/or degradation processes. Sediment risk assessments of products take into account this phenomenon by recognizing the unique fate and effects of individual ingredients, their by-products and metabolites. Results of whole (intact) product fate-and-effects testing rarely if ever can provide a realistic assessment of the collective environmental risks of discrete ingredients. For this reason, assessments of products are typically conducted on an ingredient-by-ingredient basis.

Exposure data (either estimated or measured environmental concentrations) are typically compared to various effects endpoints to determine whether a sufficient margin of safety exists (see Section 4.5.2). The EEC should be considered as a range of values rather than a single number. The EEC should consider not only the presence of chemical in the compartment, but also its availability to the organisms of concern. For sediment- (and soil-) borne materials, the issue of bioavailability is particularly critical in defining the exposure to the organism tested in the laboratory as well as the organisms in the field.

Methods for estimating/measuring exposure for the purpose of developing the exposure characterization include the following:

- Exposure models based on a variety of inputs including physical-chemical properties, environmental conditions, degradation routes (photolysis, hydrolysis, biodegradation), and sorption characteristics
- Estimates of model sensitivity and uncertainty in the exposure scenarios
- Refined modeling approaches, *e.g.*, the use of more sophisticated or data-intensive models, and/or the verification of modeling assumptions and parameters

- Initial environmental monitoring, if relevant for retrospective assessments, including data on similar products
- Confirmatory environmental monitoring

Various models address emission routes from publicly owned treatment works (POTWs), from manufacturing facilities, and from agricultural fields and could be used to assist in evaluating exposure scenarios to sediment. These models, combined with input from a benthic biologist on the behavior and habits of benthic organisms, should be used in establishing the routes of exposure that lead to definitive effects testing. These models vary from general use (*e.g.*, Mackay Fugacity Models - Levels I and II [Mackay 1991]) to more specialized cases used for POTWs (Cowan *et al.* 1995), agricultural products (Baker *et al.* 1994), or site-specific models that may have been developed for a given river or estuary (Chapter 15). All of these models are based on a combination of current mechanistic assumptions about partitioning and chemical degradation. In combination with field data describing the range of environmental parameters (*e.g.*, percent organic matter, particle size distribution, partition coefficients, water depth), they can be used in a deterministic fashion to address site-specific issues as well as to establish a range of exposures that may occur in the environment. These ranges of exposure might then be used in establishing the dose for a single-concentration effects study or for studies employing a range of concentrations for a full effects characterization. While models are useful in guiding the exposure analysis leading to the risk assessment, it is also important to recognize their limitations and be sure that the assumptions are clearly stated in the final exposure assessment presented to the risk manager.

Distribution models such as the Mackay fugacity models (Mackay 1991) have been used to estimate the temporal distribution of a material via various emission points. Key input parameters include log P, biodegradation rate constants, and Henry's Law constant as an index of volatility. Distribution between water and sediment is calculated, as well as partitioning of a material into biomass. The models can be used to develop local, regional, and global exposure levels. Di Toro *et al.* (1991) have employed an equilibrium partitioning model to predict chemical concentrations in pore water on the basis of K_{oc}. The equilibrium partitioning approach has been recommended for screening purposes for nonionized, organic materials by a recent OECD document on aquatic effects assessment (OECD 1995).

Surface water models such as Exposure Analysis Modeling System (EXAMS), Water Quality Analysis Simulation Program (WASP 5) are used to estimate the distribution of a material among the various compartments of an aquatic system (*e.g.*, water, sediment, biota). These models have been applied to both household consumer product chemicals (Cowan *et al.* 1995) and agricultural products (Baker *et al.* 1994).

Runoff models such as Erosion Productivity Impact Calculator (EPIC), Plant Root Zone Model (PRZM 2), Groundwater Loading Effects of Agricultural Management Systems (GLEAMS), Simulator for Water Resources in Rural Basins (SWRRB), and others have been used in the United States to predict the potential levels of product which may leave a field during storm-driven runoff events (Baker *et al.* 1994). These models require additional inputs to describe the product's use pattern (rate of application, application equipment method, frequency of application, time of year, *etc.*), the crop (canopy interception and uptake, crop degradation, water budget), regional conditions of use (rainfall, water budget, soil characteristics, *etc.*) and degradation parameters (soil degradation rates). The edge of field loads (both water- and soil-borne runoff) developed from these models are then entered into the surface water models to determine the distribution of the product in the aquatic system and the exposures in the various aquatic compartments, including sediments.

In addition to runoff modeling, agricultural products may move off-target at application. Models used to estimate what those levels may be include the aerial spray drift model AGDISP and the DowElanco Spray Drift Model (DESDM) (Baker *et al.* 1994). This information can then be used in a surface water model to determine the product's distribution in an aquatic system.

4.5.2 Exposure characterization: dose considerations

An area of significant uncertainty in a sediment risk assessment is the definition and accurate measurement of the dose experienced by the organism. In either laboratory or field sediments, the actual route of exposure (pore water, whole sediment, sediment particle fractions, *etc.*) is not known, and the dose to an organism is often difficult to measure analytically. Di Toro *et al.* (1991) have proposed the use of equilibrium partitioning and normalization of exposure based on OC in whole sediment for nonionic chemicals. For some products and some organisms, exposure may better correlate with the concentration related to total (whole) sediment, a particular particle size fraction, or cation exchange capacity. This is especially true for polar and ionic chemicals. For cationic metals, sediment bioavailability/exposure may be related to the acid volatile sulfide (AVS) content of the sediments (Di Toro *et al.* 1990). When the molar concentration of the AVS exceeds the molar concentration of the simultaneously extracted metal (SEM), the bioavailable metal concentration has been shown to be low and below toxic effects levels. In addition to the bioavailability measures mentioned above, other factors to consider in defining dose include habitat characteristics (substrate, and percent and type of organic matter), organism function (shredders, gatherers, filterers), and exposure through the food web (dietary exposure and trophic transfer).

Chemical loads, distribution, and compartment information can be used to set an upper level of exposure for comparison to existing data on biological effects for a wide variety of species. Setting a load (dose) for a "limit test" based upon this comparison with effects data and on some percentage of likelihood for that load to occur allows for a first-level or screening-level assessment of possible biological effects. Similarly, a range of exposure estimates can be used for comparative purposes with effects data and for setting concentrations in a full dose-response effects study to be used in a more thorough sediment risk analysis. Both direct and indirect routes of exposure need to be considered when exposure characterization is performed and doses for toxicity tests are selected.

4.5.3 Tiered exposure characterization

A tiered approach for collecting data to characterize exposure is often optimal to develop cost-effective and scientific data. A suggested three-tier approach for characterizing exposure in a sequential manner in conjunction with effects data is depicted in Table 4-2. Results of the exposure estimates at each tier are used in conjunction with effects data to determine whether or not there is need for further refinement of the estimates. If an adequate margin of safety exists, no further data may be required. This decision is often made at the discretion of the risk manager with input from the risk assessor. Generally, if a reasonable worst-case exposure estimate does not exceed the measured or estimated chronic effect level for the most sensitive species tested, it is presumed that an adequate safety margin exists and no further exposure assessment is needed.

Tier 1 estimates of exposure concentrations are based upon volume of production, use and anticipated release rates, and are usually obtained using fairly simple partitioning models based on physical-chemical properties. Alternatively, they may be based upon other compounds with similar properties and uses. Tier 2 estimates of exposure are typically obtained by improving the quality and/or quantity of data used as inputs to computer models. For example, site-specific data on stream flow, sediment type, or release rate may be incorporated. Additional fate data may be collected on rates, *e.g.*, rates of hydrolysis, biodegradation, or photolysis. The models used may sometimes be calibrated with another dataset for a similar chemical for which there are environmental monitoring data. Tier 3 estimates of exposure may require field-collected data to validate models and to determine actual exposures following the use of a product. Often this is done when there is a potential for exposure to exceed the threshold effects levels, *i.e.*, when the margin of safety is small. Environmental monitoring may, in some cases, follow a period of test application or limited commercial usage of a product.

4.6 Effects analyses

Effects analyses for benthic organisms should be carried out in conjunction with exposure analyses to ensure that similar assumptions are used and questions are addressed in parallel. Both should be clearly focused upon the problem formulated and commensurate with the level of data required in each tier. The effects characterization should consider the duration of exposure (acute or chronic) and the potential for bioaccumulation. Chronic assessments typically include survival, growth, and reproduction endpoints. Additional studies or measurements may be needed to assess bioaccumulation endpoints. A major challenge in sediment effects analyses is identifying and quantifying the relevant route of exposure to a test material, which may differ widely across benthic taxa (Adams *et al.* 1985). Measurement of exposure requires detailed sampling and analytical protocols to obtain reliable data for correlating effects with exposure concentrations in a given matrix.

Like the exposure analysis, a three-tiered approach that enables decisions to be made at successive stages and precludes unnecessary testing is presented.

Table 4-2   Tiered approach to collecting data for risk characterization

Tier	Exposure assessment	Decision alternatives
1	Screening-level EECs (computer estimations)	a) Margin of safety is large; exposure characterization is complete. b) Margin of safety is small; refine exposure estimate.
2	Predictive EECs; estimates based on improved model simulations	a) Margin of safety is large; exposure characterization is complete. b) Margin of safety is small; refine exposure estimate.
3	Confirmed EECs; EEC confirmed by environmental monitoring	a) Margin of safety is large; exposure characterization is complete. b) Margin of safety is small; collect additional data or consider risk mitigation options.

4.6.1 Tier 1

The first tier in effects analysis is a screening step, which summarizes existing effects data (measurement endpoints) for aquatic species, both benthic and pelagic, and which generates predicted effects data from relevant QSARs. These data are used in conjunction with PECs from a Tier 1 exposure assessment. The porewater concentration is usually compared to the effects endpoint measured for the most sensitive species in the aquatic effects assessment (*e.g.*, an acute measurement endpoint [LCx, ECx] in a daphnid, fish, or benthic invertebrate study). Additionally, the porewater concentration could be used to estimate tissue residue levels in species of interest by using the chemical's log P.

Due to the relatively high uncertainty associated with screening-level assessments, assessment factors of 1, 10, or 100 are often applied by regulatory authorities in order to account for extrapolations from acute to chronic data, laboratory to field exposure scenarios, and extreme organism sensitivities. The adoption and magnitude of assessment factors used for screening-level assessments reflects the regulatory policies of particular countries or regional regulatory authorities (*e.g.*, the European Union). At the present time, there is no broad international consensus on the use of assessment factors.

This method of screening applies to nonionic organic chemicals. Greater uncertainty is associated with the use of predictive models for metals and ionic organic compounds than for nonionic organics. In these cases, measured sediment partition coefficient (K_d) values should be used to estimate porewater concentrations. In all cases, the magnitude of the application factor should vary according to the uncertainty of the exposure estimate.

4.6.2 Tier 2

For Tier 2 effects analyses, direct measurement of sediment toxicity (and bioaccumulation) in benthic organisms is often used for assessment purposes (see Table 4-1). Greater emphasis is placed on chronic sediment studies because these tests better simulate the chemical equilibria established between sediment, water, and tissue. However, chronic toxicity and bioaccumulation test methods for sediments are still in a developmental phase.

Where possible, standard test methods with associated validity or performance criteria are employed. Criteria for species selection have been outlined in a SETAC-Europe guidance document on sediment toxicity tests (Hill *et al.* 1993) and by the American Petroleum Institute (API 1994).

Recommended test species include 10-d lethality and growth tests with chironomids and amphipods for freshwater applications. Amphipods, polychaetes, and bivalves are often tested for marine/estuarine applications. Standard methods have been published or are in preparation for a number of test protocols (Table 3-2). Ideally, test species with different feeding strategies should be used, but this may be dictated by the availability of standard methods and the scope of the assessment. Water column toxicity tests of sediment extracts provide an additional approach for some sediment assessment applications. Due to additional uncertainties associated with extracts preparation and interpretation of results, however, whole sediment tests are generally preferred.

Both formulated and natural sediments have been used for effects analyses. Issues of standardization within and between product assessments yet need to be addressed. If natural sediments are used, key physical-chemical properties such as OC, particle size distribution, ammonia, and pH should be measured. The sediments should be collected, handled, stored, and dosed according to existing standard methods (see Table 3-2). The sediment should be dosed with a concentration range from which a dose-response may be observed. The concentration range may be predicted from existing aquatic effects data or preliminary range-finding tests.

Replication, reference substances, and control treatments should be used according to the standard test method. Measurement endpoints and treatment of data should be consistent with methods used in aquatic risk assessment. Ideally, a sufficient amount of effects data can be obtained or generated to allow for a probabilistic assessment approach. The USEPA water quality criteria employ a probabilistic approach that protects 95% of the species. SETAC (Baker *et al.* 1994) has recommended the use of a somewhat similar approach based upon 90th percentile toxicity data. When there is a very limited dataset available, the most sensitive measure should be used for risk characterization.

For nonpolar organics, effect concentrations should be normalized to sediment OC content. This would allow comparison with PECs also calculated on the basis of OC. For metals and polar organics, models for calculating PECs should be parameterized using the same physical-chemical properties and parameters of the test sediment for both the effects and exposure characterization. In many cases, test sediments representing worst case and typical case (*e.g.*, low and average OC content, respectively; low and average cation exchange capacity; low and average AVS concentration) canvass the full range of potential effects. These tests should be guided by the PEC for sediments and by the need for dose-response data.

4.6.3 Tier 3

Tier 3 effects analyses include higher level or simulation tests that may be used to assess application-specific risks. Such tests often include microcosm, mesocosm, or field studies. The need for such tests would, in part, be dependent upon the potential for the product to remain in the environment for extended periods of time. The tests are designed to address particular toxicity, bioaccumulation, or exposure questions identified in prior testing. They may include alternative approaches for application of products (*e.g.*, overspray for pesticides, addition of slurry or suspended solids for drilling fluids), exposure of specific organisms of concern, alternative exposure regimes such as *in situ* or multispecies testing, or assessment of field effects. Any unique fate pathways, toxicological concerns or mechanisms, or action identified in problem identification should be reflected in the test design. For this level of testing, sediment test methods are largely in a developmental phase.

4.7 Risk characterization

Risk characterization is the final phase of sediment assessments for products. Recommendations and considerations for risk characterization in ERA have been reviewed by Hoffman *et al.* (1995) and are fully relevant to product assessments of potential sediment risks. These are briefly summarized below, with particular emphasis on how they may be tailored specifically to products and sediments.

In the risk characterization, the exposure and effects assessments are integrated to yield an expression of the likelihood of adverse effects to benthic species, populations, or communities, depending on the nature and scope of the risk assessment. If an additional or alternative purpose of the assessment is to evaluate the potential for bioaccumulation in benthic species or biomagnification through associated food webs, the risk characterization applies the exposure estimate to derive an estimated steady-state tissue concentration in the species of interest. In either case, the objective of the risk characterization is to describe the risk in terms of the assessment endpoint identified in the problem formulation phase. In addition, the risk assessor provides an interpretation of the ecological significance of the identified risks to benthic populations. This may include consideration of the integrity of the benthic population or community as well as possible implications to other (non-benthic) components of the ecosystem. However, extrapolations of risks beyond the population level tend to be speculative when only single-species effects data are available.

Both qualitative and quantitative descriptors are appropriate and necessary in characterizing risks. To the extent feasible, the full range of risks should be communicated, including those likely to occur in representative scenarios of sediment-mediated exposure as well as those pertaining to highly exposed or highly susceptible benthic populations. In the context of tiered assessments for products, the scope of the risk assessment must be clearly identified. Screening-level assessments typically result in deterministic expressions of risk where only one or a few point estimates of risk are provided. Higher level tiers of assessment may allow estimation and characterization of probabilistic distributions of exposure and effects, as summarized by SETAC (Baker *et al.* 1994).

Possible outcomes of the risk assessor–risk manager interface include the following:

- When risk probability is low and there is an adequate margin of safety, the process stops.
- When risk probability is intermediate or high or when margins of safety are unclear or unknown, more assessment is generally needed and risk mitigation action may ultimately be needed.
- When additional assessment is needed, it may take the form of additional refinements of the risk assessment or verification of the effectiveness of proposed mitigation actions.

In cases where multiple and independent estimates of sediment exposure or effects exist (*e.g.*, results from several models or toxicity data on multiple species), and in order to account for seemingly conflicting data due to variability or uncertainty, a weight-of-evidence approach may be necessary. The relevance of the measurement endpoints to the assessment endpoint should be described in the risk characterization. This may include a ranking or prioritization of measurement endpoints. For example, toxicity data gained from direct sediment testing of a benthic species may be considered more relevant than toxicity estimates inferred from extrapolations from pelagic test species.

In retrospective assessments of existing products, evidence for causal relationships between product and effects to benthic communities should be clearly delineated where possible. When the causal relationship is not clear, this must be communicated in the sediment risk characterization.

Spatial and temporal distributions of the risk (*e.g.*, constant versus intermittent exposure, homogeneous versus contagious distribution of organisms, seasonal occurrence of particularly sensitive benthic life forms) should be identified. The expression of the risks to benthic populations is largely dictated by the type and form of sediment exposure and effects data available. Exposure may be expressed in various forms, such as total (whole) sediment concentration, particle size fraction concentration, porewater concentration, or OC-normalized whole sediment concentration. It is difficult or often invalid to attempt extrapolations between or among different expressions of exposure due to the potential moderating effects of key environmental parameters upon exposure or bioavailability. These limitations in extrapolation should be clearly communicated in the risk characterization.

Risk characterizations should clearly relate to and communicate the problem formulated and the scope of the risk assessment. Relevant questions include these:

- Was the initial sediment problem addressed?
- What aspects could not be addressed, and do these limit or preclude the estimation of risks or limit the scope of the assessment?
- On the basis of the data and information obtained in the effects and exposure assessment, was the initial problem accurately formulated?
- Can the risks to benthic populations be expressed in a quantitative manner?
- Can risk management or mitigation options be identified to reduce potential risk associated with product usage, if necessary?

Default assumptions used in the effects and exposure assessment should be clearly communicated. Extrapolations of the data (*e.g.*, endpoint to endpoint, chemical to chemical, species to species, population to community) should be identified. Moderating effects of physical-chemical characteristics of the benthic environment (*e.g.*, OC content, particle size), limited bioavailability or distribution of the product, or biological factors (life history stage sensitivity, behavioral mechanisms or avoidance) should be described.

The limitations of the sediment assessment, including sampling biases and model extrapolations of the exposure assessment, should be clearly identified. Uncertainties of the effects and exposure assessments due to data and knowledge gaps, and any assumptions used to bridge these gaps, should be described.

For sediment risk assessments of products, complete characterization of the chemical technology should be described. Key aspects include the name and structure of the chemical, chemical mixtures, and impurities; formulation, use, or disposal processes that may affect environmental fate, distribution or bioavailability; the stage of the product's life cycle being addressed (*e.g.*, sourcing, discharges from manufacturing facilities, both intended and unintended; consumer use and disposal); usage volume, both by individual corporations and industry-wide; expected degradation and transformation processes and metabolites; availability of QSAR data; and regulatory considerations or requirements that influence the design of the study design.

Safety factors (*i.e.*, application factors) have been conventionally used in ERA as an approach for establishing toxicity threshold concentrations above which concern or risk may exist. This is usually done when there is concern that the available effects data are not truly representative of the ecosystem or species at risk or when there is uncertainty associated with the available effects data. Safety factors have also been used with limited effects datasets to estimate threshold concentrations for the purpose of determining the need for higher tiered testing or to assist in the selection of a particular risk management option. Safety factors can be used in considering risks of products to benthic populations or communities. Safety factors may be applied in the risk characterization for products or in subsequent discussions between the risk assessor and risk manager, in order to ensure organism protection and to provide the risk manager with a basis for decision-making. Because the rationale for and magnitude of safety factors vary from one risk assessment to another (and often from country to country), it is important to document the rationale and method used in applying safety factors. It should also be noted that safety factors are not always needed, depending upon the extent of the effects data collected.

The risk characterization forms the basis for discussions with the risk manager, whether within the private or public sector. For this reason, it is important that risks be identified in terms relevant to the assessment endpoints and that the assumptions, uncertainties, and confidence in the assessment be clearly presented.

4.8   Application of product risk assessment for sediments: integration of risk assessment and risk management

The prospective risk assessment process for chemical products is perhaps best viewed as the application of science as part of an informed risk management program. This view is supported by emerging risk-based approaches to environmental regulations, which promote increased integration of societal values, science, and risk mitigation practices for environmental decision-making. This integrated decision-making process goes beyond the traditional scientific boundaries of risk assessment to include risk mitigation and risk management. Within this context, the following definitions apply:

- Risk assessment is a science-based process that consists of effects and exposure analyses which are ultimately integrated into a risk characterization.

- Risk mitigation is an activity that involves remediation or mitigation measures to reduce or eliminate adverse environmental impact.

- Risk management is a policy-based approach that identifies risk assessment questions and assessment endpoints to protect human health and the environment. It utilizes the risk characterization decisions and incorporates social, economic, political, and legal factors that impinge on or influence the final decision and selects regulatory actions.

The integrated decision-making process represents a fundamental shift in thinking that focuses attention on making timely, environmentally protective decisions through better integration of risk assessment, risk management, and risk mitigation options. For products that enter the marketplace with the potential to sorb to sediments, risk characterization can be integrated with risk policy decisions in a number of ways. This can include a decision of no risk and no need to consider mitigation; conversely, a decision of moderate or significant risk could be reached, which would indicate the need to consider the magnitude of the risk and ways to mitigate the potential for exposure. There are a number of options available which can limit the potential for a product to reach the environment or which can influence the potential for sediment concentrations reaching levels of concern. For additional details on product mitigation, the reader is referred to SETAC (Baker *et al.* 1994).

While scientists are variously engaged in basic and applied issues concerning the development and refinement of ERA concepts and practices, they rarely serve as the final decision-makers in risk management actions. Rather, scientists provide technical opinions and support to risk managers who must weigh the scientific risk against societal values dictated by economic benefits and constraints, existing precedents of law and policy, and often conflicting opinions of different special interest groups. In some respects, risk mitigation options act as a bridge between risk assessment and risk management, requiring practical solutions and corrective actions to environmental problems. Recognizing and understanding that risk assessors and risk managers have different roles and responsibilities should go a long way toward improving the decision-making process.

4.9 Research needs

There are several areas in the sediment product risk assessment framework that would be improved by additional research and development. General consensus areas include these:

- Development of predictive models for estimating exposure/bioavailability for metals and ionic substances sorbed to sediments
- A better understanding of the routes of exposure of benthic invertebrates to sediment-sorbed chemicals
- Development and standardization of additional chronic sediment toxicity tests
- Identification and standardization of an appropriate formulated sediment
- Determination of limits of acceptability for natural sediments
- Standardization of sediment spiking techniques
- Development and standardization of higher level tests (*e.g.*, microcosms and mesocosms)
- International harmonization of test methodologies and sediment characterizations

4.10 References

Adams WJ, Kimerle RA, Mosher RG. 1985. Aquatic safety assessment of chemicals sorbed to sediments. In: Cardwell RD, Purdy R, Bahner RC, editors. Aquatic Toxicology and Hazard Assessment: 7th Symposium. Philadelphia: American Soc for Testing and Materials (ASTM). STP 854. p 429–453.

[API] American Petroleum Institute. 1994. User's guide and technical resource document: evaluation of sediment toxicity tests for biomonitoring programs. Washington DC: API. API Publication No. 4607.

Baker JL, Barefoot AC, Beasley LE, Burns LA, Caulkins PP, Clark JE, Feulner RL, Giesy JP, Graney RL, Griggs RH, Jacoby HM, Laskowski DA, Maciorowski AF, Mihaich EM, Nelson Jr HP, Parrish PR, Siefert RE, Solomon KR, van der Schalie WH, editors. 1994. Aquatic dialogue group: pesticide risk assessment and mitigation. Pensacola FL: SETAC Pr. 220 p.

Barnthouse LW, Suter II GW. 1986. User's manual for ecological risk assessment. Oak Ridge TN: Oak Ridge National Laboratory (ORNL), Environmental Sciences Division. Publication No. 2679, ORNL 6251.

Cowan CE, Versteeg DJ, Larson RJ, Kloepper-Sams PJ. 1995. Integrated approach for environmental assessment of new and existing substances. *Regulatory Toxicol Pharmacol* 21:3–31.

Di Toro DM, Mahony JD, Hansen DJ, Scott KJ, Hicks MB, Mayr SM, Redmond MS. 1990. Toxicity of cadmium in sediments: the role of acid volatile sulfide. *Environ Toxicol Chem* 9:1487–1502.

Di Toro DM, Zarba CS, Hansen DJ, Berry WJ, Swartz RC, Cowan CE, Pavlou SP, Allen HE, Thomas NA, Paquin PR. 1991. Technical basis for establishing sediment quality criteria for nonionic organic chemicals by using equilibrium partitioning. *Environ Toxicol Chem* 10:1541–1583.

[EC] European Commission. 1994. Technical guidance on risk assessment of existing substances in context of Commission Regulation (EC) No. 1488/94 in accordance with Council Regulation (EEC) No. 793/93 on the Evaluation and control of existing substances. Draft report. Brussels: EC.

Environment Canada. 1996. Ecological risk assessments of priority substances under the Canadian Environmental Protection Act: guidance manual. Draft 2.0, March 1996. Hull Quebec: Environment Canada, Chemicals Evaluation Division, Commercial Chemicals Evaluation Branch, 351 St. Joseph Blvd., K1A OH3.

Hill IR, Matthiessen P, Heimbach F. 1993. Guidance document on sediment toxicity tests and bioassays for freshwater and marine environments. Proceedings from the Workshop on Sediment Toxicity Assessment; the Netherlands. Brussels: SETAC-Europe.

Hoffman DJ, Rattner BA, Burton Jr GA, Cairns Jr J. 1995. Handbook of ecotoxicology. Ann Arbor MI: CRC Press, Lewis.

Larson RJ, Cowan CE. 1995. Quantitative application of biodegradation data to environmental exposure assessments. *Environ Toxicol Chem* 14:1433–1442.

Mackay D. 1991. Multimedia environmental fate models: the fugacity approach. Chelsea MI: Lewis.

[OECD] Organization for Economic and Cooperative Development. 1995. Guidance document on sediment aquatic effects assessment. Paris: Organization for Economic and Cooperative Development Environment Monographs No. 92.

[USEPA] U.S. Environmental Protection Agency. 1992. Framework for ecological risk assessment. Washington DC: USEPA Risk Assessment Forum. EPA/630/R-92/001, 41 p.

SESSION 3
NAVIGATIONAL DREDGING

Chapter 5

Workgroup summary report on navigational dredging

Richard Peddicord, Tom Chase, Tom Dillon, Jim McGrath,
Wayne R. Munns, Kees van de Gucthe, William van der Schalie

5.1 Introduction

Dredging is the process of excavating sediment from a waterway, often involving transportation of the excavated dredged material to another site. Dredging occurs most often to maintain or increase the depth of waterways to provide safe passage for vessels, thus the term *navigation dredging*. Because sediments are the ultimate reservoir for many contaminants, sediments can be the subjects of environmental concerns. These concerns can focus on navigation dredging when contaminated sediments occur in a navigation channel or can lead to remediation dredging of sediments not associated with navigation needs. Remediation dredging is not specifically discussed in this section on navigation dredging, although the fundamental risk assessment process would be similar but would be conducted in a different risk management context.

A dredged material manager must make decisions regarding the most appropriate location for placement of dredged material. Such decisions are often constrained by the urgency for dredging and by the cost and environmental consequences of various placement alternatives. These constraints result from international treaties, national laws, regulations and policies, and pressures from various stakeholder groups. This section provides a brief environmental overview of navigation dredging and illustrates two different regulatory approaches taken in the United States and the Netherlands, followed by an examination of the application of ERA methodology to dredged material evaluations.

5.2 Existing approaches for evaluating dredged material

5.2.1 Overview

Contaminants enter aquatic systems through a variety of point and nonpoint sources and in time may become associated with sediments by a variety of chemical or physical mechanisms. Dredged material encompasses a wide range of sediment types that vary in physical and chemical properties and amount of both anthropogenic pollutants and naturally occurring toxicants (*e.g.*, NH_4, H_2S that may also be enriched in polluted sediments). Sediment–contaminant interactions are both complex and dynamic. Geochemical processes such as partitioning and organic complexation directly affect the biological availability of sediment-associated contaminants. Only the biologically available (bioavailable) fraction is relevant to the actual contaminant exposure that may ultimately elicit adverse ecosystem impacts.

When sediment contamination is an issue in navigation dredging, it is rare for there to be only a single contaminant of concern; typically, there is a complex mixture of several classes of contaminants (*e.g.*, metals, polycyclic aromatic hydrocarbons [PAHs], chlorinated compounds) that interact with the sediment matrix in complex ways. The components and relative concentrations in the mixture, and the interactions with the sediment matrix, differ from sediment to sediment. Toxicity reference values and other quantifiers of effect in risk assessments have almost always been developed on a single chemical basis: the effect of each chemical acting alone is evaluated, and these effects are usually either assumed to be independent or they are combined by some mathematical approach. At the same time, sediment toxicity and other bioeffects tests in most dredged material evaluations have typically been whole sediment tests that measured the total effect of the entire "black box" of contaminants and sediment matrix interactions, with no ability to attribute effect to a specific cause other than to say "this sediment caused this effect in this test" (however, see Chapter 16).

When navigation dredging is considered, it is typically true that the sediment under evaluation will be dredged regardless of its contamination status; economic factors usually force dredging, and the issue becomes one of environmentally responsible management of the dredged material. The dredged material could theoretically be placed either on land or at another location in the water. The physical-chemical conditions affecting bioavailability of contaminants and the receptors of concern are quite different on land and in the water, resulting in different kinds and degrees of risk from the same dredged material, depending on whether it were to be placed in a land or water environment. Even though such cross-media risk considerations are crucial to environmentally sound dredged material management, they substantially broaden the issues to be considered and are beyond the scope of this chapter, which considers the application of risk assessment to dredged material placed in an aquatic environment.

The entire issue of risk management is fundamental to environmental protection and is the primary reason risk assessments are being considered in the context of navigation dredging. However, risk management is a subject unto itself and is not included in this discussion, except to note the necessity of involving risk managers and all appropriate stakeholders in the problem formulation process (Section 5.3.2) to avoid conducting a risk assessment of little practical use due to improper focus.

This chapter reviews the application of scientific understanding and technology in dredged material decision-making within the regulatory and management constraints of the United States and the Netherlands. This review is used as a context for applying the ERA framework to dredged material operations and in making conclusions and recommendations. The remainder of this section summarizes the existing approach to evaluating dredged material for open-water disposal in the U.S. and the Netherlands, especially focusing on the relationship between management constraints and scientific understanding in dredged material decision-making.

5.2.2 United States approach

Ports and harbors and their associated waterways play a vital role in U.S. economic, defense, and recreational interests. In 1992, U.S. ports handled 2.9 billion metric tons of cargo for a total import/export value of $487 billion and supported approximately 15 million jobs, while more than 94 million Americans participated in some form of recreational boating or fishing activity (American Association of Port Authorities 1994). Maintaining adequate navigation depths in ports and waterways is fundamental to national interests.

As sediments are transported and settle in channels and basins, periodic maintenance dredging is required to insure safe passage for shipping. Excavated sediments or dredged materials are removed from channels and transported to another location. There are more than 4000 km of navigation channels and over 400 harbors in the United States that may require maintenance dredging in any one year. This activity generates approximately 300 million m^3 of dredged material that must be relocated annually. About 80% of dredged material is moved to other designated sites in the aquatic environment.

5.2.2.1 Regulatory context

Dredging and placement of dredged material are regulated in accordance with a number of environmental statutes including the National Environmental Policy Act of 1969 (NEPA), the Clean Water Act (CWA), and the Marine Protection, Research, and Sanctuaries Act of 1972 (MPRSA). In addition, the United States is signatory to the London Convention, which governs the disposal of material in ocean waters. The U.S. Army Corps of Engineers (USACE) has primary responsibility for all dredged material permitting activities as well as responsibility for all aspects of federal navigation dredging activities. The USACE, using federal funds, performs the dredging necessary to maintain commercial waterways throughout the United States. Any party (port authority, industrial facility, marina, *etc.*) that wishes to access the federal channel must obtain a permit from the USACE to dredge their own berths, turning basins, and access channels. Many dredged areas experience rapid sedimentation and require relatively frequent dredging to maintain safe depths for navigation. Most dredging projects require environmental evaluation, public notice and comment, opportunity for a public hearing, and permit issuance. The USEPA is responsible for developing environmental criteria and guidelines used by the USACE for permit evaluation under the CWA and MPRSA, the two major statutes governing the disposal of dredged material. USEPA is also responsible for environmental oversight of dredging projects and has the authority to veto projects it determines not to be in compliance with environmental requirements.

The environmental policy goals for dredged material management are established in the CWA and the MPRSA. While the language in these laws differs slightly, the underlying concept in both laws is that placement of dredged material result in "no unacceptable adverse affects." The general policy considerations in the two laws provide neither unambiguous statements of the elements of the ecosystem that are to be protected nor quantitative or qualitative descriptions of the nature of protection that is to be afforded those elements.

NEPA, CWA, and MPRSA and their implementing regulations (40 U.S. Code of Federal Regulations [CFR] 1500–1508, 40 CFR 230 and 40 CFR 220–228, respectively) each requires evaluation of human health and environmental impacts resulting from proposed projects, comparison with other alternatives, and an opportunity for public review and comment. In 1992, the USEPA and USACE issued a generalized, consistent, technical framework document (USEPA and USACE 1992). This document is used to evaluate environmental impacts of dredged material placement in open water, confined disposal facilities, or beneficial uses listed above. This document describes the exposure pathways for each option that should be assessed to determine the potential for contaminants in dredged material to cause unacceptable adverse impacts to human health or the environment. In general, the evaluations are concerned with the potential for toxicity and for long-term or secondary effects due to bioaccumulation of contaminants in the dredged material.

The three principal components of dredged material management embodied in the regulations are 1) site selection and designation, 2) dredged material evaluation (permitting), and 3) site monitoring. The site designation process consists of baseline studies of the environs of the proposed site. The site with the least potential for unacceptable adverse impact is designated as an acceptable location for placement of dredged material that is judged acceptable for aquatic placement in the dredged material evaluation or permitting process. The permitting process requires a detailed evaluation of the specific dredged material under consideration in accordance with regulatory criteria or guidelines and with technical evaluative procedures developed jointly by USEPA and USACE (USEPA and USACE 1991, 1994). Site monitoring serves as a feedback mechanism to ensure that the evaluative procedures are appropriate to maintain the environmental integrity of the site. These three components can be thought of as a management continuum; they act together to ensure that dredged material from multiple projects is evaluated and managed in an environmentally acceptable manner.

USEPA and USACE have jointly developed evaluative procedures for determining the suitability of dredged material for placement at aquatic sites. Because of the complex nature of sediment–contaminant interactions and the fact that contaminated dredged materials can contain complex mixtures of a multitude of toxicants, the CWA, MPRSA, and USEPA and USACE (1991, 1994) guidances focus on effects-based evaluations. The primary evaluative endpoints are toxicity and bioaccumulation potential of sediment-associated contaminants.

The current guidance manuals (USEPA and USACE 1991, 1994) utilize a tiered approach designed to proceed from simple, cost-effective evaluations, which take advantage of available information, to more complex and costly assessments that provide more detailed answers. An evaluation proceeds through the tiers until the necessary and sufficient information is developed to make a permit decision. The tiered hierarchy may be entered at any point, providing the required information is available.

Tier I is primarily an evaluation of existing data. However, in most cases, a more complete chemical characterization of the dredged material will have to be generated. In many cases, a permit decision can be made in Tier I, thus providing a timely and cost-effective regulatory decision. However, in dredged material evaluations involving concerns about contaminants, Tier I will typically indicate that further testing in subsequent tiers is warranted.

Tier II, III, and IV test procedures and evaluations depend on use of a reference sediment. The reference sediment concept implements the regulatory requirement that there be "no unacceptable adverse impact." Reference sediment is defined as a sediment that is substantially free of contaminants, is as similar to the grain size of the dredged material and the sediment at the site as possible, and reflects conditions at the environs of the site in the absence of dredged material placement (USEPA and USACE 1991). Both the reference sediment and the dredged material are tested, and the dredged material is evaluated in relation to the reference material. In concept, if the dredged material does not produce greater response than the reference material, "no unacceptable adverse impacts" are expected.

Tier II is designed to take advantage of predictive assessment models. When sediment quality criteria are fully developed and adopted for regulatory use they may be applied in Tier II as part of the permit decision process. Currently a number of modeling approaches (USEPA and USACE 1991) are recommended for use in Tier II, including the Automated Dredging and Disposal Alternatives Management System (ADDAMS) collection of models, which provide predictive assessment of physical behavior and potential water quality impacts, and the Theoretical Bioaccumulation Potential (TBP) model, which is used to assess bioaccumulation potential of nonpolar organic contaminants of concern associated with the dredged material. Permitting decisions can be reached on the basis of comparisons to reference sediment in Tier II, or further evaluation in subsequent tiers may be necessary.

The Tier III protocols consist of water column and benthic toxicity tests and whole sediment bioaccumulation tests. Tier III water-column toxicity testing addresses the acute toxicity of both suspended and dissolved fractions of the dredged material remaining in the water column following a 4-h period to allow for initial mixing. Tier III benthic tests are conducted with appropriate, sensitive, benthic marine organisms to determine the potential for toxicity and bioaccumulation of contaminants from the dredged material. Dredged material is considered unsuitable for unrestricted placement at aquatic sites if the dredged material toxicity is statistically greater than the reference sediment toxicity and exceeds the reference sediment toxicity by at least 20%. Bioaccumulation potential of contaminants associated with dredged material is evaluated by comparing contaminant concentrations in organisms exposed to reference sediment with concentrations in the same species exposed to dredged material following 28 days of exposure. If the dredged material produces concentrations below U.S. Food and Drug Administration (FDA) Action Levels for Deleterious Substances and below reference sediment results, the dredged material satisfies the bioaccumulation endpoint (USEPA and USACE 1991). If the dredged material results exceed the FDA values, the dredged material is considered unacceptable for unrestricted placement at an aquatic site. When dredged material results are below FDA values but above reference results, interpretation becomes more problematic. While the regulations are concerned about the potential for bioaccumulation, the permit may be denied only if the bioaccumulation is likely to cause an unacceptable adverse effect. Therefore, case-specific criteria that reflect local conditions are used to enhance the interpretative framework for bioaccumulation test results. Tier III is intended to be sufficient for most dredged material evaluations.

Tier IV is intended for use only in those instances where decisions cannot be made in Tier III due to lack of adequate data. Tier IV tests consist of chronic sublethal sediment toxicity tests and steady-state bioaccumulation tests which account for long-term effects of exposure to dredged material. Tier IV methodology is primarily a research and development activity at this time, and the guidance requires that case-specific testing and interpretive procedures be agreed upon by USEPA and USACE before any Tier IV testing is used in the regulatory program.

5.2.2.2 Risk-based techniques in current U.S. regulatory evaluations

Risk assessment concepts and terms have seldom been consciously used in dredged material regulatory evaluations. Risk assessment approaches are clearly within the scope of the regulations and may offer the best opportunity for refinement of dredged material regulatory evaluations. While seldom recognized as such, the present procedure for evaluating water column impacts (USEPA and USACE 1991) is an application of the generic ERA paradigm to a complex mixture of chemicals in the matrix of a specific sediment. The effects assessment consists of determining the toxicity of a whole sediment suspension, which generally approximates the components of the dredged material that might remain in suspension around an aquatic discharge. Responses are observed periodically throughout the test, and LC50 values are calculated at the end of each time period. The LC50 concentrations are plotted through time, producing a time-mortality curve (Figure 5-1). The equivalent of exposure assessment in the risk assessment paradigm is evaluated using a numerical model to calculate short-term spatial-temporal distributions of dredged material in the water column. Model output is plotted on the same graph as the LC50 values, producing a time-concentration curve (Figure 5-1). Risk characterization consists of comparing the mortality curve to the concentration curve to determine whether exposure concentrations are expected to exceed effects concentrations. Thus, risk assessment using direct measurement of biological effects has been applied in dredged material evaluations since 1977.

No equivalent, straightforward application of the risk assessment paradigm has been used for evaluating ecological risk to benthic biota as a result of exposure to deposited dredged material (however, see Chapter 16 for a possible approach to this issue). This has

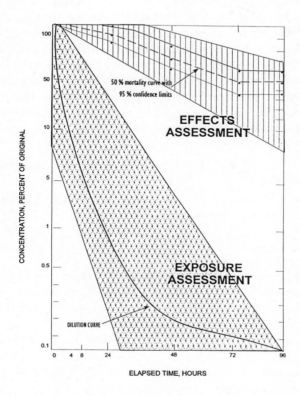

Figure 5-1 Illustration of risk characterization based on toxicity of a complex mixture of chemicals. The effects measurement endpoint (mortality curve) is compared to the exposure measurement endpoint (dilution curve) to determine risk. Reproduced from USEPA and USACE (1977).

been a severe gap because water column impacts associated with dredged material discharges typically are ephemeral and minor relative to potential benthic impacts associated with long-term exposure to deposited sediment.

The water column procedure notwithstanding, most regulatory evaluations of dredged material in the U.S. have not taken advantage of the risk assessment concept. They generally consist largely of effects assessment to the virtual exclusion of exposure assessment and risk characterization. Yet risk assessment is inherently compatible with the regulatory requirements, as demonstrated by the longstanding use of the risk assessment concept in water column evaluations. An approach by which the entire dredged material evaluation process could be conducted on a risk assessment basis is presented and discussed in Section 5.3.

5.2.3 The Netherlands approach

Dredged material management policies in the Netherlands grew out of regulatory and social backgrounds and political pressures different from those in the United States and therefore approach the problem differently. Geographical differences have had a major influence: most dredging in the Netherlands involves sediments associated with the Rhine River estuary, while the U.S. regulations must be applicable to a wide variety of sediments and circumstances encountered throughout the country. The general aims of the dredged material management policies in the Netherlands include these:

- Protection of the structure and function of the ecosystem, at least theoretically protecting 95% of the species, including endangered species and other species of concern
- Application of the "stand still" principle, *i.e.*, no further deterioration of receiving environments
- Minimization of the amount of dredged material placed at aquatic sites, regardless of the degree of contamination
- Reduction of contaminant inputs based on international agreements

5.2.3.1 Short-term policies

Short-term policies are based mainly on chemical analyses of about 40 priority pollutants relevant to sediment contamination in the Netherlands. Based on extensive research on biological responses to contaminated sediment, sediment chemistry-based regulatory values have been established. Dredged material that exceeds upper contaminant levels is placed in diked containment areas. Dredged materials that do not exceed lower thresholds may be placed at aquatic sites. Dredged materials with contaminant concentrations between the upper and lower levels may be used for beneficial purposes or placed at aquatic sites as long as the "stand still" principle is adhered to.

Dredged material in the intermediate category is evaluated in relation to more detailed criteria in order to further reduce the aquatic placement of moderately polluted sediments. In addition to the chemical analyses-based quality assessment, the use of toxicity tests is being considered to identify dredged materials that are toxic even though they do not exceed the lower threshold chemical concentrations. Such materials would be placed in diked containment areas. A cost-effective, stepwise approach is being followed in implementing the toxicity-based program. In the Netherlands' regulatory program, toxicity tests are useful only when the results might change a decision that was based upon chemical analyses of all dredged materials. Monitoring confirms that contaminants of concern are properly identified and that potential contaminants are considered in future dredged material evaluations.

5.2.3.2 Long-term policies

Long-term policies call for further reduction of the placement of dredged material at aquatic sites due to concerns about possible implications of contaminants and international agreements on dredged material management. In the meantime, point and nonpoint source control should lead to less contaminated dredged materials, preferably to the point that placement in diked containment areas would not be required.

5.2.3.3 Regulatory approach

Chemical analyses of whole sediment are performed for about 40 contaminants of concern, and results are compared to the sediment quality values established for each contaminant of concern. Dredged material is considered acceptable for placement at aquatic sites if none of the sediment quality values is exceeded. The sediment quality values are assumed to cover both direct and indirect (bioaccumulation) exposure-effect estimates.

Toxicity tests are conducted on dredged materials with chemical concentrations in the intermediate category. If severe toxicity is revealed in any test, placement at aquatic sites is not allowed. If necessary, bioaccumulation experiments are performed for specific compounds. Tissue levels are compared to values derived from sediment quality guidelines. The bioaccumulation assays are also used to check deviations from the EqP theory used to derive sediment quality values.

Field surveys of benthic communities at aquatic sites are carried out through time. The monitoring does not influence dredged material placement decisions but merely indicates needs for further identification of contaminants of concern.

In order to set new sediment regulatory values, risk-based evaluations of selected contaminants of concern are carried out, using sophisticated exposure models and effects estimates.

Sediment quality values for dissolved, suspended, and deposited contaminants are based upon ecotoxicological data and apply to freshwater as well as to saltwater environments. Sediment quality values are in principle equal to soil values, but compartment-specific risk evaluations are carried out. Criteria are normalized for OC content and particle size distribution. Sediment quality values are derived from spiked whole-sediment toxicity tests and tests of dissolved contaminants. Data from dissolved contaminant tests are translated into sediment values using the EqP theory. For metals, field-derived partitioning constants are used.

If necessary, sediment samples are treated to fulfill criteria on confounding factors like ammonia, pH, and oxygen, which represent the tolerance limits of the specific test species. Responses in effect parameters are compared to those in control and in reference sediments from near the site, resembling the characteristics of the dredged material of concern. However, reference sediments often appear not to be free from contaminants and effects. Responses are classified as no effect, moderate effect, and severe effect based upon the most sensitive parameter per species (survival, growth, reproduction, luminescence, morphological abnormalities).

5.2.3.4 Linking bioassay responses to sediment quality objectives

Short-term acute toxicity testing cannot be a basis for classifying a dredged material as acceptable for placement at an aquatic site, but it can be a basis for declaring a sediment unacceptable. If chronic toxicity testing shows toxicity to only one appropriate species, the dredged material may presently be placed at an aquatic site, but such placement will become unlikely in the future as policies become more restrictive. Dredged material that shows no adverse biological effects is regarded as acceptable for placement at aquatic sites and will remain so in the future. For these sediments, monitoring at the placement site is an important tool to confirm the dredged material management decision-making process.

5.3 Application of risk assessment methodology to dredged material evaluations

5.3.1 Overview

The basic risk assessment concept includes fundamental components that can be summarized very briefly as follows. Each will be discussed in more detail in the context of navigation dredging in the remaining sections of this description of the application of risk assessment methodology to dredged material evaluations.

- Problem formulation (Section 5.3.2) in which the problem is defined, potential stressors are identified, and the application of the risk assessment framework to the specific situation is structured
- Effects assessment (Section 5.3.3) in which the nature of the potential effects of the stressors and dose-response relationships are determined
- Exposure assessment (Section 5.3.4) in which the nature of the exposure to the stressors is determined
- Risk characterization (Section 5.3.5) in which expected exposure conditions are compared with conditions necessary to cause effects, to determine whether effects would be likely to occur

In the past, most evaluations of dredged material for management purposes focused on the dredged material itself. Evaluation of the placement site was often a secondary issue, usually treated more or less separately from evaluations of the individual materials to be placed at the site. In reality, characteristics of both the site and the materials placed there are inextricably involved in determining the environmental impacts that could occur.

Both exposure and effect must exist before risk can occur. In relation to navigation dredging, exposure occurs at the placement site (except in evaluations of the "no action alternative"). Therefore, the potential risks of a particular dredged material can be determined only in the context of the proposed placement site (or its present location in evaluations of the "no action alternative") and the exposure conditions existing at that site. Effects tests of the material are a necessary but insufficient component; risk can be determined only when the effects information is considered in conjunction with the exposure conditions at the placement site. Therefore, exposure assessment at the placement site is essential if risk assessment is to be applied to navigation dredging. This principle has not been commonly recognized in dredged material evaluation and underlies the remainder of this discussion of the application of risk assessment to navigation dredging.

Although risk management discussions are beyond the scope of this document, a corollary of the above principle must be mentioned here for completeness. Exposure at the placement site is key, not only for risk assessment but also for risk management. The potential toxicity and other effects of a particular dredged material are characteristic of the material and cannot be easily altered. Exposure, however, can be controlled by a variety of proven techniques, including use of a different placement site, various controls on the placement process, and engineering techniques such as capping to reduce exposure.

The value of risk assessment cannot be realized if it is conducted in a vacuum. It is imperative that the risk assessment be fully integrated with risk management needs from the very earliest stages. Only in this way can the entire process lead to optimal environmental decisions. The remaining discussion relates to application of risk assessment to evaluation of dredged material proposed for placement at an aquatic (subaqueous) site, which may also receive other dredged materials from other locations before, while, or after the material in question is placed there (a multi-user site). Dredged material is described in Section 5.2 as a "complex mixture" of sediments and chemicals. Dredged material placed at a multi-user site in fact becomes part of a "complex mixture of complex mixtures." Environmental exposure at a multi-user site is the most important issue for application of risk assessment to dredged material evaluations and is the focus of the following discussion. Therefore, the discussion does not address evaluation of the "no action alternative" or placement at upland or other sites. Neither does this discussion consider evaluation of the dredging operation itself, as distinct from the placement of the dredged material. However, the basic principles are applicable to all dredged material evaluations and can readily be modified as appropriate for use in a wide variety of contexts. Similarly, this SETAC Special Publication focuses on ERA, but many of the basic principles are equally appropriate to HHRA.

5.3.2 Problem formulation
Dredged material disposed into an aquatic system can potentially affect the environment through a number of different pathways. This section examines the stressors associated with ecological risk, identifies the types of potential ecological receptors that risk assessment should focus on, and indicates the types of assessment endpoints that should be used to analyze ecological risk. Two things should be kept in mind. First, pathways to human receptors are not addressed in this discussion; the issue is important but beyond the scope of this effort. Second, the examples provided are nothing more or less than examples; any application of these principles must provide comprehensive analysis of the appropriate parameters for the system in question.

5.3.2.1 Stressors
Placement of dredged material at an aquatic site can involve physical and/or chemical stressors that can act in the short or long term, either near or far from the site. Physical

stressors are common to all aquatic placement of dredged material and can include such things as the initial discharge that can bury the benthos and may physically alter the substrate at the site, thereby affecting recolonization. Dredged materials may or may not contain chemical stressors, which can be present singly or as a wide variety of chemical types and forms, depending on the source of the dredged material and the sources of contamination to it. These chemical stressors include not only the chemicals that are measured but also those that are unmeasured. Major kinds of stressors that may or may not be associated with any particular dredged material are summarized with their most likely field of influence in Table 5-1 and discussed in the context of the conceptual site model in Section 5.3.2.4. Physical stressors can be important in some circumstances, but for the sake of brevity, the remainder of this discussion is focused on chemical stressors.

5.3.2.2 Ecological receptors and pathways

Multi-user aquatic placement sites, which are the focus of this discussion, can be found on all sorts of bottom types in fresh or salt water at depths of a few meters to a few thousand meters, with wind and/or current energies causing negligible to rapid sediment erosion and transport. Across such a variety of environmental conditions, few specifics can be offered about the exposure pathways or potential receptors that might be of concern for any particular dredging project. However, some general concepts are provided to guide the identification of pathways and receptors for consideration on individual projects.

Ecological risk assessment of aquatic placement of dredged material will often focus on four primary categories of receptors, whose relevance and importance will differ depending on the circumstances of each operation. These primary receptor groups are 1) the infaunal benthic community, 2) the epibenthic community, 3) the demersal fish community, and 4) the pelagic fish and invertebrate communities. A fifth group, birds and wildlife that rely on fish and invertebrates as prey, could also be potentially affected in situations where a complete exposure route from the dredged material exists and, in such cases, should be considered where food chain transfer of contaminants to these receptors is a potential concern.

The selection of receptors follows directly from an understanding of the exposure pathways for contaminants from dredged material to the environment (Figure 5-2). The primary pathways of concern at the placement site (*i.e.*, below the dashed line in Figure 5-2) are these:

1) Entrainment and dispersion as the material settles to the bottom. Dissolved and particulate stressors are directly exposed to the water column communities, leading to identification of the demersal and pelagic fish and invertebrate communities as potential receptors.

2) Direct exposure from the mound of dredged material deposited on the bottom. Sediments are directly ingested by species within the epibenthic and benthic infaunal communities, leading to identification of these communities as potential receptors. Exposure can also occur due to dermal contact with

Table 5-1 Summary of major types of potential stressors that may exist at an aquatic dredged material placement site

Type of stressor	Area of potential influence		Time of potential influence	
	Near field	Far field	Short term	Long term
Physical				
Burial	X		X	
Alteration of substrate type	primary, during placement	secondary, due to subsequent sediment transport	primary, during placement	secondary, due to subsequent sediment transport
Chemical				
Dissolved contaminants	X		primary, during placement	secondary, due to desorption
Sediment-sorbed contaminants	X			X
Contaminants entering food web	X	X		X

sediments or contact with pore water, again potentially affecting the epifaunal and infaunal benthic communities.

3) Movement from the deposited mound. Sediments may move from their original point of deposition through resuspension, transport, and redeposition. These sediment movements could increase the intensity of exposure and/or the area over which exposure occurs for the benthic and epibenthic communities. Subsequent disposal at multi-user sites can substantially alter the intensity and/or spatial nature of exposure to one dredged material by displacing and/or covering it entirely or in part with other dredged material.

4) Trophic transfer. Uptake into the food web and transfer to higher trophic levels is a possible exposure pathway for all of the above receptor communities. This pathway might warrant particular consideration with DDT, DDE, PCB, toxaphene, arsenic, methyl mercury, and total mercury, which have been shown to have the potential to biomagnify through aquatic food webs (Suedel *et al.* 1994).

5.3.2.3 Assessment endpoints and measurement endpoints

Assessment endpoints must be developed for each individual site in keeping with the characteristics of the site and the potential receptors that might be present. This discussion is intended to illustrate the nature of analysis entailed in reaching a site-specific list of assessment endpoints.

Under the structure suggested by Suter (1993), establishing assessment endpoints is the first of several steps that translate general policy objectives, such as those found in the international treaties or national laws, regulations, and policies, into measurement end-

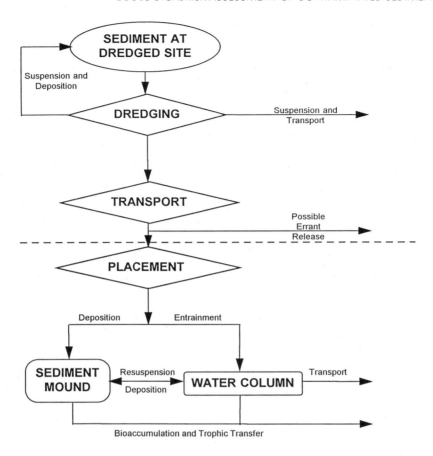

Figure 5-2 Possible exposure pathways during dredging and dredged material placement at an aquatic site

points that can be evaluated in risk assessment framework. Suter's (1993) intermediate step between assessment endpoint and measurement endpoint was to select indicators of effects (*e.g.*, changes in abundance). A slightly more elaborate series of steps is used here to select assessment and measurement endpoints. The first step is to identify the receptors (*i.e.*, ecological elements) of concern. For example, a species such as flounder may be the basis of an important fishery or food for other valued species. The second step is to identify the receptor attribute that might be affected. For example, the abundance of flounder might be affected by physical and/or chemical stressors associated with placement of the dredged material at the site. The third step is to identify a quantitative assessment endpoint that reflects the importance of the receptor and specifies both the attribute of interest and the level of effect. In addition, a time component is often important because dredged material effects may be of finite duration. In the flounder example, an assessment endpoint might be that the abundance of flounder in the immediate vicinity of the site be within 20% of the abundance in a nearby reference area within one year after the operation is completed.

Assessment endpoints are the basis for establishing measurement endpoints that allow assessment of the risk to the receptor from placement of the dredged material at the site. Measurement endpoints must consider the potential exposure pathways from the dredged material to the receptors of concern. In the case of the flounder, measurement endpoints might include toxicity of the dredged material to juvenile flounder or to amphipods, polychaetes, or other surrogates for important flounder food in the site vicinity. Such endpoints can be measured in laboratory toxicity tests, which if appropriate could be designed to measure other endpoints related to sublethal effects of the dredged material on growth and reproduction, or even the potential for bioaccumulation and trophic transfer through prey to flounder. The latter could also be "measured" using verified and calibrated mathematical models.

Measurement endpoints provide information needed to evaluate assessment endpoints and might be considered surrogate measures or indicators of the latter. In this regard, measurement endpoints relate to ecological effects (USEPA 1992). The rationale for identification and selection of measurement endpoints includes their relevance to the assessment endpoints, the practicality of approaches for their quantification, their sensitivity and response characteristics, and their consistency with assessment endpoint exposure scenarios (USEPA 1992). Research in the field of sediment assessment has identified and evaluated several tools that are potentially useful as measurement endpoints in risk assessment of dredged sediment. Although not intended to be an exhaustive list, several classes of these are identified in Table 5-2.

The utility of any of these within a particular assessment should be evaluated against the rationale listed above. For example, measurement endpoints appropriate for the assessment endpoints addressing declines in flounder abundance include sediment and water quality criteria and guidelines; toxicity of deposited and suspended dredged material to life stages of flounder, surrogate species, and infaunal prey species; and perhaps summary statistics derived from an appropriately developed model of flounder population dynamics. Some of these (*e.g.*, toxicity to life stages of flounder) relate to the direct effects of chemical stressors associated with the dredged material, whereas others (*e.g.*, toxicity to infaunal species) are appropriate in evaluations of indirect effects. Each varies with respect to its degree of association to the assessment endpoint and therefore with respect to the level of uncertainty associated with its extrapolation.

The process of identifying the important ecosystem characteristics concerning a particular dredged material placement operation, determining the appropriate corresponding receptors of concern, establishing assessment endpoints, and determining the appropriate measurement endpoints should be applied for each ecosystem characteristic of concern for a particular project until assessment and measurement endpoints have been established for all ecosystem characteristics of concern.

Table 5-2 Example assessment endpoint and relevant measurement endpoints and exposure measures that might be useful in ecological risk assessment of navigational dredging

Assessment endpoint	Measurement endpoints	Exposure measures
Ecologically important decline in flounder abundance	Sediment quality criteria/guidelines	Dredged sediment volume
	Water quality criteria/guidelines	Dredged and site sediment geotechnical attributes (*e.g.*, granulometry, porosity, specific gravity)
	Endpoints from acute and chronic toxicity tests of whole sediment, pore water, and elutriate water	Dredged material and site sediment bulk, bioavailability, and elutriate chemistry
	Abundance of prey species	Dredged and site sediment geochemistry (AVS, TOC, pH, ammonia)
	Inputs to ecological models	Modeled and measured contaminant concentrations in prey species
	Demographic and population statistics of unimpacted (reference) populations	

5.3.2.4 Conceptual model

The final step of problem formulation involves development of the conceptual model. The purpose of this model is to summarize available information concerning stressors, ecological receptors, and assessment and measurement endpoints important to the assessment. Developed as a set of working hypotheses, this model describes the environmental processes and exposure pathways relating stressors to receptors, identifies points along those pathways where important exposure information should be obtained, and indicates possible mechanisms of impact to ecological systems. The conceptual model also identifies the spatial and temporal scales and boundaries of the assessment, and it identifies and relates measurement endpoints to the assessment endpoints. This is the appropriate time to select the risk characterization approach to be used, as discussed in Section 5.3.5.

A generalized conceptual model that addresses several spatial and temporal scales can be developed. At the broadest scale is a description of potential exposure to ecological receptors during all phases of dredging and dredged material placement (Figure 5-2). As described previously, potential release of dredged material and possible associated contaminants can occur at several points along a sequence from the dredging operation to the dredged material mound deposited at the site. Since these "releases" occur in different parts of the water body and on different temporal scales, the ecological risks associated with each will vary with respect to stressors, ecological receptors, and routes of exposure. In keeping with the previously established scope of this paper, the remaining discussion of conceptual models will focus on the ecological risks of a specific dredged material placed at a multi-user aquatic site (illustrated below the dashed line in Figure 5-2). The spatial boundaries of the assessment therefore include the site itself, as well as areas surrounding the site, which might be influenced by resuspension and subsequent transport from the mound.

Scenarios of exposure at the site include the possible entrainment of dredged material into the water column during convective descent and settlement following release from the transport vessels, exposure to deposited material on and around the mound, and resuspension and subsequent transport of dredged material particles from the mound. At a smaller scale, the dynamics of particles and chemical stressors become important. Contaminants may partition dynamically between sorbed and dissolved states and may be available for uptake by organisms in either state. Similar behavior occurs in the deposited sediment, where chemicals may partition between particle surfaces and pore water. These processes affect the transport and fate of chemical contaminants and thus the exposure of receptors.

The aquatic environment into which dredged material is placed can be partitioned into several compartments, based on biological, chemical, and physical properties. Included are the benthic and pelagic or water column compartments as well as the epibenthic compartment (Figure 5-3). The benthic compartment can be considered to consist of the newly created mound and the original or relic underlying sediments which existed at the site prior to dredged material placement. Exposure dynamics and potential ecological effects are different in all compartments and are functions of the physical, geochemical, and biological processes occurring in each. The conceptual model should recognize the existence of these various compartments and processes and should evaluate potential ecological risks within each as appropriate.

The conceptual model reflects explicit exposure pathways, from the source and nature of the environmental stressor (such as chemical contaminants in the dredged material) to the ecological receptors about which assessment endpoints are developed. Borrowed from classical exposure pathway analysis, this level of representation of the conceptual model allows not only for explicit visualization of routes of exposure but also helps to identify key *exposure points* (loci along the exposure pathway which, through direct or indirect measurement, permit quantification of exposure experienced by the receptors of concern) for exposure characterization activities. Exposure pathways for the flounder population abundance assessment endpoint are shown in Figure 5-4. Clearly, there are several possible pathways by which flounder could be exposed to stressors associated with dredged material. Included are both direct effects of contaminant exposure through direct contact and trophic transfer and indirect effects associated with possible reductions in prey availability resulting from direct stressor effects on those prey. Burial as a stress is represented in Figure 5-4, but burial of flounder, while possible, is deemed to be relatively rare and therefore is not considered further in this discussion. However, both

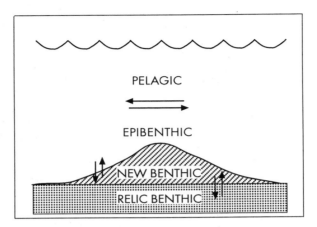

Figure 5-3 Compartments of the aquatic environment to be considered in dredged material risk assessment

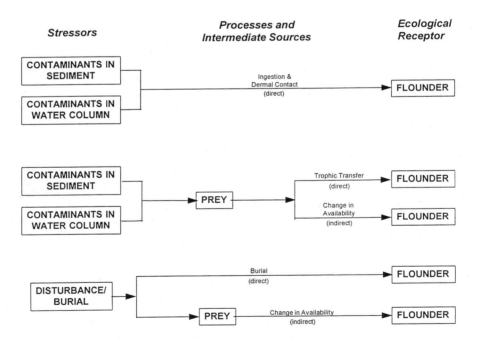

Figure 5-4 Major exposure pathways that might be relevant to a flounder population abundance assessment endpoint

the direct effects of flounder burial and the indirect effects of reductions in prey availability due to burial should be considered in actual dredged material evaluations.

Key exposure points, and therefore media for exposure measurement for the flounder assessment endpoint, include the deposited dredged material, water influenced by contaminants dissolved from the dredged material, and flounder prey. In the prospective view characteristic of dredged material risk assessments, exposure measures may be derived from physical or mathematical simulations of exposure conditions (involving such processes as sediment deposition and entrainment, hydrodynamic transport, and bioaccumulation and trophic transfer) or may be based upon data collected in field investigations. Such exposure measurements are subject to validation and are associated with some degree of uncertainty. This uncertainty should be quantified to the extent possible to aid in interpretation of risk. Several classes of exposure measures proven to be useful tools are presented as examples in Table 5-2. Selecting appropriate exposure measures involves evaluating their ability to reflect characteristics of exposure along a pathway from stressor source to ecological receptor. As importantly, they should provide information useful for predicting and interpreting changes in measurement endpoints. Examples of valuable exposure measures for the flounder assessment endpoint include the volume and spatial distribution of the dredged material deposits and the physical and chemical characteristics of the dredged material.

The conceptual model also considers the potential ecological effects expected to result from stressor presence and subsequent exposure. As identified earlier, such effects depend on the nature of the stressors as well as on the kinds and numbers of ecological receptors present within the influence of the dredged material. Potential ecological effects are summarized in the conceptual model as a function of important assessment endpoints and pathways of exposure from source to receptors. Ecological effects hypothesized for the flounder assessment endpoint include direct toxicological impact to individual flounder (perhaps ranging from subtle physiological and cytogenetic effects to gross reproductive and survival impacts) as a result of contact with contaminants in the sediment, direct impact due to burial of flounder (not followed in this discussion), as well as impacts on flounder population dynamics resulting from changes in demographics. Direct toxicological effects also might occur as a result of the transfer of chemical contaminants to flounder through trophic pathways. Indirect effects resulting from reduction in the availability of prey through toxicological impact and burial might also be considered. An understanding of the potential effects of stressor exposure on ecological receptors supports identification of measurement endpoints suitable to evaluate assessment endpoints. The conceptual model should explicitly define the activities to be conducted in the effects assessment, exposure assessment, and risk characterization components of the risk assessment. These activities derive directly from problem formulation and the fully developed conceptual model. Clearly, the product of problem formulation depends upon the aspect of navigation dredging being evaluated, that is, the component along the sequence from dredging to placement illustrated in Figure 5-2. The discussion provided above should aid in appropriate problem formulation for any aspect of navigation dredging.

5.3.3 Effects assessment

Effects assessment is the process of establishing the strength or magnitude of the potential impact dredged material could have if organisms were to be exposed to it. Impacts may be related to physical stressors (*e.g.*, burial or change in physical characteristics like grain size) or to chemical stressors (*e.g.*, dissolved or sorbed contaminants). Conduct of an effects assessment for dredged material is driven by 1) the biological receptors of concern and 2) the anticipated exposure scenarios. That is, effects are assessed in the context of affected receptors and expected environmental exposures (see exposure assessment as discussed in Section 5.3.4). One of the most frequently evaluated scenarios is the effect of sediment-associated chemicals in deposited dredged material on the benthic community. To assess these effects, toxicity tests of deposited dredged material are conducted with appropriately sensitive infaunal or epifaunal species. The potential for contaminants to bioaccumulate from deposited sediments is also usually determined as part of the effects assessment. Although bioaccumulation *per se* is a phenomenon and not an effect, the data are reviewed for potential residue-effects relationships. Effects in the water column are evaluated by conducting acute (48-h to 96-h) elutriate toxicity tests with appropriately sensitive water column species. Elutriate and deposited sediment toxicity tests evaluate response to all contaminants in the dredged material, both known and unknown. Most sediment toxicity tests available today measure survival or growth after short-term exposures (4 to 10 days). A new generation of toxicity tests involving longer exposures and sublethal endpoints (*e.g.*, growth and reproduction) are being developed (Dillon 1993; also see Chapter 18). As these chronic sublethal sediment bioassays become available, they will likely be used with increasing frequency as effects assessment tools.

If effects on receptors at higher trophic levels are of concern, a chemical-by-chemical approach may be taken. For example, the approach can begin with case-specific bioaccumulation data and can model trophic transfer of individual chemicals from benthic infauna to a predator (*e.g.*, flounder, humans). Effects can then be evaluated by examining residue-based or ingestion-based benchmark toxicity taken from the literature or generated empirically on a site-specific basis.

To improve the utility of effects assessments in dredged material risk assessments, four major issues must be addressed: 1) interpretive guidance, 2) extrapolations, 3) cumulative effects, and 4) recovery potential. The first two concern the laboratory assessments of the effects of dredged material (*i.e.*, toxicity tests) that are done today. The latter two issues refer to the field evaluation of dredged material impacts at the placement site. Although lab evaluations have historically been emphasized over field studies, it is the field impacts that are of ultimate ecological and regulatory concern ("no unacceptable adverse impacts").

5.3.3.1 Interpretive guidance

Comparison to a reference sediment is currently a major component of interpreting dredged material toxicity and bioaccumulation test results in the U.S. regulatory program for dredged material. Comparison to reference provides a relative basis for interpreting results. The technical basis for an absolute interpretation, (*i.e.*, what does 15% mortality mean ecologically versus 25% mortality?) is more elusive. Absolute interpretations become even more problematic when sublethal endpoints are examined. Demographic models may form the basis for interpreting chronic sublethal bioassays with dredged material. Demographic models integrate multiple life history information (*e.g.*, survival, growth, reproduction) and express toxicity at a population level (Hummon 1974; Gentile *et al.* 1982; Wong and Wong 1990). This source of interpretive guidance is still firmly embedded in the research and development community.

In some instances, the basis for interpretive guidance is a chemical-specific benchmark toxicity value. For example, trophic transfer modeling may predict that a contaminant of concern associated with the dredged material in question could bioaccumulate to X concentration in flounder. One would compare this predicted tissue concentration with a benchmark toxicity value. These values are occasionally reported in the scientific literature, but in practice, there often is no benchmark toxicity value for the chemical-receptor combination of interest, and one must be developed by extrapolation among species and/or contaminants. Interpretive guidance based on chemical-specific toxicity benchmarks must also always consider the potential impacts of contaminant interactions, *e.g.*, synergism and antagonism (see Chapters 16 and 18).

5.3.3.2 Extrapolations

Risk assessment is basically an exercise in extrapolations. Important extrapolations, required to relate effects assessment data to the original assessment endpoints during risk characterization (see Section 5.3.5), include the following:

- Species to species: It is seldom practical to test all species of concern.
- Response to response: One may measure acute toxicity but also may be interested in effects on survival or reproduction.
- Individual to population: Effects on individual organisms are measured in most tests, but it is usually the population of organisms that is of concern.
- Laboratory to field: Effects under controlled laboratory conditions may or may not mimic effects under actual exposure conditions at the placement site.
- Present to future: Effects observed today may or may not persist into the future.

5.3.3.3 Cumulative effects

The term *cumulative effects* has been used to refer to effects over time or to the combined effects of multiple stressors (*e.g.*, metals plus pesticides). It has also been used to refer to the combined effects of multiple events (*e.g.*, placement of dredged material from several different projects at a site over time). The term is used in the latter sense in the U.S. dredging regulations (*e.g.*, 40 CFR 230.11(g)) and in this discussion. The concern about cumulative effects focuses not on project-specific toxicity tests but on the health and viability of the benthic communities near the placement site. International dredged-material regulations address the concept of cumulative effects of all activities taking place at the site, and cumulative effects are important in complete risk assessment and risk management contexts. However, practical and appropriate ways to quantitatively predict cumulative effects have received little research attention, some of which is addressed in Chapter 16. A way of including cumulative effects in dredged material risk assessments, based on the effects assessments of each material previously placed at the site, is described in the context of cumulative exposure at the site in Section 5.3.4.

5.3.4 Exposure assessment

In the context of risk assessment applied to dredged material placed at a multi-user site, exposure assessment must address a variety of conditions, including, for example, the following:

- Suspended and deposited sediments
- Multiple projects of different size, sediment quality, and dredging method (and their different physical behaviors) placed at a single site over time
- Sedentary and mobile receptors of concern
- Direct exposure of biota and food web vectors for contaminant transfer

In addition, the regulatory context in which many risk assessments of navigation dredging are conducted imposes other constraints. For example, in the U.S., most navigation dredging risk assessments will have to contend with

- effects evaluated on the basis of whole sediment biological tests as well as on the basis of chemical analyses and
- compatibility with a regulatory context in which permit decisions must be made expeditiously.

5.3.4.1 Exposures to be considered

The following discussion of exposure assessment at a multi-user site focuses on the temporal and spatial distribution of the stressors of interest (*i.e.*, contaminants associated with suspended or deposited sediment). Also important in all cases but not addressed here is the temporal and spatial distribution of the organisms or receptors of concern in relation to the distribution of the stressor. It is important to know which species or receptors of concern are likely to encounter the stressors, the frequency and duration of contact, the proportion of the receptor population encountering the stressor, and other relevant ecological considerations in addition to the physical considerations described below.

During placement. During dredged material placement at the site, suspended sediment is introduced throughout the vertical extent of the water column as the dredged material is discharged from the transportation vessel or pipeline. This continues, often on a frequent intermittent basis, until the dredging project being evaluated is complete, then ceases until the next dredging project begins. When dredged material is released near the water surface, most of the material descends rapidly to the bottom in a mass descent phenomenon. However, a few percent of the total mass of sediment discharged may remain in suspension long enough for short-term dispersion in the water column, providing a potential exposure mechanism for water column organisms.

Exposure to deposited sediment during placement is a theoretical possibility but seldom appears to be a major focus of concern for contaminants. Organisms directly buried by the deposited material may face major physical stresses, but the burial usually overshadows any chemical concerns on a short-term basis. Because new sediment is more or less continuously being deposited throughout the placement operation, site recolonization does not begin in earnest until after placement is complete. Thus, there are limited benthos present to be exposed to deposited sediment until after placement is completed.

Post placement. After the discharge operation is completed, sediment may be resuspended from the site by wave and/or current action. This may be a mechanism of exposure for organisms on and off the site, depending on the distribution of the suspended sediment. After the project is completed, recolonization of the deposited dredged material by larvae and mobile organisms may occur. The deposited material probably provides the most direct exposure opportunity for infauna and epifauna and for the organisms that feed on them.

5.3.4.2 Exposure assessment approach
The following approach to exposure assessment is presented at the conceptual level. Clearly, many aspects remain to be worked out on a case-by-case basis, and some will require further research and development before routine implementation becomes practical. However, the conceptual approach is appropriate and provides a framework to guide development of practical implementation procedures.

During placement. The USEPA and USACE procedure (1991) for evaluation of water column impacts was designed specifically to address exposure to suspended sediments at the site during discharge. This approach uses a mathematical model developed specifically for this purpose to describe the temporal and spatial distribution of suspended sediments and associated chemicals during and immediately following a discharge event. The model can describe concentrations of dissolved contaminants and thus is compatible with chemical-based effects endpoints for effects assessment purposes. It can also describe suspended sediment concentrations and so can be used with toxicity or other effects tests based on exposure to suspended sediment.

Post placement. Resuspension of deposited dredged material after discharge is completed can be modeled mathematically or estimated by other means. Once the temporal and spatial distribution of suspended sediments or associated chemicals is known, these distributions can be used with either chemical- or biological-based measurement endpoints in effects assessment.

Long-term exposure to the deposited dredged material mound depends less on movement of the sediment than on movement of receptors onto the sediment. This can occur either through larval setting or movement of adults, including feeding on infauna at the site. Exposure to deposited sediment can be evaluated in relation to either biological or chemical effects endpoints. Exposure to deposited sediment is controlled predominately by biotic and/or physical-chemical, rather than physical, factors.

5.3.4.3 Exposure calculation concept

In the simple case, material from only one dredging project will be placed at the site. However, at multi-user sites, the complexity of dredged material from multiple projects placed at the site over time must be addressed. Exposure to deposited sediment in relation to either biological or chemical effects measurement endpoints is controlled largely by movement of receptors onto the deposit. Exposure is assessed considering the size of the deposit, frequency and duration of receptor contact with the sediment stressors, proportion of the receptor populations exposed, *etc.* When a second dredging project is placed at the same site, exposure to the new deposit is evaluated in the same way. However, total exposure is now the cumulative exposure to the first project mound plus the exposure to the second project mound. If the second project mound partially or entirely covers the first mound, the reduction in surface area of the first mound accessible to receptors must be considered in calculating the cumulative exposure to receptors as a result of the second project. The same process is applied to calculating cumulative exposure as a result of placement of subsequent mounds.

For multi-user sites, cumulative exposure is evaluated as shown in the following conceptual illustration (based on Figure 5-5). The approach is the same whether exposure is evaluated in relation to sediment chemistry or in relation to sediment toxicity or other biological effects endpoints. When placement of the first project is complete, exposure of receptors at the site, all other things being equal, is a function "f" of the surface area of "mound" 1. This "f" involves aspects of chemical bioavailability and co-occurrence of receptors at the site.

Cumulative exposure as a result of adding a particular dredged material to the placement site can be illustrated by the following equations. These equations are not intended for mathematical solution but are intended simply to illustrate the concept of quantifying cumulative exposure, assuming additivity:

$$X_1 = f(A_1)$$

where X_1 = cumulative exposure at the site after placement of project 1 and
A_1 = surface area of mound 1.

Project 2 is placed at the site in such a way that mound 2 does not overlap mound 1. After project 2 is completed, the cumulative exposure (X_2) of receptors at the site will be

$$X_2 = f[(A_2) + (A_1)].$$

The third project to be placed at the site is predicted to create a mound that partially covers both mounds 1 and 2. After project 3 is completed, the cumulative exposure (X_3) of receptors at the site will be

$$X_3 = f[(A_3) + (A_2 - A_{2c}) + (A_1 - A_{1c})]$$

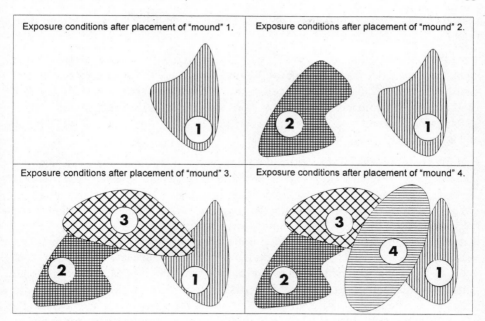

Figure 5-5 Conceptual illustration of cumulative exposure to deposited dredged material at a multi-user aquatic site as "mounds" of dredged material from different projects are sequentially added

where A_{2c} = surface area of mound 2 covered by subsequent mounds and
        A_{1c} = surface area of mound 1 covered by subsequent mounds.

The fourth project is placed at the site in such a manner that it partially overlaps mound 3 and a portion of mound 1 not covered by mound 3. The cumulative exposure (X_4) of receptors at the site after placement of the mound from project 4 will be

$$X_4 = f[(A_4) + (A_3 - A_{3c}) + (A_2 - A_{2c}) + (A_1 - A_{1c})].$$

Note that A_{1c} here has a larger value after the placement of mound 3 because mound 4 covers part of mound 1 that was not previously covered.

Subsequent projects placed at the site would be addressed by extension of this approach.

5.3.5 Risk characterization

Risk characterization is the final phase of the risk assessment process. Major elements of this phase include integrating exposure and effects information, summarizing relevant supporting information in a weight-of-evidence discussion, interpreting the ecological importance of the risks, and discussing the results of the assessment with the risk manager and other stakeholders. In this section, these principles are discussed with respect to the placement of dredged material at a multi-user aquatic site. A hypothetical characterization of potential risks to flounder abundance in the vicinity of such a site is used as an example.

The approaches used for characterizing risk associated with dredged material at an aquatic site will depend on the purpose of the assessment as well as on the resources (personnel and time) available for the assessment. Decisions concerning selection of risk characterization approaches should have been made as part of the conceptual model. In general, empirical and mechanistic approaches may be used to characterize risk (Wiegert and Bartell 1994).

Examples of empirical approaches include joint distributional analysis and the quotient method. In the quotient method (Barnthouse *et al.* 1986), a point estimate of exposure is divided by a point estimate of an effect level or toxicological benchmark concentration, *i.e.*, by an effect measurement endpoint. For example, in a preliminary evaluation of the risk to flounder of chemicals released into the water column during placement of dredged material, an estimate of a chemical concentration might be divided by the corresponding acute-toxicity water quality criterion. If the quotient value is much less than one, a lack of potential acute toxicity might be inferred; if the quotient value is much greater than one, potentially substantial effects might be inferred. Intermediate quotient values generally indicate the need for more detailed evaluation. While the quotient method offers an efficient means for risk screening, it allows qualitative rather than quantitative estimates of uncertainty. In addition, the quotient method does not facilitate management of risks through exposure reduction, since the meaning of a quotient's drop (from, *e.g.*, 5.4 to 2.1), which results from reduced exposure, cannot be readily interpreted in terms of potential ecological effects.

An alternative empirical approach to the quotient method for integrating risk is the use of joint probability analysis (*e.g.*, Cardwell *et al.* 1993). In this approach, distributions of both exposure concentrations and species sensitivity values (*i.e.*, effects measurement endpoints) are graphed. For example, consider the pathway to flounder through contaminated prey. In this case, a range of predicted exposure values for tissue-contaminant levels might be compared to a distribution of toxicity effects obtained from a feeding study with flounder. The overlap or lack thereof in the exposure and toxicity functions could be used to predict risk. A potential limitation of this approach is the need to develop or validate and calibrate an acceptable food web model. However, other applications of distributional methods that rely upon existing data have been described (*e.g.*, Cardwell *et al.* 1993).

Mechanistic (or process) models use hypotheses about causal processes to simulate the responses of biological systems to stressors. In the flounder example, if the pathway involved concerns for flounder population changes resulting from reduced prey abundance, changes predicted from population models for important prey species such as amphipods or worms might be coupled to a flounder population model. While such models can generate probabilistic data and facilitate exploration of a range of alternative exposure scenarios, necessary input data may be unavailable and associated uncertainty levels may be high.

Whatever method is used to integrate exposure and effects data, the risk should be expressed in terms of effects on an assessment endpoint. For example, if the measurement endpoint for water-column chemical effects on flounder is toxicity to a surrogate species, the risk characterization should express risk to the flounder through an evaluation of the extrapolation from the surrogate to flounder.

Major uncertainties associated with such extrapolations and with the entire risk assessment process should be aggregated and evaluated during risk characterization. If a quotient method is used, only a qualitative estimate of uncertainty may be possible. On the other hand, if mechanistic models were used, quantitative uncertainty estimates may be available. Other major uncertainties associated with risk assessment should be identified as well. Examples include uncertainties associated with a range of extrapolations (from toxicity to surrogate species to toxicity to flounder; from laboratory bioassays with whole sediments to field effects) and with assumptions regarding exposure routes. For example, when determining the risk to flounder feeding in the vicinity of the site, there may be considerable uncertainty as to the proportion of prey consumed that is contaminated from the site, relative to the proportion that is unaffected by the dredged material.

Information bearing on the risk assessment may be obtained from sources other than the analyses used in the risk integration. Such information should be summarized and evaluated as part of a weight-of-evidence discussion (USEPA 1992). Using multiple lines of evidence can be quite valuable, especially when the individual elements are not conclusive (USEPA 1994). In the example of evaluating risks to flounder from consuming contaminated prey, flounder food web models might be evaluated in conjunction with any applicable monitoring data on flounder or prey populations in the vicinity of the site.

Once the risks and uncertainties have been summarized, the ecological importance of the risk should be evaluated. While ecological importance has many facets (Harwell et al. 1994), there are several aspects particularly relevant to dredged material. Consideration of the temporal scale may be important. For example, rate of recolonization of benthic invertebrates at the site should be considered. Even if the risk manager is not concerned with dredged material effects on benthic invertebrates within the site, the rate of recolonization of the newly deposited sediments is one factor that influences how and when flounder feeding in the area may begin to encounter contaminated prey.

Temporal scale is also important if multiple dredging projects are to be placed at a particular site. In this case, the risk assessment should consider changes in exposure resulting from burial and reburial of previously applied sediments (Section 5.3.4.3). Changes in benthic invertebrate populations (as prey for flounder) that result from repeated disturbances should also be considered.

Spatial scale is another aspect of ecological importance for placement of dredged material at aquatic sites. The areal extent of bottom covered by deposits is obviously important in evaluating the severity of the effect on benthic invertebrates. If the concern is for flounder, the foraging area of the fish needs to be considered relative to the area of the site. Spatial scale is also critical in determining the nature and extent of effects on flounder. A severe effect on individual flounder in the immediate vicinity of the site may be of negligible consequence to the flounder population in the area.

When the risk assessment has been completed, the results are discussed with the risk manager. Wiegert and Bartell (1994) summarize a number of factors that can lead to effective presentations of risk assessment findings. Some important areas are highlighted below:

- Present results in terms of the assessment endpoints that were chosen in conjunction with the risk manager in the problem formulation stage. This will help ensure that the risk assessment is recognized as relevant to the needs of the risk manager. If the assessment endpoint is flounder abundance, do not present risk in terms of a surrogate test species.

- Where appropriate, describe a range of risks. If dredged material is placed at a site with little lateral dispersion, the risk to benthic invertebrates associated with that material may be relatively uniform across the site. However, if there is substantial spreading of the sediment and mixing with adjacent sediments, it may be appropriate to provide a range of risk estimates associated with the range of exposure conditions expected on different sub-areas of the site.

- Summarize major assumptions and uncertainties. While it is neither necessary nor desirable to list every possible uncertainty, the risk manager should clearly understand major limitations, data gaps, assumptions, and uncertainties. Examples might include 1) exposure/effects pathways not considered in the assessment, such as direct burial of disposal site flounder or benthic invertebrates by sediments, or 2) extrapolations from toxicity tests with a few invertebrate species to acute and chronic effects on the benthic invertebrate community.

It may also be appropriate to describe potential follow-on activities, such as monitoring, to evaluate the predictions of the risk assessment.

Recovery potential has been essentially unaddressed by either research and development or U.S. regulatory policy. This is a critical gap since it is often assumed that when stressor input stops, the ecosystem will recover in a normal and acceptable pattern and rate. This may or may not be the case (Chapter 11). Only well-designed monitoring programs based on testable hypotheses will be able to assess recovery potential and provide a basis for quantitatively incorporating consideration of recovery into future risk assessments.

5.4 Research needs

A number of topics that are important to the application of risk assessment to navigation dredging could benefit greatly from additional, carefully focused research, development, and refinement. Some of the more urgent or promising of these are identified briefly below:

- Chronic toxicity effects endpoint measurement techniques using dissolved and sediment-sorbed contaminants
- Shortcut toxicity tests, perhaps using concentration techniques, as an alternative for chronic, long-term tests
- Exposure–time relationships in chronic whole-sediment toxicity tests analogous to exposure–dilution relationships in aquatic toxicity tests
- Ecological relevance of bioassay response for *in situ* effects, that is, extrapolation of laboratory test responses to field conditions. This could involve such issues as
  - application of toxicity test results with individual organisms to population-level effects,
  - application of tests under laboratory conditions to effects under "real world" exposures, and
  - adaptation of effects test methods to take into account multiple exposures to different sediments in the field.
- Confounding factors in laboratory toxicity tests, such as effects on test species of ammonia, hydrogen sulfide, pH and salt shifts, *etc.*, in laboratory toxicity tests
- Conceptual framework for shortcut identification of sediments having negligible contaminant risks, perhaps based on lack of effect in appropriate tests
- Better methods and guidance for quantitatively predicting cumulative exposure at the placement site
- Methods for quantitatively predicting recovery potential at the placement site
- Cost-effective procedures for measuring effects and exposure endpoints

5.5 Conclusions and recommendations

- The ERA process is a useful framework for conceptualizing problems, conducting testing and evaluations, and communicating technical results to the dredged material manager and stakeholders.
- A tiered risk assessment process is an efficient approach for assessing dredged material.
- In relation to navigation dredging, assessment endpoints and their relationship to measurement endpoints have historically been poorly developed.
- Actual quantitative exposures at placement sites have been poorly documented.
- Chemical and/or biological measurements by themselves are insufficient to support management decisions about dredged material. Exposure assessment must also be incorporated into a complete risk assessment before management decisions can be properly made.
- Managing exposure at the placement site is the key to managing ecological risks associated with dredged material.

- Dredged material should be evaluated and managed in the context of the placement site ecosystem. Attention to appropriate temporal and spatial scales is essential.
- Site selection, dredged material testing and evaluation, site management, and monitoring should be viewed as a continuum, with feedback leading to refinement and improvement of the methods used for all four components.
- The scientific basis for quantifying the ecological importance of observed or predicted changes in assessment endpoints is poorly developed.
- Monitoring can be useful in the field validation of ERA.

5.6 References

American Association of Port Authorities. 1994. U.S. public port facts. Alexandria VA.

Barnthouse LW, Suter II GW, Bartell SM, Beauchamp JJ, Gardner RH, Linder E, O'Neill RV, Rosen AE. 1986. User's manual for ecological risk assessment. Oak Ridge TN: Environmental Sciences Division, Oak Ridge National Laboratory (ORNL). Publication No. 2679, ORNL-6251.

Cardwell RD, Parkhurst B, Warren-Hicks W, Volosin J. 1993. Aquatic ecological risk assessment and cleanup goals for metals arising from mining operations. In: Application of ecological risk assessment to hazardous waste site assessment. Alexandria VA: Water Environment Federation. p 61–72.

[CWA] Clean Water Act. 33 U.S.C. §1251 *et seq.* (June 30, 1948). Also titled Federal Water Pollution Control Act.

Dillon TM. 1993. Developing chronic sublethal sediment bioassays: a challenge to the scientific community. In: Gorsuch JW, Dwyer FJ, Ingersoll CG, La Point TW, editors. Volume 2, Environmental toxicology and risk assessment. Philadelphia: American Soc for Testing and Materials (ASTM). STP 1216. p 623–639.

Gentile JH, Gentile SM, Hairston Jr NG, Sullivan BK. 1982. The use of life-tables for evaluating the chronic toxicity of pollutants to *Mysidopsis bahia*. *Hydrobiologia* 93:179–187.

Harwell M, Gentile J, Norton B, Cooper W. 1994. Ecological significance. In: Ecological risk assessment issue papers. Washington DC: USEPA Office of Research and Development. EPA/630/R-94/009.

Hummon WD. 1974. Effects of DDT on longevity and reproductive rate in *lepidodermella squammata* (gastrotricha, chaetonotida). *American Midland Naturalist* 92:327–339.

[MPRSA] Marine Protection, Research, and Sanctuaries Act of 1972. 33 U.S.C. 1401 *et seq.* (October 213, 1972).

[NEPA] National Environmental Policy Act of 1969. 42 U.S.C. 4321 *et seq.* (January 1, 1970).

Suedel BC, Boraczek JA, Peddicord RK, Clifford PA, Dillon TM. 1994. Trophic transfer and biomagnification potential of contaminants in aquatic ecosystems. *Rev Env Contam Tox* 136:21–89.

Suter II GW. 1993. Ecological risk assessment. Boca Raton FL: Lewis.

[USEPA and USACE] U.S. Environmental Protection Agency and U.S. Army Corps of Engineers. 1977. Ecological evaluation of proposed discharge of dredged material into ocean waters. Vicksburg MS: Environmental Effects Lab, U.S. Army Engineer Waterways Experiment Station.

[USEPA and USACE] U.S. Environmental Protection Agency and U.S. Army Corps of Engineers. 1991. Evaluation of dredged material proposed for ocean disposal: testing manual. Washington DC: Office of Water (WH-556F). EPA-503/8-91/001.

[USEPA and USACE] U.S. Environmental Protection Agency and U.S. Army Corps of Engineers. 1992. Evaluating environmental effects of dredged material management alternatives: a technical framework. Washington DC. EPA 842-B-92-008.

[USEPA and USACE] U.S. Environmental Protection Agency and U.S. Army Corps of Engineers. 1994. Evaluation of dredged material proposed for discharge in waters of the U.S.: testing manual (draft). Washington DC: Office of Water (4305). EPA-823-B-94-002.

[USEPA] U.S. Environmental Protection Agency. 1992. Framework for ecological risk assessment. Washington DC: Risk Assessment Forum. EPA/630/R-92/001.

[USEPA] U.S. Environmental Protection Agency. 1994. A review of ecological assessment case studies from a risk assessment perspective. Volume II. Washington DC: Risk Assessment Forum. EPA/630/R-94/003.

Wiegert RG, Bartell SM. 1994. Issue paper on risk integration methods. In: Ecological risk assessment issue papers. Chapter 9. Washington DC: Risk Assessment Forum. EPA/630/R-94/009.

Wong CK, Wong PK. 1990. Life table evaluation of the effects of cadmium exposure on the freshwater cladoceran, Moina macrocopa. *Bull Environ Contam Toxicol* 44:135–141.

SESSION 4
CONTAMINATED SITE CLEANUP DECISIONS

Chapter 6

Perspectives on sediment ecological risk assessment for hazardous waste sites

Charles A. Menzie

6.1 Introduction

This chapter provides some perspectives concerning the application of ERA to sediment contamination at hazardous waste sites. It is not a comprehensive review but rather is intended to foster discussion. In some places, reference is made to other chapters where additional detail can be found. Human health risk issues are not addressed, although both direct contact and transfer of chemicals to food items are recognized pathways of exposure to humans from sediments at hazardous waste sites.

Sediment risk analyses for hazardous waste sites can be both retrospective and predictive in nature, and they should focus on specific questions. Examples of common questions that sediment risk analyses at hazardous waste sites serve to address include these:

1) Are chemical levels in sediments toxic to species living or foraging in stream or wetland sediments?

2) Are chemicals present in sediments at levels that may affect the abundance or species composition of benthic invertebrates?

3) Are chemicals in sediments being bioaccumulated and transferred via food webs to fish and wildlife species at levels that may affect the survival or reproductive success of these species?

4) What is the likelihood that chemicals in sediments may be released to the water column as a result of future hydrological events? Could the resulting exposure levels cause unacceptable environmental effects?

5) What is the likelihood that chemicals will be buried and removed as sources of exposure?

6) What are the relative risk reductions associated with alternative remedial options?

7) What are the risks associated with implementing a specific remedial action?

6.2 Historical perspective

Sediment risk assessment at hazardous waste sites has a history rooted in three general areas: 1) the development of chemical and biological tools for assessing sediment conditions, 2) the conduct of a number of high-profile site-specific assessments in particular areas of the country, and 3) the emergence of ERA as an organizing framework.

Over the past few decades, numerous tools have been developed and applied to gain a better understanding of sediment conditions. To a large extent, these tools are designed to determine the status of the sediments either in terms of toxicity, geochemistry, physical conditions, structure of the benthic invertebrate community, habitat suitability, or health of fish populations using the area. Recent tools include methods for measuring bioavailable fractions of organic chemicals and metals. With this expanding tool kit, scientists are able to examine the general status of sediment conditions and examine trends either spatially or temporally. The U.S. National Oceanic Atmospheric Administration's (NOAA's) National Status and Trends Program (NS&T) and the USEPA's Environmental Monitoring and Assessment Program (EMAP) are examples of large-scale efforts that use a battery of tools to assess conditions (NOAA 1991; Reifsteck et al. 1992). While these programs — and many on smaller scales — are useful for examining conditions and formulating questions concerning risks associated with various stressors, the programs themselves are not examples of risk analyses.

A number of high-profile sediment contamination problems in the U.S. and elsewhere led to the development of area or regional approaches for evaluating risks associated with contaminated sediments during the 1970s and 1980s, prior to publication of national HHRA or ERA paradigms or frameworks. Most of the high-profile cases involved retrospective assessments associated with historical releases of contaminants such as PCBs in the Hudson River and New Bedford Harbor, Kepone in the James River, and numerous organic and inorganic contaminants in embayments of Puget Sound. These and other high-profile cases demanded that decisions be made concerning the need for remediation and the nature and extent of remediation. The analyses that were undertaken to inform those decisions brought together many of the tools mentioned above into frameworks that could be used to "get a handle" on problems and determine the need for action. During the 1970s and 1980s, there were also a number of high-profile predictive assessments including ocean dumping and offshore oil and gas exploration (International Ocean Disposal Symposium [IODS] Special Symposium 1983; Duedall et al. 1985; Nocito et al. 1989). These retrospective and predictive programs had elements of risk analyses, although they were not explicitly stated as such.

In the late 1980s and early 1990s, hazardous waste site programs in the U.S., Canada, and elsewhere identified risk assessment as a critical tool for evaluating the need for and efficacy of cleanup options. Most attention was initially given to assessing human health risks and formal guidance was developed in the mid-1980s to support the U.S. Superfund Program. Ecological risk assessments have now become a key component of the remedial investigation and feasibility studies (RI/FS) at Superfund sites (USEPA 1994), state sites, and sites in Canadian provinces (Gaudet et al. 1995). While risk analyses certainly preceded the development of frameworks, guidance, and terminology, the process is now more explicitly defined so that it can be discussed in scientific and policy forums.

One of the major benefits of risk assessment is that the process provides an organizing framework for approaching problems wherein tools and experience can be brought together in an appropriate way to address the questions related to the decisions at hand.

One of the most important aspects of the process is formulating the questions the assessments are intended to address. Questions set the direction for the technical aspects of both human health and ecological risk analyses. An assessment that does not begin with a clear question is like a ship without a rudder. No matter how sophisticated the equipment aboard, the ship will probably never reach a satisfactory destination.

6.3 Guidance

Ecological risk assessment guidance is being developed by the USEPA for Superfund sites, by Environment Canada, and by many states and provinces. Some states — Wisconsin, for example — have developed guidance specific for sediments. For the most part, guidance is based on the USEPA's Framework for Ecological Risk Assessment (USEPA 1992), often using the guidance as a starting point (Menzie-Cura & Associates 1996).

The USEPA's Superfund guidance illustrates the steps that may be involved in assessing risks associated with contaminated sediments at hazardous waste sites (Table 6-1). The approach incorporates the concepts of ERA articulated in the USEPA's framework into a procedure.

As indicated in Table 6-1, a stepwise approach is suggested, beginning with preliminary problem formulation. Guidance regarding phased or tiered approaches can be found in Environment Canada's guidance (Canadian Council of the Ministers of the Environment [CCME] 1995) and in guidelines developed by a number of states. For example, the Massachusetts Department of Environmental Protection (MA DEP) Ecological Risk Assessment Guidance identifies two stages, Stage I: Environmental screening and Stage II: Risk characterization (quantitative process), and has developed specific guidance on how each of these phases would be implemented. Phased or tiered approaches permit the assessment to fit the size, scope, and implication of the decisions. A possible constraint is the schedule imposed upon the investigation process; a set clock for completion of a study may not be compatible with a phased study unless the phases are identified at the onset.

6.4 Sediment risk assessment tools for hazardous waste sites

A broad range of tools has been developed to evaluate physical, chemical, or biological aspects of sediments. As already noted, these tools may be applied for various purposes including the analysis of risks at hazardous waste sites. To make effective use of tools in a sediment risk assessment for a hazardous waste site, the questions and potential decisions should be understood at the beginning of the process. The amount of information needed to answer the questions or reach a decision may vary depending on the scope of the problem and the certainty to which an answer is desired. Thus, risk analyses and tools may be employed in a tiered fashion, leading from simpler (generally conservative) to more sophisticated approaches.

A few of the common biological assessment tools that have been used for risk assessment of contaminated sediments are described in Table 6-2 with reference to their use in risk assessment (see Chapter 18 for additional detail).

Table 6-1   Steps in ecological risk assessment for Superfund sites from draft guidance

Step	Description
1	Preliminary problem formulation, ecological effects evaluation
2	Preliminary exposure estimate and risk calculation
3	Problem formulation: assessment endpoint selection
4	Problem formulation: conceptual model, measurement endpoint selection, study design
5	Site assessment for sampling feasibility
6	Site investigation
7	Risk characterization (integration of exposure and effects)
8	Risk management

Source: USEPA 1994.

In addition to the common biological assessment tools in Table 6-2, several other categories of tools are worth noting:

1) Sediment fate and transport studies (laboratory and field) including deposition, burial, and scouring studies. Information is used to evaluate exposures (Chapter 17).
2) Equilibrium and other partitioning methods developed to determine the bioavailable fractions of chemicals in sediments. Such information is commonly used in exposure assessments (Chapter 18).
3) Models of fate, transport, and bioaccumulation (Chapter 17)

6.5 Psychological "dilemmas"

There are a number of psychological dilemmas that risk assessors encounter during the scoping, analysis, and characterization phases of sediment risk assessments. In some cases, these can limit the information content of the assessment or yield results in a form that does not capture the information adequately. For discussion purposes, I have identified these dilemmas as 1) safety in numbers, 2) the testing hot potato, 3) the comfort zone, and 4) so what does it mean?

6.5.1 Safety in numbers

Numerical criteria or guidelines — derived either on a generic or site-specific basis — appear to be highly desirable to risk managers seeking a bright line against which to compare a chemical measurement. Research on risk perception and communication has shown that people relate most easily to the question, "Is it safe?," if clear benchmarks are provided. This desire led to the development of ambient WQC and is the basis for ongoing efforts to develop SQC. These are described further in Chapter 18.

Table 6-2 Examples of biological assessment tools

Assessment endpoint	Measurement endpoints	Strengths and limitations
Toxicity of sediments		
• Is sediment toxicity associated with contaminants affecting abundance and composition of benthic invertebrates?[1]	Compare sediment chemistry values to sediment quality criteria or other benchmarks; usually conducted as part of preliminary assessment	Easy to do; does not require more expensive testing; not site-specific and usually overestimates toxicity (*Recommended for all sites*)
• Is sediment toxicity associated with contaminants affecting abundance, composition, or health of benthic invertebrates, fish, or other biota?	Conduct sediment toxicity tests with one or more species of invertebrates or fish; regulatory agencies often want these data if chemicals are present; tests include acute toxicity as well as sublethal effects; tests available for evaluating reproductive effects on some species	Becoming widely accepted by regulatory agencies; provides direct measure of toxicity to a standard test species; test may not reflect *in situ* conditions because of sediment handling and type of organism used
• Is sediment toxicity associated with contaminants affecting abundance, composition, or health of benthic invertebrates, fish, or other biota?	Conduct benthic biological survey; agencies often want these data as part of their evaluations Conduct field assessments for fish, including use of habitats, and environmental epidemiological assessments related to tumors or reproductive impairment	Accepted by regulatory agencies for many years; provides direct information on organisms naturally present in sediments or utilizing sediment as habitat; other environmental factors may complicate data interpretation
Bioaccumulation in food webs[2]		
• Do sediment contaminants bio-accumulate in food webs at levels that could affect fish or wildlife survival?	Calculate bioaccumulation using published BAF or BSAF values; often-used approach; used in exposure assessment	Easy to do; requires limited site-specific data; subject to great uncertainty
• Do sediment contaminants bio-accumulate in food webs at levels that could affect fish or wildlife survival?	Place organisms in the field for specified time periods and measure tissue levels; being done at a number of sites; used in exposure assessment	Has some acceptance (*e.g.*, use of mussels, invertebrate plate samplers, and fish); can provide comparative spatial picture of accumulation; may not reflect actual levels of accumulation in the field
• Do sediment contaminants bio-accumulate in food webs at levels that could affect fish or wildlife survival?	Collect indigenous fauna, measure body burdens of chemicals; best method to evaluate accumulation in local fish/shellfish; used in exposure assessment	Most widely accepted method; provides direct measurement of bioaccumulation; in some cases, difficult to relate body burdens to particular exposure regimes

[1] Invertebrate and other biological groups identified would be ones that are considered to be either ecologically, recreationally, or commercially important.
[2] Usually considered when compounds have potential for bioaccumulation.

While sediment criteria or guidelines (*i.e.*, numbers) are desirable to risk managers and can be useful for screening sites (especially where conditions are simple), they should be used with caution in sediment risk assessment. The criteria or guideline may not relate to the question that is being asked in the risk assessment, and thus there may be a mismatch between the question and assessment tool (the criterion). Because experience has shown that sediment contamination problems can be quite complex and variable from one place to the next, criteria and guidelines can be overprotective or underprotective depending on circumstances. Equilibrium partitioning methods apply to individual chemicals in simple systems and may not be appropriate for evaluating complex mixtures, especially where aging has occurred. The challenge for the risk assessor and risk manager is in developing an understanding of the strengths and limitations of chemical-specific numbers within the context of the risk assessment and risk management decisions. Further, where numerical values fall short, the risk assessor and risk manager need to identify appropriate risk communication tools for use in management decisions (Chapter 18).

6.5.2 The testing hot potato

This might best be described as "fear of the unknown." This fear is usually expressed by a potentially responsible party (PRP) or their attorneys during planning stages of an assessment. While chemical measurements in media are familiar to PRPs, ecological testing — especially toxicity testing and tissue analyses — can be unfamiliar territory. As a result, there are concerns about what such testing will yield as well as the relevance of the testing for the decision. To some degree, these fears are well founded in that there does not exist a clear framework for incorporating such information into assessment of risks, and our understanding about how to relate measurements to assessment endpoints is still evolving. A result of these concerns may be debate about the scopes of the assessments when these are cast simply as measurement programs.

6.5.3 The comfort zone

Many scientists and managers who participate in risk assessments bring with them a knowledge and experience base that forms their personal "comfort zone" regarding what is important and what works when assessing risks associated with contaminated sediments. As a result, there is often a desire to apply familiar tools or approaches from past experience to new situations. These may or may not be the best tools or approaches to use. Because people do not share common values or a common knowledge or experience base, conflicts often arise concerning what is important and how the assessment should be conducted. These can be best handled in the problem formulation and planning stages of the project, during which participants — especially the risk assessors and risk managers — should be open minded.

6.5.4 So what does it mean?

Communication between the stakeholders, risk assessor, and risk manager are vitally important during the planning and implementation of the assessment in order to avoid a huge SO WHAT? question at the end of the analyses. A common misconception is that the ecologist or environmental scientist (risk assessor) is the one who knows what is important to address in an ERA and should, therefore, formulate the questions. Where this occurs, risk assessors assume the roles of risk assessor, risk manager, and stakeholder, bringing to the problem formulation their own personal set of values. "What is important?" is not solely an ecological question. It is also a question of "What is important to people?" If the assessment is disconnected from what stakeholders and risk managers consider important or if the importance of the assessment cannot be established in a meaningful way, then the analysis is likely to fail in providing information useful for decision-making.

6.6 Recommendations for sediment risk assessment at hazardous waste sites

There are two recommendations for future risk assessment of sediments that I consider especially important: 1) develop guidance for the problem formulation phase of sediment risk analyses and 2) utilize a weight-of-evidence approach within a tiered or phased strategy.

6.6.1 Guidance for problem formulation

This is the most critical stage of the risk assessment. Based on a review of many case studies, it is clear that the strengths and limitations of ERAs stem from how well the problem was formulated from the onset. The most important thing to occur in problem formulation is defining the questions the assessment is intended to address. This process involves communication among the stakeholders, risk manager, and risk assessor. These form the assessment endpoints to be evaluated with regard to sediments. At present, there is very little guidance on this process but its importance can be underscored by the adage, "It is more important to do the right thing, than to do the thing right." The balance of the assessment — how to do it right — begins with asking the right questions.

6.6.2 Utilize weight-of-evidence approach within tiered or phased strategy

Most risk analyses for sediments at hazardous waste sites will involve a *weight-of-evidence approach*. However, there is no consensus on exactly what this means or how it should be carried out. Published definitions include the following:

> Each risk estimate will have its own assumptions and associated uncertainties and these may not be expressed equivalently. The separate lines of evidence must be evaluated, organized in some coherent fashion, and explained to the risk manager so that a weight-of-evidence evaluation can be made (Suter 1993).

> Risk description has two primary elements. The first is the ecological risk summary, which summarizes the results of the risk estimation and uncertainty analysis and assesses confidence in the risk estimates through a discussion of the weight of evidence (USEPA 1992).

> For many Superfund ecological risk assessments, a weight-of-evidence approach will be used. This frequently will require that different types of data are evalu-

ated together. These types of data may include toxicity test results, assessments of existing impacts on-site, or true risk calculations comparing estimated exposure doses with toxicity values from the literature Balancing and interpreting the different types of data can be a major task...the strength of evidence provided by different types of tests and the precedence that one type of study has over another should already have been determined.... This will insure that data interpretation is objective and not designed (*i.e.*, biased) to support a preconceived answer (USEPA 1994).

A multi-agency/consultant workgroup established in Massachusetts to examine the question of how to conduct a weight-of-evidence approach defined the process as follows:

The weight-of-evidence approach is the process by which measurement endpoint(s) are related to an assessment endpoint to evaluate if there is a significant risk of harm to the environment. The approach is planned and initiated at the Problem Formulation Stage and results are integrated at the Risk Characterization Stage (MA DEP Workgroup 1995; Menzie *et al.* 1996).

This definition provides an explicit link between risk characterization and the questions developed — assessment endpoints — during the problem formulation phase. The Massachusetts workgroup has developed a qualitative and quantitative procedure based on the following considerations:

1) Degree of confidence or weight placed in each measurement endpoint
2) Response of the measurement endpoint
3) Extent of agreement or divergence among measurement endpoints judged with respect to the initial weight or confidence

The Massachusetts workgroup has applied the approach to case studies involving sediment contamination and found that it works well as a planning and communication tool. It has the advantage of making transparent the risk assessors' understanding of the strengths and limitations of specific lines of evidence as they relate to a particular assessment endpoint. The workgroup has reached general agreement on the attributes to consider in developing a level of confidence or weight associated with each measurement endpoint (Table 6-3). In the weight-of-evidence procedure, ten attributes of each measurement endpoint are evaluated. For a given assessment endpoint, the quality of each measurement endpoint is compared with respect to these attributes; those measurement endpoints with the highest quality for the most attributes are given the greatest weight in the overall characterization of risk.

Table 6-3   Attributes for judging measurement endpoints

I. Attributes related to strength of association between assessment and measurement endpoints	II. Attributes related to data quality	III. Attributes related to study design and execution in relation to the assessment endpoint
• Biological linkage between measurement endpoint and assessment endpoint	• Extent to which data quality objectives are met	• Site specificity
• Utility of measure for judging environmental harm		• Sensitivity of the measurement endpoint for detecting changes
• Correlation of stressor to response		• Spatial representativeness
		• Temporal representativeness
		• Quantitativeness
		• Use of a standard method

Source: MA DEP 1995.

6.7 References

[CCME] Canadian Council of the Ministers of the Environment. 1995. A framework for ecological risk assessment: general guidance. CCME Subcommittee on Environmental Quality Criteria for Contaminated Sites.

Duedall IW, Kester DR, Park PK, Ketchum BH. 1985. Wastes in the ocean. Volume 4, Energy wastes in the ocean. New York: J Wiley. 818 p.

Gaudet CL, Milne DA, Nason TGE, Wong MP. 1995. A framework for ecological risk assessment at contaminated sites in Canada, Part II: Recommendations. *Human Ecol Risk Assess* 1:207–230.

[IODS] International Ocean Disposal Symposium. 1983. Ocean waste management: policy and strategies. An International Ocean Disposal Symposium Series Special Symposium; 1983 May 2–6; Whispering Pines Conference Center, The University of Rhode Island.

[MA DEP] Massachusetts Department of Environmental Protection Agency Workgroup on Weight-of-Evidence for Ecological Risk Assessment. 1995. A weight-of-evidence approach for ecological risk assessment. Boston: MA DEP.

Menzie CA, Henning MH, Cura JJ, Finkelstein K, Gentile J, Maughan J, Mitchell D, Petron S, Potocki BM, Svirsky S, Tyler P. 1996. Special report of the Massachusetts Weight-of-Evidence Workgroup: a weight-of evidence approach for evaluating ecological risks. *Human Ecol Risk Assess* 2:277–304.

Menzie-Cura & Associates Inc. 1996. An assessment of the risk assessment paradigm for ecological risk assessment. In: Risk assessment and risk management in regulatory decision-making. Washington DC: Commission on Risk Assessment and Risk Management.

[NOAA] U.S. National Oceanic and Atmospheric Administration. 1991. National Status and Trends Program for marine environmental quality. Rockville MD: NOAA.

Nocito JA, Walker HA, Paul JF, Menzie CA. 1989. Application of a risk assessment framework for marine disposal of sewage sludge at midshelf and offshelf sites. *Aquat Toxicol* 11:101–120.

Reifsteck DM, Stobel CJ, Schimmel SC. 1992. Environmental monitoring and assessment program near coastal component, 1992 Virginia Province effort, field operations and safety manual. Narragansett RI: Environmental Research Laboratory, Office of Research and Development, USEPA.

Suter GW. 1993. Ecological risk assessment. Boca Raton FL: Lewis. 538 p.

[USEPA] U.S. Environmental Protection Agency. 1992. Framework for ecological risk assessment. Washington DC: Risk Assessment Forum. EPA/630/R 92/001.

[USEPA] U.S. Environmental Protection Agency. 1994. Ecological risk assessment guidance for Superfund: process for designing and conducting ecological risk assessments. Review draft. Edison NJ: USEPA.

SESSION 4
CONTAMINATED SITE CLEANUP DECISIONS

_____Chapter 7

Workgroup summary report on contaminated site cleanup decisions

Peter M. Chapman, Manuel Cano, Alyce T. Fritz, Connie Gaudet,
Charles A. Menzie, Mark Sprenger, William A. Stubblefield

7.1 Introduction

Risk assessments are an integral part of the RI/FS conducted as part of any waste-contaminated site cleanup effort. Ecological risk assessments provide a basis for evaluating existing risks (*i.e.*, baseline conditions) and provide a framework for evaluating management alternatives, including the potential effects associated with remedial alternatives under consideration for the site. As described in Chapter 1, several documents have previously outlined and offered guidance for conducting ERAs (*e.g.*, USEPA 1992). Additional forthcoming guidance includes the draft USEPA document, Ecological Risk Assessment Guidance for Superfund: Process for Designing and Conducting Ecological Risk Assessments (USEPA 1994a) and the Canadian approach (Gaudet *et al.* 1995). However, existing or forthcoming guidance documents do not specifically address the issue of assessing the risks associated with sediments at contaminated sites.

The four basic steps involved in conducting any ERA are applicable to contaminated site cleanups: 1) problem formulation and site characterization, 2) exposure assessment, 3) effects assessment, and 4) risk characterization. Each risk assessment varies as to the level of detail required in the above steps. The need for detailed information is determined by regulatory site status, the goals of the remedial actions, the complexity of the site, the potential magnitude of remedial options, and the difficulty in adequately describing exposure, toxicity, and the properties of potential concern. In all cases, SERA should make optimal use of all available data. Ideally, additional empirical data should be collected only to fill information needs, in particular substituting site-specific information for assumptions, and to reduce uncertainties in critical factors that strongly influence risk estimates.

An SERA should be conducted in a tiered or phased approach (USEPA 1992). Sediment ecological risk assessments may be conducted as part of an overall site risk assessment and may need to be integrated with the overall ERA. An initial analysis (Tier 1) is conducted, using very conservative assumptions about exposure and toxicity, then a decision is made whether and how to proceed. Typically, this initial screening-level risk assessment is conducted using available data and conservative assumptions about exposure and toxicity or using existing effects-based benchmark values or guidelines (*e.g.*, Smith *et al.* in press). On the basis of this screening-level assessment, conclusions may be reached suggesting that ecological risks which require management or reduction do not exist or are insignificant. Alternatively, when the screening-level SERA indicates that there is a potential risk, the subsequent risk assessment may be used to prioritize areas, chemicals, and species at the site and to decide what kinds of additional data are necessary to develop a more definitive risk estimate (*e.g.*, Tiers 1 and 2). In later tiers, additional data may be collected that will better define the exposure-response relationship at the site. At each tier, a pass/fail decision is made whether to proceed to advanced tiers and, if so, how best to proceed. The criteria for tier advancement or for reaching a conclusion of no significant risk should be specified to the extent possible during risk assessment planning. Data needs for subsequent tiers are refined after the results of preceding tiers are analyzed.

A number of issues differentiate SERAs conducted for waste-contaminated sites from those conducted for other reasons and detailed in other chapters (*e.g.*, evaluation of new products in Chapters 6 and 7, or evaluation of sediments for dredge spoils relocation in Chapter 5). These issues include the following:

- Assessments for waste-contaminated sites are typically retrospective, being conducted after wastes have been deposited and often after ecological effects have occurred. However, the evaluation of cleanup or remediation alternatives requires predictive or prospective evaluation techniques.

- Assessments for waste-contaminated sites are interactive in nature, requiring ongoing interaction and communication between the risk assessor and the risk manager. This requires that the risk manager be an integral part of the assessment process from the outset.

- Because assessments at waste-contaminated sites require interaction with a variety of disciplines and groups other than environmental toxicologists and chemists, including but not restricted to remediation engineers, lawyers, hydrogeologists, and the public, it is imperative that the approach used and interpretation of results and conclusions be easily understandable and thoroughly presented.

This chapter provides guidance on conducting SERAs for waste-contaminated sites, including refinement or modification of generic ERA procedures (Chapter 1). This guidance is not intended to be inflexible; to the contrary, flexibility is encouraged given the site- and situation-specificity of contaminated sites. However, we caution investigators to remember that sediment evaluations (or other investigations) typically focus on only a snapshot in time, and both temporal and spatial shifts must be considered as they may affect redistribution of contaminants or result in modifications to biological communities which may or may not be related to contaminants in sediments.

Further, note that the current state-of-the-art for evaluating sediment effects is in its infancy. This is particularly true for mechanistic understanding of sediment toxicity. Thus, presently, the only possible approach is primarily empirical. With the development of the science and of our knowledge of mechanistic relationships, this will no doubt change, but an empirical approach will probably always play a major role in SERA.

7.2 Problem formulation

7.2.1 Initial problem/issue definition

Background information and preliminary sampling data are used to identify the problem and define the issues that need to be addressed at a contaminated site. At this stage of the sediment assessment, a number of aspects should be considered in formulating the problem and developing the conceptual model for the site. These include source definition, consideration of the potential effects of the identified contaminants, and initial identification and characterization of potential receptors.

7.2.1.1 Source definition

Source definition at contaminated sites is important because

- from a legal perspective, it establishes linkages to potential contributors (*i.e.*, to the problem) who might participate in the investigation or remediation of the site; and

- from a scientific perspective, it identifies the form of discharge and hence the fate and transport processes linking land-based contamination to aquatic systems, including sediments at one location with other areas where exposure could occur. It is therefore the first step in the conceptual model. Note that areal and temporal dimensions are both important.

In addition, source identification may help to focus the analyses for contaminants or limit the number of contaminants of concern if historical information about the site is available.

The identification of sources of sediment contamination generally involves an examination of the operational histories of private or public facilities that may have contributed directly to contaminated sediments or to land-based contamination that has or might migrate to the sediments. The process can become complicated by the presence of many different historical or current sources. In such cases, the site investigators and managers must decide how to define sources. One option is to define sources in terms of a sediment contamination problem within the aquatic system. In such cases, a number of sources may be identified. Another option is to differentiate a particular source from ubiquitous or upstream contamination associated with other point and nonpoint sources (*i.e.*, background).

An additional set of considerations involves fate and transport processes that may link land-based contamination or in-place sediments with other areas where exposure may occur. The following processes are important to consider at this stage:

- Surface erosion and bulk transport of contaminants present in surface soils at a site

- Seep discharge of non-aqueous phase liquids (NAPL) present at a site or along the bank of a water body which may be denser than water (D-NAPL) or lighter than water (L-NAPL), which could enter the aquatic environment by migrating along geological features or on ground water and contaminate sediments

- Contaminated groundwater from a site which could discharge to an adjacent water body and result in exposure to benthic organisms or result in contamination of sediments
- Resuspension of contaminated sediments at or below the sediment-water interface (sometimes by infrequent storm or flood events), such that they become exposed to the surface or are transported to other areas

Each of these considerations can affect how the problem and sources are defined at a contaminated site and how the risk assessment is implemented. Each may lead to different remedial strategies.

7.2.1.2 Potential effects of contaminants

Contaminants vary in their environmental behavior and effects. These characteristics will affect how an SERA is structured. At the problem formulation stage, it is important to consider characteristics of individual chemicals or classes of chemicals such as persistence, hydrophobicity, potential for food chain transfer, and toxic effects on receptors. Most chemicals that tend to sorb to sediments can be placed into one of the general categories shown in Table 7-1.

7.2.1.3 Identification and characterization of potential receptors

Identification of potential receptors depends, in part, on the habitat and contaminant. An SERA should consider all receptors potentially exposed to sediment contaminants either directly or indirectly. For sediment assessments, the benthic invertebrate community is often identified as a receptor along with demersal fish species. If chemicals (*e.g.*, chlorinated pesticides, PCBs, certain mercury compounds) have a potential for transfer from sediments via food webs, then pelagic fish species (*e.g.*, lake trout) and other wildlife (diving birds and fish-eating mammals) may be considered in the risk assessment. Typically, wading birds and mammals that feed on benthic invertebrates or may be exposed directly to sediments are considered for shallow water sediments. Plants may also be considered as potential receptors.

The process of characterizing receptors usually involves a combination of reconnaissance survey work (site visits) together with natural history information. Depending on the needs of the risk assessment, additional data on receptors may be gathered at later stages of the evaluation.

7.2.1.4 Summary

The product of the initial problem/issue definition should be a clear statement of the nature of the sources at the site, potential issues concerning migration, potential contaminants of concern at the site, potential effects associated with these chemicals, and receptors that should be considered. This information may be focused further as the assessment proceeds. Alternatively, additional sources, contaminants, or receptors may be added as additional information is gathered. However, the primary objective should be to gather and organize information in a manner that can lead to the development of

Table 7-1   General categories of chemicals sorbing to sediments (matrix would be filled in for specific studies)

Effects	Bioaccumulation		
	Limited bioaccumulation in invertebrates	Bioaccumulates in invertebrates but limited food chain transfer	Bioaccumulates and can be transferred via food chains
Lethal and sublethal effects upon direct contact			[Relatively most serious]
Lethal and sublethal effects when bioaccumulated			
Sublethal effects when bioaccumulated			

clearly stated assessment endpoints. All parties involved should be in agreement with the clearly stated problem definition.

7.2.2 Conceptual model development

Development of the conceptual model as part of the problem formulation process is an iterative process. Initially, one conceptualizes how the contaminants may have moved both physically and biologically, their exposure pathways, receptors that might be exposed, and mechanisms for toxic effects in order to determine the primary concerns or priorities for what needs to be protected (*i.e.*, assessment endpoints). As one progresses through problem formulation, the conceptual model is refined to establish potential exposure pathways and receptors, appropriate measurement endpoints, and a study design to meet data needs. The finalized conceptual model serves as input to the analysis phase of the assessment.

The outcome of the problem formulation process is a conceptual model describing how a stressor might affect ecological components at the site. The conceptual model identifies the exposure pathways and specific adverse effects (ecological effects) which will be evaluated in the SERA. The model traces contaminant physical-chemical fate and transport, including food chain transmission, and defines potential adverse effects. It allows the risk assessor to evaluate the exposure pathway to potential receptors, particularly related to the assessment endpoints, to ensure that the exposure pathway is complete. Further, it describes the relationship among assessment and measurement endpoints and is used to confirm that the selected measurement endpoints are in the same exposure pathway as the assessment endpoints.

The conceptual model's underlying principle is that it is representative of the critical exposure pathways and the ecotoxicity threats posed to specific trophic levels or other ecosystem components. Identification of these exposure pathways is analogous to food chain energy transfer models. For particular species some food sources are of greater value and some food items are particularly high in energy value. These same concepts are true for contaminant transfers in ecosystems. Sources are identified, physical-chemical transport mechanisms are identified, and exposures to receptors are evaluated based upon the rate of utilization (duration of exposure), modes of toxicity, and contaminant transfer efficiency (bioavailability).

Thus, the site conceptual model "represents" the exposure pathways and receptors (assessment endpoints) which are to be evaluated in the SERA. At any given site, the appropriate assessment endpoints may involve individual endangered species, local populations of particular species, community-level integrity, or habitat preservation. The site conceptual model must encompass the level of biological organization appropriate for the assessment endpoints for the site. It may or may not be completely realistic relative to the measurement endpoints; strict "reality" in the model relative to the measurement endpoints can undermine the utility of the model for evaluation of risk to the assessment endpoint.

Development of a conceptual model also identifies data requirements, methodologies needed to analyze the data (study design), and points or assumptions including the greatest degree of uncertainty. By identifying these conservative assumptions, effort can be focused in the study design to address information gaps or sources of uncertainty, thereby utilizing site-specific information to minimize over- or underestimation of the actual ecological risks. Typically, this translates into direct field evaluation of contaminant concentrations at exposure points and of bioavailability or toxicity of contaminants in sediments.

7.2.3 Selecting and defining assessment endpoints

Assessment endpoints for sediments at contaminated sites establish the overall direction and focus of the risk assessment. Assessment endpoints are established through discussions among the risk assessor, risk manager, and others (*e.g.*, the public) who may be interested in, or affected by, the outcome of the decisions. In some cases — where the problems are simple or well understood — it may be possible to proceed based on past experience without extensive discussions.

Assessment endpoints reflect what human beings are concerned with or care about, expressed in a manner that can be evaluated through an objective scientific process. Such endpoints generally include but are certainly not restricted to ecological concerns. Assessment endpoints are most useful when expressed in terms of a specific receptor (species, habitat, system) and a specific function or quality that is to be maintained or protected. An overly broad assessment endpoint (*e.g.*, "health of the environment") can be difficult to evaluate. In such cases, the assessor may find it helpful to disaggregate the broad assessment endpoint into multiple assessment endpoints that consider more specific characteristics of the environment or species under consideration. In general, this process involves professional judgment and an appreciation of what is needed to communicate the analysis effectively to the risk manager and to other interested parties.

At contaminated sites, more than one assessment endpoint is typically considered. The exact number depends on the chemicals and receptors present.

7.2.3.1 Selecting assessment endpoints for sediments

The selection of assessment endpoints is a critical communication step among the risk assessor, risk manager, and others interested in the outcome of the assessment (*e.g.*, stakeholders). The objective is to reach a consensus on what the assessment endpoints will be. The site manager, with assistance from the risk assessor/scientist, is typically responsible for resolving issues that may arise among various involved parties.

The process of selecting assessment endpoints (once the endpoints have been defined) usually starts with the risk assessor/scientist and comprises the initial site conceptual model. Based on a general knowledge of the chemicals, receptors, and site, the assessor/scientist usually develops an initial set of assessment endpoints for consideration by the risk manager and other interested parties. For simple or well-understood sites, this typically involves a discussion between the risk assessor and site manager.

This initial set of endpoints forms the basis for discussions that may result in additions, deletions, or modifications to the list of assessment endpoints. The process is iterative. The risk assessor/scientist often plays a key role in informing other parties about the ecological or human health "relevance" of various assessment endpoints, and helps focus the assessment. Other assessment endpoints or modifications to endpoints which may be suggested by the site manager or others need to be discussed and a decision made by the site manager (not risk assessor) regarding inclusion, exclusion, or modification.

7.2.3.2 Defining assessment endpoints

Assessment endpoints should be defined in terms of
- a receptor (species, community, other level of organization) and
- a characteristic or function (*e.g.*, survival, maintenance, reproduction) to be protected.

With regard to contaminated sediments, there are several categories of assessment endpoints that might be considered depending on the receptor and chemicals. Examples of assessment endpoints for sites with contaminated sediments are given in Table 7-2.

7.2.4 Defining and selecting measurement endpoints

Measurement endpoints are components of assessment endpoints which are nonsubjective and quantal (*i.e.*, measurable). There can be, and often are, multiple measurement endpoint options associated with an assessment endpoint. Measurement endpoints are linked to assessment endpoints by the mechanism of toxicity and by exposure (*e.g.*, reproductive success and tissue concentrations of PCBs).

Table 7-2 Example assessment endpoints for sediment ecological risk assessment

Receptors	Assessment endpoints
Benthic invertebrates	• community structure and function as a reflection of an aquatic environment typical for the water body • survival and maintenance of selected (keystone) species • support of fish and wildlife species
Fish	• community structure and function as a reflection of an aquatic environment typical for the water body • maintenance of recreational or commercial fisheries • survival • reproductive success

Birds
  Feeding on invertebrates
  Feeding on fish
	• maintenance
	• survival
	• reproductive success

Mammals
  Feeding on invertebrates
  Feeding on fish
	• maintenance
	• survival
	• reproductive success

Humans (usually a separate assessment)
  Ingesting contaminated fish or shellfish
  Contacting contaminated sediments
	• carcinogenic risks
	• non-carcinogenic risks
	• exceedance of public health values (loss of recreational or commercial fishery)

Vegetation
  Wetlands
  Submerged aquatic vegetation
	• stressed or lost vegetation
	• survival and maintenance of selected (keystone) species
	• alteration of community structure and function
	• support of fish and wildlife species

Threatened or endangered species	• survival of individuals • reproductive success of individuals

7.2.4.1 Defining measurement endpoints

A *measurement endpoint* is a measurable ecological characteristic that is related to the assessment endpoint. Because measurement endpoints are the bases for structuring the analysis phase of the sediment risk assessment, and because they will ultimately be used to estimate risk, they should have certain features:

- They should be related explicitly, either directly or indirectly, to specific assessment endpoints.
- They should include metrics that can be used for estimating risks; these metrics incorporate both test or study "endpoints" (*e.g.*, toxicity results, tissue levels, and community structure) with nonsubjective scaler functions or values that will be used to judge the response at the risk characterization stage (*i.e.*, the response variable).

One or more of the following scales might be included as metrics for the measurement endpoint:

- A change or difference in the response variable that is considered potentially ecologically relevant (*e.g.*, degree of mortality in a test, or change in abundance or biomass)
- Spatial scale of the change or difference as related to the assessment endpoint (*e.g.*, hectares, fraction of foraging area, fraction of area utilized by a local population)
- Temporal scale of the change or difference as related to the assessment endpoint (duration; changes with time, with and without major events such as storms or floods; rate of recovery)

The measurement endpoint might also include some statement of statistical confidence, although this is not always possible.

A hypothetical example of the relationships among an assessment endpoint, measurement endpoints, and measurement tools/results is given in Table 7-3. The statements and values used in the assessment and measurement endpoint descriptions are used only for illustration and do not reflect any scientific or managerial consensus. In a real-world situation, careful consideration would be given to how the measurement endpoints should be developed and stated.

7.2.4.2 Selecting measurement endpoints

In selecting measurement endpoints for evaluating contaminated sediments at sites, a number of issues should be addressed, including the following:

1) The relationship between the assessment and measurement endpoint
   a) the degree of association between measurement and assessment endpoints can vary from direct to indirect, more confidence is given to measurement endpoints that are more closely related to or provide an indication of response tied to the assessment endpoint (for instance, exposure mechanism and mechanism of toxicity)
   b) the degree to which the measurement endpoint can be linked to the chemical stressors of concern in a stress-response relationship
   c) the availability of objective criteria for judging or evaluating the ecological relevance of changes in the measurement endpoint
2) The quality of the data that can be developed for the measurement endpoint
   a) the need to minimize both Type I and Type II errors in the selected measurement endpoint
3) Study design considerations for measurement endpoint selection
   a) the extent to which the measurement endpoint (the test as well as the scaler components) relies upon site-specific information

Table 7-3   Example assessment and measurement endpoints

Assessment endpoint	Measurement endpoints[1]	Measurement tools and test endpoints (results)
Maintenance of keystone benthic amphipod species important in diet of local groundfish	• Abundance and biomass of the species as compared to that at appropriate reference areas; a "reduction" or difference of __% that appears related to presence of contaminants is considered ecologically relevant at spatial scales of ____ to ____ hectares for a period of ____ years; desired level of statistical confidence is 0.10	• Benthic community studies at the site and in one or more reference areas; results include species identifications, enumeration, and biomass
	• Mortality of amphipods in a 10-d whole-sediment test; mortality >25% related to presence of contaminants is considered potentially ecologically relevant at spatial scales of ____ hectares	• 10-d toxicity test with amphipod; results expressed as % mortality

[1] The specificity of this example is intended to be illustrative, not constrictive.

b) the sensitivity of the measurement endpoint for detecting changes or effects of interest; involves a consideration of test variability and natural variability

c) the spatial representativeness of the measurement endpoint relative to the assessment endpoint

d) the temporal representativeness of the measurement endpoint relative to the assessment endpoint

e) the ability of the measurement endpoint to provide quantitative information (measurement endpoints can be qualitative)

f) the extent to which the measurement endpoint is based on standard methods (Measurement endpoints can include procedures that are in the developmental or research arena but there may be less confidence in such methods.)

g) the need for multiple measurement endpoints

4) Other factors

a) site schedule/phasing: Overall schedules for the site investigation and feasibility study may dictate the types of measurements that can be made and when they can be made; external schedule constraints must be considered when planning the SERA and selecting measurement endpoints for which site data are desired; in some cases, there may not be an opportunity to conduct the assessment in phases, and the risk assessor may need to select more measurement endpoints up front than might be the case if the assessment were phased.

b) resources: Measurement endpoints vary in implementation costs; assessments are often resource constrained, and in such cases, measurement endpoints must be selected carefully to insure the most useful information for the available budgets; measurement endpoints should also be appropriate for the nature and size of the problem.

c) imminent hazard evaluations: Relative to severity of impact or potential impact, at some sites there may be concern regarding imminent hazards; in such cases, some measurement endpoints may be selected to determine whether such a hazard exists; these endpoints typically permit rapid collection and evaluation of data.

d) physical limitations: Hard-bottom streams, *e.g.*, obviate sediment toxicity tests.

7.2.5 Study design

The study design incorporates the selected measurement endpoint tools and the site conceptual model within the data rigor needed for extrapolation from the measurement endpoints to the assessment endpoints. As such, it is basically the plan for the collection of field samples which serves to generate the measurement endpoints. In addition, study design incorporates elements of the analysis phase, in particular the risk characterization.

The number of sample locations and number of replicate samples which must be taken to answer the questions developed through the problem formulation phase are specified in the study design. This inherently requires an evaluation of statistical and logistical issues, for example, Can replicate samples really be obtained? or How many observations will be require to meet the data requirements? Consultation with an experienced biostatistician is highly recommended, particularly to determine the necessary statistical power to resolve differences between sampling locations or exposure levels. Note that one of the major benefits of involving a biostatistician in the early stages of study design is that certain measurements may be shown to be worthless due to the excessively high number of samples required to show a difference, while others may be shown to be highly effective and worthy of substantive additional effort (Chapter 18).

If, for instance, it is determined that the number of samples required to answer a specific question is not obtainable, either due to physical sampling limitations or resource limitations, then the measurement endpoint and the measurement endpoint tools must be reevaluated. Collection of samples which are insufficient to answer the questions arising from problem formulation is wasted effort, since the data will not be interpretable in the manner required to reach a management decision. If sample size alterations occur due to field conditions, the risk manager must be consulted since failure to collect sufficient data may translate into an inability to obtain a measurement endpoint and hence can affect the evaluation of the assessment endpoint.

Either the sampling locations or the mechanism by which sampling locations will be identified is specified in the study design. For example, if a study requires samples to be collected along a concentration gradient to develop an exposure-response curve and detailed contaminant distribution data exist, specific locations could be targets for sample collection. Alternatively, if little information exists on contaminant distribution but there is a field screening technique for the contaminant of interest or a co-distributed material, the study design could specify the concentrations required such that the sampling locations can then be determined in the field.

It may be advantageous to develop an exposure-response relationship in association with the measurement endpoint tools. While it is frequently difficult at best to determine the actual dose received by organisms, it may be possible to estimate exposure levels (exposure point concentrations) which can then be related to measurement tool observations. For example, if the measurement tool being employed is the tissue concentration of resident species, then representative media, sediment, or water samples could be collected within the home range of the organisms sampled for tissue contaminant levels. This can be repeated at several media concentration levels. From these data, an exposure point (media concentration) to tissue concentration relationship could be drawn.

Sediment sampling requires an evaluation of the depth of sediment to be collected, which may vary based upon the assessment endpoints and thereby the measurement endpoints. For example, if the measurement endpoint is the laboratory toxicity associated with the aerobic sediment at the sediment-water interface, then only the uppermost layer of the sediment should be collected. Alternately, if the measurement endpoint is the laboratory toxicity of the biologically active layer of the sediment, then the upper 10 cm may be the appropriate sampling depth. Knowledge of the contaminant distribution in the sediment should always be incorporated into the decision of the appropriate sampling depth. For example, if information indicates that the contaminated portion of the sediments is below 5 cm, then inclusion of the upper 5 cm in the sample could dilute the sample, potentially resulting in erroneous conclusions.

Synoptic sampling is required to provide a temporal link between all data collected. Synoptic sampling can also provide for cost savings, in that data can be used for multiple purposes. For most media (e.g., sediment), chemical data generated can be used for all other tools being employed.

The study design must be critically evaluated to ensure that data generated will fully answer all of the specific questions developed through problem formulation. In particular, the measurement endpoint tools must be relevant to the concerns associated with the assessment endpoint.

7.3 Analysis phase

7.3.1 Exposure assessment

The goal of any exposure assessment is to estimate, to the degree of accuracy needed, the zone of influence of potential concern in the environment or the bioavailability of chemicals to organisms. Exposure assessments include analysis of the magnitude, duration, and frequency of exposure, based on data for 1) chemical sources, 2) chemical distributions in media including transformations, and 3) spatial-temporal distributions of key receptors. Available data are used initially for an exposure assessment, and focused studies may be conducted to collect data for subsequent tier assessments. A variety of methods or tools exist to allow the assessor to estimate potential exposures. These tools each have varying degrees of accuracy, ecological relevance, and inherent uncertainty (Chapter 18). For example, a mechanistic model can provide estimates of water concentrations; however, its accuracy depends on the inherent understanding of the environmental processes on which the model is based and the validity of the assumptions used in the model scenario. Alternately, empirical measurement of the chemicals of concern at a given site is perhaps a less uncertain method for evaluating exposure, because it permits the direct evaluation of exposure at a given location. However, it has the drawback of being labor intensive and costly, and bulk concentration does not directly translate to exposure.

The evaluation of exposure is perhaps the most unique, and the most uncertain, aspect of conducting ERAs for contaminated sediments. A wealth of information is available suggesting that evaluation of contaminant concentrations in whole sediments may not be sufficient to address the question of bioavailability and, thus, risk of adverse effects. This is reflected in recent regulatory strategies for developing sediment criteria for both nonpolar organic materials (USEPA 1993) and metals (USEPA 1994b). In both cases, empirical data suggest that porewater contaminant concentrations more accurately predict observed toxicity and community-level effects than do whole sediment concentrations. This is chiefly due to mitigating or modifying factors (*e.g.*, OC, AVS) that affect the bioavailability of sediment-associated contaminants. On the other hand, depending on the degree of rigor necessary (*i.e.*, the tier of the assessment), the use of less rigorous, more uncertain methods may be all that is needed.

A variety of tools are available for estimating exposures in sediments. These tools and a brief description of their advantages and disadvantages are provided in Table 7-4.

The decision to use a particular tool in assessing exposure must be based on a number of factors, including these:
- Type of contaminants present
- Inherent effects or concerns associated with the contaminants-of-concern (*e.g.*, direct acting toxicants versus bioconcentratable materials)
- Level of assessment (*i.e.*, screening-level assessment versus advanced tier)
- Quantity and quality of available data

In most instances, a battery of tools must be employed in conducting the assessment; however, the degree to which any or all tools will be used may vary.

7.3.2 Effects assessment
The goal of the effects assessment is to provide information on toxicity or other effects that form integral components of the measurement endpoint. Effects assessment tools

Table 7-4 Exposure assessment tools

Exposure assessment method	Advantages	Limitations
Empirical estimates		
Whole sediment concentrations	Extremely valuable for retrospective screening-level assessment. Permits one of most direct procedures for determining distribution and quantification of sediment contaminants. Valuable for evaluation of bioaccumulative contaminants.	Requires that assessor dictate chemicals of concern up front and requires that a sampling design of sufficient statistical rigor exist. Can be costly; *a priori* information should be used to quantify number of samples needed.
Porewater concentrations	May provide more direct estimation of actual exposure than whole sediment concentrations.	Requires more difficult sampling procedures that can be limited in certain types of sediments and can be more costly.
Overlying water concentrations	Can prove valuable for retrospective screening-level assessments, especially, in cases where sediments are "coarse grained", *i.e.*, overlying waters may appreciably affect porewater concentrations or in those cases where it is anticipated that sediments may serve as a "source" for water contamination.	May not be appropriate for all types of sediments. Water flow rates may result in dilution to levels below detection limits.
SEM concentrations	Provides most direct empirically derived estimation of actual exposure for metals in contaminated sediments. Only applicable in those cases where metals are a contaminant of concern. See pore water.	Requires difficult analytical procedures. Samples must be obtained using rigorous procedures and can be costly. Relates only to divalent metals. See pore water.
Tissue concentrations	Can be most direct method for estimating tissue bioaccumulation. Will allow separation of tissue concentrations to edible and inedible tissues. Permits species-specific evaluation of tissue concentrations and allows direct evaluation of food chain transfer.	Can be difficult or impossible for species that are difficult to capture or cannot be practically obtained. Generally, sample numbers are small and can limit statistical rigor. Prediction of adverse effects from these data must have its basis in controlled experiments where a clear exposure-response relationship has been established; few such data currently exist.
Biomarkers	Can be useful in addressing question of exposure and bioavailability. Can be extremely sensitive.	Generally, nonspecific for contaminants. May not provide an exposure-response relationship. Requires understanding of mechanisms and must be related to a quantifiable adverse effect.

Table 7-4   continued

Exposure assessment method	Advantages	Limitations
Modeled / calculated estimates		
TOC/ AVS-corrected concentrations	Normalization of whole or SEM sediment concentrations on the basis of TOC or AVS can be useful in providing sediment-specific estimate of bioavailability and receptor exposure.	There is error in these estimates.
Fate and transfer mechanistic models	Can be important tools in conducting predictive ERAs. May provide one of the few tools available for estimating spatial or temporal shifts in contaminants.	Require accurate understanding of various fate-and-transport processes that affect contaminant concentrations. Outputs are only as accurate as assumptions used to derive them. Accuracy of modeled results may be directly correlated with quantity and quality of available data.
Fate and transfer/ probabilistic models	Can be important tools in conducting predictive ERAs and may provide one of few tools available for estimating spatial or temporal shifts in contaminants. When sufficient data are available, can provide powerful tool based on site-specific empirical data, reducing uncertainty associated with estimation of contaminant concentrations and spatial distribution.	Accuracy is directly related to available data and may require great deal of data to be very robust. Predictability often depends on quality of extant data.
Food chain models	Provide only method for estimating tissue concentrations in certain types of single organisms, communities, or trophic groups. Are most accurate when rooted in empirical data but can provide estimations of bioaccumulation or biomagnification based on conservative assumptions. In cases where it is not practical or possible to directly measure tissue concentrations in species (*e.g.*, threatened or endangered species), these models are only way to estimate concentrations.	As with all models, a high degree of mechanistic understanding or empirical data will improve model predictions. Prediction of adverse effects from these data must have its basis in controlled experiments where a clear exposure-response relationship has been established. Few data currently exist that would provide this type of relationship.

should be appropriate for the contaminants and assessment and measurement endpoints under consideration. As described below, tools range from numerical sediment criteria or values to exposure-response curves or relationships.

Effects assessment tools employed at earlier tiers or phases are usually simpler than for those that may be used at later stages. Various tools exist to allow the assessor to estimate potential effects. As with exposure assessment, these have varying degrees of accuracy, ecological relevance, and inherent uncertainty (Chapter 18).

Effects assessment tools and a brief description of their advantages and disadvantages are provided in Table 7-5. Biomarkers, with the exception of histopathology studies, are not included because they are considered measures of exposure, not effects.

In general, the following tools are used to evaluate the effects of sediments on benthic invertebrates:
- Benchmark values (numerical criteria)
- Toxicity studies
- Field studies of abundance and community structure

The following tools are used to evaluate effects on fish:
- Benchmark values (numerical criteria)
- Bioaccumulation measurements or estimates
- Toxicity studies
- Field studies (abundance and community structure, possibly also histopathology)

The following tools are used to evaluate effects on birds and mammals:
- Bioaccumulation measurements or estimates in prey species
- Literature values on toxic effects
- Field studies of effects (environmental epidemiology)

7.4 Risk characterization

Risk characterization for hazardous waste sites should not be done in isolation but rather should consider other compartments that may be part of the environment as a whole. Specifically, consideration should be given to the environment outside the boundaries of the waste site and to potential cumulative impacts from other sources or effects related to ambient background concentrations of contaminants which may not be attributable directly to the waste site. Risk characterization involves bringing together exposure and exposure-response information to estimate the extent and nature of the risk at a contaminated sediment site and the level of uncertainty associated with the estimate of risk. The study design (Section 7.2.5) and the assessment and measurement endpoints (Sections 7.2.3 and 7.2.4) will determine the nature of the risk characterization and the level of uncertainty associated with it (Chapter 18). There are several key considerations that must be incorporated into the risk characterization for effective management decisions to be made in remediating contaminated sediments.

Table 7-5 Effects assessment tools[1]

Effects assessment method	Advantages	Limitations
Empirical estimates		
Benchmarks (*e.g.*, sediment contaminant numbers)	Provide simple approaches from comparing sediment concentrations (measures of exposure) to reference values (surrogate measures of potential effects). Considerable use of this approach around the world; results are easy to communicate to risk managers. Also requires limited evaluation, usually involving measures of chemical concentrations. Employed for measurement of both bulk chemicals and bioavailable fractions of chemicals.	Technical bases of values are often not well established and not included as part of publishing the numbers. Makes it difficult to determine if values are appropriate for a particular assessment endpoint. Most numbers are based on either background values or protection of benthic invertebrates and may not be appropriate for other assessment endpoints. Numbers based on bulk chemical measurements also do not consider effects of matrix on bioavailability. Even values that explicitly consider certain aspects of bioavailability (*e.g.*, EqP) may miss others (*e.g.*, physical sequestration).
Toxicity tests (experimental) • laboratory • field	Primary advantage in lab or field is that it is possible to control or account for many variables that may exist in the field and to focus on a specific attribute of the sediment, *i.e.*, its toxicity. Information is quantitative and can be interpreted within context of what is known about toxicity of other sediments or chemicals to organism used in test.	Major disadvantage is that they are a model of the real system and do not include many factors that affect populations. Use of toxicity tests introduces uncertainties related to extrapolating results. Potential disadvantage is that toxicity tests typically consider one or a few life stages of a few species.
Bioaccumulation tests (experimental) • laboratory • field	Bioaccumulation is used both to estimate exposure and effects. Direct measurements of bioaccumulation provide information on potential effects where the measurement endpoint of concern is a concentration.	May not be specific to particular target organ. Uncertainty exists concerning relationships between tissue levels and effects. In some cases, effects are inferred from tissue levels. In such cases, there may be uncertainties associated with relationships among tissue levels and effects. Bioaccumulation is a phenomenon, not an effect and is primarily an effect tool, except where there is a target level that is actually the measurement endpoint.

Table 7-5   continued

Effects assessment method	Advantages	Limitations
Field studies of communities and populations	Primary advantage is that they provide direct information on populations and communities of interest.	Major limitation is occasional difficulty in detecting ecologically important changes from natural variability. In addition, it is sometimes difficult to sort out patterns related to contaminants as compared to other environmental variables.
Histopathology studies	Serve as indicators of effects in certain species. Effects can be measured directly and quantified in terms of frequency of incidence.	Population significance not well understood.
Development of exposure-response relationships [applicable to all above]	Provide basis for evaluating implications of exposure concentrations to receptors of interest.	Unless relationships have been developed for species at site of interest, adjustments are typically made to account for interspecies extrapolations. Can be considerable uncertainty associated with these adjustment factors.

[1] All tools have varying degrees of difficulty in determining causality.

7.4.1 Multiple lines of evidence

In sediment assessment, multiple lines of evidence are typically available for the final risk characterization (*e.g.*, results of toxicity tests, benthic community surveys); thus, a weight-of-evidence approach is used to evaluate and characterize risk. The study design will normally prescribe the lines of evidence required to support evaluation of risk to a particular assessment endpoint, though the process may be tiered or iterative (Section 7.1).

Characterization of risk relative to a broad assessment endpoint (*e.g.*, a healthy benthic community which poses no risk to wildlife consumers) and the associated confidence in the estimate may be expressed in simple narrative terms such as low, moderate, or high risk to the benthic community. Such statements represent the integration and interpretation of multiple lines of evidence which are supported by various mathematical and statistical analyses. As emphasized by Suter (1993), "...each risk estimate will have its own assumptions and associated uncertainties. The separate lines of evidence must be evaluated, organized in some coherent fashion, and explained to the risk manager so that a weight-of-evidence evaluation can be made." A weight-of-evidence approach is described more fully below.

Though measurement endpoints can be directly characterized in terms of risk, the goal of the risk characterization is to estimate risk relative to the assessment endpoints. Measurement endpoints must be clearly linked back to the original assessment endpoints when interpreting and communicating results of the risk characterization.

The weight-of-evidence approach requires best professional judgment and is the process by which measurement endpoints are related to assessment endpoints to evaluate if there is a significant risk of harm to the environment. The approach is planned and initiated in the problem formulation stage, and results are integrated at the risk characterization stage (Menzie *et al.* 1996).

In using a weight-of-evidence approach, there may be competing or contradictory lines of evidence and potential for inconsistency in interpretation of the significance of results. Thus, a coherent framework is needed for evaluating and weighting measurement endpoints in terms of their importance in the final risk characterization. For example, in sediment assessment using a triad approach (Chapman 1990, 1996), community indices, analytical chemistry data, and results of toxicity tests may not all similarly indicate risk to the benthic community. Weighting of measurement endpoints based on the factors described below will focus decisions and risk characterization.

7.4.1.1 Confidence in measurement endpoints
In interpreting measurement endpoints as part of the overall risk characterization, more confidence may be placed on nonsubjective endpoints (*e.g.*, lethality, growth, or reproduction) where the nature of the adverse effect and mechanistic links with the assessment endpoint can be clearly made, than on a measurement endpoint where the mechanistic link with the assessment endpoint may not be clearly understood (*e.g.*, biomarkers, histopathology). The latter may be used to corroborate or support interpretation of risk but would have less weight in the characterization. Confidence may also be linked to statistical uncertainty, how the value was derived (*e.g.*, direct observation, modeled with default parameters). Factors affecting confidence in measurement endpoints are described in Sections 7.2.4 and in Chapter 18.

7.4.1.2 Magnitude of response relative to the metric
The greater the magnitude of the response, the more confidence or weight that will normally be placed on it. However, the magnitude of the response can only be evaluated relative to an accepted metric or yardstick. For example, judging whether a 20% or a 50% reduction in growth is a "high" level of effect will be based on precedent and established metrics for that endpoint.

7.4.1.3 Multiple indications of effect
The greater the number of lines of evidence supporting a particular indication of risk, the greater the confidence in that endpoint (*e.g.*, low confidence may be placed on a line of evidence or measurement endpoint which is not supported by any other evidence). Extreme caution should be exercised in interpreting such results where the outlier is indicative of an adverse effect and fits either or both of the above considerations. Such a result may point to a need for further investigation.

Ideally, a weight-of-evidence approach will express not only the absolute value of the measurement endpoint, but the degree of confidence (weighting) in the endpoint based on the above considerations. Results may be descriptively presented in tabular or graphical form to aid in interpretation.

7.4.2 Retrospective versus prospective risk characterization

The information needed to reach a management decision and consideration of remediation options may require both a retrospective risk characterization (What is the existing risk of harm to the environment?) and a prospective risk characterization (Given the probability of certain events as a result of a remedial alternative — *e.g.*, mobilization of contaminated sediments as a result of dredging activities, biological changes such as methylation of mercury — what is the prospective risk of harm to the environment?). For example, a prospective evaluation of risk is important for buried sediments which may be exposed through a storm event or other perturbation. A "snapshot" retrospective assessment may indicate low risk, whereas a prospective assessment may indicate high probability of risk and lead to a different management decision. Other prospective considerations can include transport and redistribution of contaminated sediment off-site, seasonal pulse events (lake bottom turnover), *etc.* Recommendation of appropriate and technically sound models for such prospective analyses is part of the problem formulation and study design phase.

A further consideration is prospective evaluation of remedial options which could involve modeling. Whatever the means, there will be uncertainty which needs to be quantified (Chapter 18).

7.4.3 Risk estimation methods

There is a rapidly expanding toolbox for risk estimation methods amenable to sediment assessment. However, it is important to remember that these are tools to characterize or interpret effects and exposure data in terms of risk.

Primary tools for estimating risk and their application include these:
- The quotient method for assessments (described briefly below)
- Joint probability analysis to estimate risk (described briefly below)
- Ecological models to extrapolate measurement endpoints to assessment endpoints, with model uncertainty analysis (*e.g.*, Monte Carlo) for risk characterization
- Chemical analyses of sediment and water in combination with toxicity testing and community analysis for assessing existing impacts to sediment communities and for developing field-based exposure-response relationships that can be used for risk predictions
- Quantification of uncertainty estimates by deriving probability functions (or ranges where data are limited in screening assessments) for risk estimates

7.4.3.1 Quotient method

The quotient method is frequently used for a screening or preliminary risk calculation for contaminated sediments. By itself, the quotient method does not provide a complete characterization of the magnitude of risk and uncertainties. Such approaches provide an approximate risk index and the confidence/uncertainty associated with this expression of risk rests largely in the confidence/uncertainty in the measurement endpoints that were entered into the calculation. Hazard quotients are used to compare single effect and exposure values and can be used whenever there is sufficient information to estimate the expected environmental concentration in the media of concern, or when it has been measured, and where there are adequate studies in the literature or from directed laboratory and field toxicity tests to determine a toxicological benchmark or threshold concentration (such as the no-observed-effects concentration [NOEC] for acute or chronic toxicity). The hazard quotient method compares the estimated or measured exposure levels to the measured or predicted threshold value for effect. Quotients are simple to interpret (*i.e.*, quotients less than 1 typically imply low risk, and quotients greater than 1 imply that there is a potential risk). However, in weighing the value of hazard quotients in the overall risk characterization, weaknesses such as the lack of indirect effects evaluation or of incremental dose impacts must be considered. Hazard quotients may be particularly useful for determining priority contaminants when the site is grossly contaminated by many chemicals (Chapter 17).

In deciding what threshold or benchmark value is used in the hazard quotient, it is important to note that, though standardized toxicity tests often have prescribed methods for interpreting results, they do not provide information for interpreting the ecological significance of the effect. For example, a 15% reduction in growth may be statistically different from a control but may have little meaning in terms of actual risk to field populations. In some cases, there are "rules of thumb" (*e.g.*, an observed 20% to 25% mortality effect in laboratory toxicity studies as the threshold concentration). Given identical measurement endpoints and statistical rigor, characterization of risk can vary widely dependent on the threshold value that is accepted as representing an ecologically rather than merely a statistically significant effect. In characterizing risk, it is important to agree, at the study design phase, what the threshold effect level will be (*e.g.*, lowest-observed-effect concentration [LOEC], 50% effect) to avoid inconsistent interpretation of results.

Comparability of risk characterizations is enhanced if the effect level refers to a quantile such as LC10 or EC20 rather than NOEC, maximum acceptable toxicant concentration (MATC), or LOEC. If quantile responses are routinely determined, it is relatively simple to proceed to the next level of complexity: continuous exposure-response relationships. Similarly, where a standardized battery of tests is used, the problem formulation stage should consider and prescribe the expectations for acceptable risk (*e.g.*, all tests must "pass").

Though traditionally used with toxicological endpoints, a quotient method may be based on virtually any of the quantified effects-based measurement endpoints identified in Section 7.2.4.1. Endpoints such as community structure indices for benthic invertebrates can be compared to threshold values for toxicity; an estimated chemical dose to a wildlife receptor (*e.g.*, estimated using a food chain model) can be compared to a toxic dose estimate such as an NOEC or LOEC.

The use of generic effects-based benchmark concentrations such as SQC or sediment quality guideline (SQG) values to calculate a quotient is perhaps the simplest form of risk characterization (also known as toxic units [TU]). Though this approach can be a useful coarse screening tool that minimizes the need for site-specific toxicity testing, interpretation must be made in light of the limitations of generic criteria as surrogates for site-specific exposure-response data. The typically conservative nature of most, but not all criteria does minimize false negatives in evaluating risk at a site and this approach is useful in focusing the efforts of further study to priority contaminants.

Population and community endpoints must be interpreted with some caution if characterized using a quotient method. For example, for data from population surveys to be directly relevant to the exposure at a site, the numbers of organisms must be largely controlled by survival and reproduction within the site, rather than by immigration and emigration.

7.4.3.2 Probabilistic risk characterization
There is a growing trend towards probabilistic risk characterization in ERA. Some commonly used techniques are described briefly below. Again, the data required to drive these methods must be clearly specified in the study design if such approaches are to be used to characterize risk.

7.4.3.3 Joint probability analyses
This approach can be used to estimate the risk of chemical concentrations exceeding toxicity thresholds, including sediment quality guidelines for a given time period in a given area (Long *et al.* 1995). A cumulative probability distribution of chemical concentrations is developed, and then the cumulative probability distribution is used to determine the risk (probability) of exceeding the toxicity threshold (Chapter 18).

7.4.3.4 Continuous exposure-response model
Data at the individual or population level are used to generate probabilistic estimates of risk. The use of population-level models in risk characterization is increasing, especially Monte Carlo simulations which provide a single estimate of risk for a single exposure concentration based on a series of model simulations using randomly selected sets of parameters. The entire procedure is repeated for a series of exposure concentrations and the estimate of risk can be plotted as a function of the exposure concentration (Bartell *et al.* 1992). Standardized computer programs are available to conduct Monte Carlo simulations, but these generally work best with real data. Disadvantages include the fact that use of computer simulations may overlook the implicit assumptions that contribute to the uncertainty of the ERA and that, due to the ease of running the model, inappropriate use and interpretation can occur among practitioners not familiar with the ecological basis of the parameters entered into the simulation.

7.4.4 Spatial characterization of risk
On-site contamination gradients (*e.g.*, depth, breadth) can be used to demonstrate on-site exposure-response functions. Where such data have been collected, they should be used as part of the risk characterization. The paired comparison/mapping of effects data from, for example, toxicity testing or community survey data with environmental concentra-

tion (analytical chemistry) data provides an important spatial dimension to risk characterization needed in final management decisions.

7.4.5 Temporal characterization of risk
Risk may also be temporally characterized in terms of seasonal or long-term trends to account for varying exposures and changes in species susceptibility to chemical effects. The use of chemical data from different times and locations to derive estimates of exposure concentrations for risk characterization must be matched to the distribution, life stages, and activities of the selected receptors and to specifics for sediments (*e.g.*, turnover, storm events, bioturbation).

7.4.6 Uncertainty analysis
Risk characterization must include the uncertainty in the risk estimate and discuss the way in which it might affect interpretation and decision-making. Sources of uncertainty include natural variation, missing information, and error associated with estimates and extrapolations. Uncertainty associated with various endpoints is discussed in Chapter 18.

7.4.7 Other considerations
In applying ERA to sediment cleanup issues, additional factors that may influence management decisions need to be considered in characterizing the site.

7.4.7.1 Comparison to a reference site
Risk may be characterized relative to an "uncontaminated" reference site. Often, the reference site does not represent pristine conditions but is representative of both the ambient or background conditions or sediment type (grain size, TOC, *etc.*) for the area and historical contamination. Multiple reference stations may be needed to cover the range of conditions. The reference site is often not a perfect control and will add to the uncertainty of the analyses.

7.4.7.2 Characterization of sensitive areas/species
A toxicologically based estimate of risk may lead to the removal or perturbation of contaminated sediments and the benthic or other communities they support. The estimate of risk must be carefully weighed against the disturbance to sensitive habitat/species that can result from any remediation. Consideration of such effects as loss of habitat and recolonization times of benthic communities will become equally important in decision-making as risk assessment that focuses on the effects of contaminants.

7.5   Decision-making
Decision-making responsibility rests with the risk manager after the risk assessor provides an objective and technically defensible evaluation of the present risks and of the risks associated with various remedial options. The first steps of the process for sediment risk assessment (see Section 7.2) should be educational, where risk assessors and risk managers achieve a common understanding of the management objectives and the nature and objectives of the site-specific assessment. Risk managers need to understand

that use of multiple assessment and measurement endpoints and a combination of assessment approaches will be necessary to determine existing and potential ecological risk at a site with contaminated sediments. Several concepts should be addressed in consultation with the risk assessor before the decision-maker or risk manager initiates preliminary screening, selection of assessment endpoints, and development of the conceptual model:

- Ecological risk assessment may evaluate one or many stressors and ecological components and thus must be flexible, *i.e.*, involve prospective, retrospective, or a combination of approaches.

- Field studies are essential for reducing uncertainty in exposure and effects assessment and for validating assumptions or predictions.

- A weight-of-evidence approach is recommended for use for risk characterization, along with explanations of uncertainty. In particular, professional judgment is key, and decision-making must involve the consensus of a team of chemists, toxicologists, ecologists, and risk assessment experts. Professional judgment involves the integration of technical expertise in ecological sciences and is important particularly in problem formulation. It must be objective, and involves developing iterative processes for assessment and analysis, evaluating sources of uncertainty, and interpreting and communicating the ecological significance of predicted or observed effects to decision-makers.

- Ecological risk is expressed in a variety of ways, ranging from probabilistic to qualitative comparison of effects and exposure.

Risk management is a distinctly different process from risk assessment. The risk assessment establishes that a risk is present and defines a range or magnitude for the risk. The risk assessors can recommend a target cleanup goal to the risk manager that will achieve the goal of protection of the assessment endpoint, and also express the risk associated with the remedial options and the resulting net risk reduction to the assessment endpoints for the remedial options.

7.5.1 Preliminary screening assessment

The risk manager and assessor need to agree on a preliminary list of assessment endpoints. The risk manager uses the results of the risk assessor's preliminary screening to decide if little or no risk exists or if ERA should be continued in order to develop a site-specific cleanup goal or to reduce the uncertainty in the evaluation of risk.

7.5.2 Iterative assessment

Subsequent iterations of the risk assessment involve a more complex problem formulation stage and may require refinement of assessment endpoints. The risk assessors and technical consulting or expert teams decide on the need for iteration for data acquisition/verification or possible modification of assessment and measurement endpoints.

The risk assessor and risk manager should coordinate with appropriate professionals in making decisions related to assumptions, models, *etc.* Risk assessors and the technical team ideally should work together in developing recommendations for specific target cleanup goals.

The risk assessors characterizing risk using weight of evidence can recommend a target cleanup goal to the risk manager. This is where risk management is a distinctly different process from risk assessment. The risk manager is the one who, through evaluation of remedial options, target cleanup goals, and the magnitude of the risk, as well as management/regulatory issues, makes the decision on the cleanup levels and actions.

In evaluating remedial options, risk managers will require additional professional assistance to interpret the implications of baseline ERA and remedial options. Because of the nature of SERA and remedial options available, it will be implicit as part of decision making that risk assessors will evaluate the risk resulting from disturbances of sediment habitat and also recovery rates under different scenarios.

The potential for adverse effects varies with each remedial alternative. For each option, ERA can be used to evaluate risk during and after remediation for both action-specific risk (*e.g.*, likelihood of physical destruction of habitat during dredging) and chemical-specific risk. Adverse effects of remedial actions that involve excavation (*e.g.*, in wetlands), dredging, redisposal, or containment may include the following:

- Physical (local) destruction of habitat and dominant aquatic species (may also involve destruction of sediment and redox strata resulting in potentially increased bioavailability of toxic contaminants during the remedial action; may result in pulse releases to vicinity and downstream areas)
- Physical removal of vegetation and animals
- Increased toxicity by transformation of chemicals or release from sediment during treatment
- Disturbance during human activity at a site, such as annoying noise to marine mammals, spawning harassment, alienation of some organisms

The comparative risk process evaluating residual chemical risk versus action-specific risk (necessarily predictive) requires information concerning rates of community recovery. Decisions may result in a less protective cleanup level resulting in short-term risk but resulting in eventual reduced risk through natural recovery.

7.6 Post-remediation monitoring

An additional iteration of any SERA is a long-term review of remediation success or of natural recovery. Also, if there will be residual risk or contaminants left in place post-remediation, a monitoring plan may be required. The magnitude and complexity of such monitoring are dependent on the remedy, residual risk, and contaminant class, as well as other factors.

Post-remediation monitoring is the final iteration of the SERA. Its purpose is to validate the results of preceding tiers in the ERA and either to evaluate the effectiveness of the remediation selected based upon the remediation goals or to confirm that acceptable decisions were made. Evaluations are typically based on one or more of the assessment end-points selected in the SERA.

The tools used to make the evaluation are typically the same tools as those used in the SERA, unless new techniques become available for the same measurement endpoint. However, if the same tools are used, they may be more specific to the measurement end-point, more easily obtained, or more cost effective.

Sampling stations or transects can be targeted specifically based on extent of remedial action and spatial extent of adverse effects or areas of concern. The monitoring time frame and sampling episodes should be based upon the contaminants remaining in the system and their characteristics, the magnitude of the contamination remaining, other point sources (for recontamination), the anticipated time of recovery for the system, and data needs for the evaluation.

Ideally, data needs for post-remediation monitoring should be evaluated during the SERA, resulting in triggers for time-dependent sampling events to evaluate trends. The intensity of post-remediation monitoring should decrease with time, reflecting the con-fidence of the risk assessor in the ERA predictions. Accordingly, some mechanism or trig-ger for sunsetting such monitoring is also required.

7.7   Research needs

The ERA process for a contaminated sediment site encompasses many sediment-specific issues due to the complex nature of the sediment environment. Due to this complexity, there are some limitations to SERA which can only be addressed by further research. Research areas and needs specific to contaminated sites are outlined below, categorized according to function in the SERA process: 1) exposure and effects analysis, 2) risk char-acterization and assessment, 3) cleanup and monitoring of the contaminated site, and 4) issues specific to wetlands.

7.7.1   Exposure and effects analysis

Exposure and effects analysis are the two processes which must be combined in order to develop any risk assessment. For contaminated sediment sites, the following research areas and needs exist:

- Improved dose-response models are needed to effectively link exposure mea-surements to effects. For example, for measurement endpoints such as tissue concentrations to be useful, a mechanistic relationship with an adverse effect must be established (*i.e.*, What impact does a concentration of a contaminant in an organism have on that organism, particularly for congener PCBs, mercury, PCDDs, and PCDFs?). More information is needed for different species on levels of persistent (bioaccumulative/bioconcentrating) compounds in tissue which cause adverse effects (*e.g.*, decrease in egg production, reduced viability, reproductive impairment). This is important for sediment risk assessment because of the variety of exposure pathways resulting from contaminated sediments.

- There is a need to develop new toxicity tests which focus on chronic endpoints (*i.e.*, growth and reproduction as well as survival), which can be related to population-level endpoints. There is also a need to refine existing tests so that factors other than contaminants can be evaluated (*e.g.*, habitat including grain size and TOC, feeding rates, water quality).

- A better understanding is needed of bioavailability in sediments that results in exposure. Specifically, are there other models/factors in addition to EqP and AVS (USEPA 1993, 1994b) which may be able to more accurately predict exposure and any resulting toxicity from contaminated sediments?

- The environmental fate and toxicity of various compounds should be investigated. Are they bioavailable? What concentrations in sediment are protective of species and habitats? A classification system for the fate and effects of compound classes encountered in SERA is needed.

- Further investigation of the factors affecting the bioavailability of contaminants often found at waste sites (*e.g.*, metals, PCBs, and PAHs) and of useful indicator organisms is needed. In particular, what environmental and biological factors increase/decrease the bioavailability of metals in wetland (freshwater, estuarine, and marine) sediments?

- There is a need to develop and evaluate environmental investigation strategies and bioassessment techniques in order to learn more about the sensitivities of various bioassays/organisms (*i.e.*, validation studies). This will allow the risk assessor to make sound recommendations for using one test over another.

- Methods of sediment toxicity identification evaluation (TIE) should be improved so that a wider range of organisms can be used and detection of potential causative agents improved.

- Existing bioassessment tools that are not routinely used at sites with contaminated sediments should be applied to test/demonstrate their utility (*e.g.*, subcellular biomarkers). Biomarkers may be a relevant endpoint, but a cause and effect relationship must first be demonstrated.

- A better understanding is required of biological interactions to determine which indirect effects are most likely to translate to direct effects. This may include the development of population models to synthesize exposure and effects for multiple species.

- A better understanding is required of benthic community structure and function. It is particularly important to determine and define what should be present in which habitats and what is functionally necessary in those habitats. A general improvement in evaluations is required for systems other than small/medium pool-run streams, in particular for big river systems.

- There is a need for agreement/standardization of the appropriate level of statistical significance for field measurement endpoints.

- There is a need for quality assurance/quality control (QA/QC) guidance for field collections of biological samples.

7.7.2 Risk characterization and assessment

Risk characterization and risk assessment synthesize the exposure and effects results. Areas and needs for future research include these:

- Developing a system for tracking what has been done and what is being done at contaminated sites with regard to the sediment risk assessment process (lessons learned). For example, cases studies including data from RI/FS unpublished literature, negative data, and other examples could be summarized and made available to others. A database of completed risk assessments could be one way of summarizing much of this information.

- Conducting sediment-based synoptic chemistry and bioassessment studies with a variety of endpoints (*e.g.*, the triad, Chapman 1990, 1996) to provide more data for field verification of benchmarks. For example, laboratory testing indicates toxicity, yet a field evaluation of the benthic community indicates no adverse effects. How are these two experimental results reconciled in the context of an SERA?

- Critically evaluating sediment benchmark values — *e.g.*, effects range low (ERLs) (Long *et al.* 1995) and SQC — relative to the technical basis of their derivation

- Summarizing commonly used toxicological benchmarks for risk assessments. For example, 20% eggshell thinning (for birds) is typically taken as an indication of an adverse effect. These benchmarks should be summarized and standardized if possible.

- Determining the relevance of behavioral responses. Can the cause-and-effect connection due to the presence of a contaminant be demonstrated?

- Determining the best way to deal with mixtures of contaminants in determining risk. For instance, is the additivity approach adequate and appropriate for joint action toxicity?

- Developing watershed approaches for single watersheds affected by more than one contaminated site and for multiple watersheds affected by one large single site

- Developing a landscape approach for measurement endpoints and study design, and for conducting the risk assessment. This approach integrates a spatial component as part of the risk assessment process.

- Determining whether volatile organic compounds (VOCs) that pass through sediment from groundwater discharges are a threat to benthic communities

7.7.3 Cleanup and monitoring

Even if a substantive risk is determined to exist at a contaminated sediment site, it may be better to leave the contamination undisturbed if the cleanup or remediation will result in greater ecological risk. In this case or if a site is cleaned up, a monitoring program may be required to determine the effectiveness of any cleanup. Several research areas and needs exist in this regard:

- Investigating the time required for the return of habitat function following a hazardous materials release through monitoring at sites after cleanup and mitigation is completed
- Determining what parameters should be monitored and how a monitoring program should be designed to determine the success of remediation
- Quantifying the risk of remediation. For instance, remediation of contaminated sediments may result in resuspension of contaminants. What tests and models can be used to predict and evaluate this risk?
- Evaluating the effectiveness of remediation technology, in particular assessing the use of capping as a remediation technology for contaminated sediments. What is the effectiveness of capping for different sediment contaminants in different environments and what is the best cap design? Does groundwater flow alter the effectiveness of the cap? What is the actual lifetime of a cap?

7.7.4 Wetland issues

Wetland environments are extremely complex ecological systems that may contain contaminated sediments. Because of this complexity, several specific research areas and needs have been identified:

- Deriving a methodology for determining when to clean up contaminated sediments in a generic wetland and when to leave them in place. Studies are needed on the effects of leaving a wetland contaminated versus the effects of cleaning it up. What is the time frame of recovery of wetland species from both food chain contamination and physical disturbance of cleanup? A cost-benefit type of analysis may be appropriate.
- Researching methods of enhancing and restoring specific contaminated wetlands that minimize disturbance. What type of wetlands have a greater success with restoration in different regions of the country? What is the risk of remediation when using different remediation technologies (*e.g.*, bioremediation or in-place remediation)?
- Determining the importance of contaminants (*e.g.*, mercury, PCBs) in wetland sediments as sources of contamination for habitats and biota outside of wetland areas

Specific application of ERA to wetlands has been addressed in a recent SETAC workshop (Lewis *et al.*, in press).

7.5   Management issues

7.5.1 Communication

There is a need for increased communication throughout the SERA process to assist in site management. Open discussion of issues which lead to the selection of initial assessment endpoints, and subsequently to clearly defined final assessment endpoints, should resolve misunderstandings and alternate interpretations as to what is being evaluated and why. Communication resulting in total transparency (to the extent legally possible) should defuse much of the apparent conflict which currently exists related to contaminated sites' SERA.

At present, site management suffers less from a lack of tools than from a lack of application of a clearly defined and stated process. This results in inadequate communication between not only the risk manager and the risk assessors but also between the other parties involved with a site. For example, if assessment endpoints are not clearly defined and stated, it is likely that one if not all of the parties involved will view the assessment point differently. When this occurs, conflicts will arise at the decision stage.

For all parties involved, the stakes are high at contaminated sediment sites. Various agencies have legal mandates related to sediment management. Remediation of sediment contamination includes financial interests, and the general public have their own interests. Because of these issues, over- or underestimating the ecological risk posed by a contaminated sediment can have far-reaching ramifications. The balancing or lack of balancing of these positions can translate into delays in the site management schedule and into overall increased cost of assessment in addition to the conflicts which result. Increased costs can result from redundant sampling efforts, increased management/oversight costs, multiple efforts to analyze data and characterize the risks, and in the worst cases, incorrect decisions for managing the sediment risks.

In addition to increased risk assessment communication, all parties must understand all responsibilities (mandates) and limitations. For example, there may be impediments to inclusion of all potential sources of ecological risk from contaminated sediments based upon what the statutes regulate. For example, USEPA Superfund does not regulate oil and related materials, as well as a variety of other materials which are not specified as "hazardous substances," nor does it regulate physical disturbances.

Through this document and other efforts, agreement must be reached on what is to be accomplished (and what can be accomplished) through the management of contaminated sediments. In particular, there must be agreement on what level of residual ecological risk is acceptable and what sediment-associated risk is ecologically relevant.

The potentially confrontational nature of the risk management decisions should be diminished by these actions:

- Ensuring a defined framework for the development of the conceptual model and selection of the assessment and measurement endpoints such that there is a clear, scientifically defensible justification for the data which need to be collected
- Developing how the exposure and effects will be evaluated and how a scientifically defensible study will be designed to answer the questions identified through problem formulation, such that a level playing field is established for the characterization of the sediment risk
- Developing monitoring plans for validation of the effectiveness of the actions taken for the particular site, which are founded in the risk assessment process

and only implemented with a clear statement as to how the data will be interpreted

7.5.2 Residual risk

The remediation of contaminated sediment sites is unlikely to remove all ecological risks, unless the remedy is to remove all of the contamination. This option is generally technically and financially unrealistic and may not be scientifically justifiable. It follows that residual ecological risk will remain after the contaminated site is remediated. Currently, this is an uncomfortable position for regulators, the regulated party, as well as those with other interests in the assessment endpoints. Regulations, including those not directly applicable to the management of contaminated sites, must be considered and an agreement reached as to how these post-remediation issues can be addressed. For example, natural resources damage assessments (NRDAs) are significant considerations in post-remediation liability to financially accountable parties. Statements of the residual risk currently have important implications to both the natural resource trustees and the financially accountable parties.

7.6 Recommendations

- The current risk paradigm is appropriate for ERA at contaminated sediment sites (Chapter 1).
- A tiered, iterative approach to sediment risk assessment should be used. This will maximize efficiency of effort and resources in reaching decisions. At the end of each tier, the information generated should be critically evaluated to determine if an additional tier is needed, if a decision can be made, or if additional information is needed to reach a decision.
- Communication between risk assessor, risk manager, and all involved parties is essential at all stages of SERA. In particular, all parties should agree to the assessment and measurement endpoints which will drive the risk assessment, understand the basis of the risk assessment process, and understand the need for flexibility in accommodating unique site characteristics.
- A weight-of-evidence approach using best professional judgment to the interpretation of data should be used in conducting an SERA.
- The ecological risk associated with contaminated sites should not be evaluated out of context of the environmental system of which it is part, including the cumulative risk at the watershed or landscape level.
- Ecological risk assessment of contaminated sediments needs to consider both spatial and temporal components.
- Assessment endpoints need to be associated with measurement endpoints through the route of exposure and the mechanism of effect. The measurement endpoints must be technically sound and defensible, and whenever possible, at least one measurement endpoint should be field-based.

- While resources, legal issues, and site management schedules are important, they are not justification for poor science.
- The initial risk assessment will characterize the existing risk at the site in a retrospective fashion. In considering remedial options and reaching final decisions, a prospective risk assessment will be required to evaluate the efficacy of the remedial options in achieving risk reduction goals.

7.7 References

Bartell SM, Gardner RH, O'Neill RP. 1992. Ecological risk estimation. Chelsea MI: Lewis.

Chapman PM. 1990. The sediment quality triad approach to determining pollution-induced degradation. *Sci Total Environ* 97:815–825.

Chapman PM. 1996. Presentation and interpretation of sediment quality triad data. *Ecotoxicology* 5:1–13.

Gaudet C, Nason GE, Milne D. 1995. A framework for ecological risk assessment at contaminated sites in Canada. Part II: Recommendations. *Human Ecol Risk Assess* 1:207–230.

Long ER, MacDonald DD, Smith SL, Calder FD. 1995. Incidence of adverse biological effects within ranges of chemical concentrations in marine and estuarine sediments. *Environ Management* 19:81-97.

Menzie C, Hemming MH, Cura J, Finkelstein K, Gentile J, Maughan J, Mitchell D, Petron S, Potocki B, Svirsky S, Tyler P. 1996. Special report of the Massachusetts Weight-of-Evidence Workgroup: a weight-of-evidence approach for evaluating ecological risks. *Human Ecol Risk Assess* 2:277–304.

Lewis M, Henry M, Nelson M, Powell R, editors. In press. Ecotoxicology and risk assessment for wetlands. Society for Environmental Toxicology and Chemistry (SETAC) Pellston Workshop; 1995 Jul 30–Aug 3; Gregson MT. Pensacola FL: SETAC Pr.

Smith SL, MacDonald DD, Keenleyside KA, Gaudet CL. In press. The development and implementation of Canadian sediment quality guidelines. Ecovision World Monograph Series.

Suter GA. 1993. Ecological risk assessment. Boca Raton FL: Lewis.

[USEPA] U.S. Environmental Protection Agency. 1992. Framework for ecological risk assessment. Washington DC: USEPA. EPA/630/R-92/001.

[USEPA] U.S. Environmental Protection Agency. 1993. Technical basis for deriving sediment quality criteria (SQC) for nonionic organic contaminants for the protection of benthic organisms by using equilibrium partitioning. Washington DC: USEPA. EPA/822/R-93/011.

[USEPA] U.S. Environmental Protection Agency. 1994a. Ecological risk assessment guidance for Superfund: process for designing and conducting ecological risk assessments. Review Draft. Edison NJ: USEPA.

[USEPA] U.S. Environmental Protection Agency. 1994b. Briefing report to the EPA Science Advisory Board: equilibrium partitioning approach to predicting metal availability in sediments and the derivation of sediment quality criteria for metals. Washington DC: Office of Water, Office of Research and Development.

SESSION 5
CRITICAL ISSUES IN ECOLOGICAL RELEVANCE

Chapter 8

Laboratory vs. field measurement endpoints: a contaminated sediment perspective

Gerald T. Ankley

8.1 Introduction

Due to the wide range of questions that may arise as part of a sediment assessment, methods that have been used in the laboratory and field to evaluate potential biological effects of sediment-associated contaminants are as diverse as those for testing any type of environmental media. A complete summary of these methods is beyond the scope of this short paper; however, comprehensive reviews on the subject are available elsewhere (Giesy and Hoke 1989; Burton 1991; Ingersoll 1995). The purpose of this essay is to identify measurement endpoints commonly used for evaluating contaminated sediments, in particular from a laboratory versus field perspective. My focus will be upon effects related to toxicity of sediment contaminants as opposed, for example, to physical disturbance of sediments and associated benthic habitats.

8.2 Measurement endpoint used in sediment assessments

From a generic point of view, the major assessment endpoint with respect to the direct toxicity (_i.e._, effects not manifested at higher trophic levels through bioaccumulation) of sediment-associated contaminants is related to the presence of benthic communities that exhibit functional and structural characteristics that would exist in the absence of impacts associated with sediment contaminants. From a practical standpoint, assessment endpoints usually are expressed in terms of a "use" statement, _e.g._, the desire to support a sustainable fishery (and hence, a viable benthic food web). Unfortunately, the linkage between assessment and measurement endpoints is at best uncertain. Part of this is related to the fact that with few exceptions, natural structure or function is unknowable, and thus typically is approximated through the use of comparative toxicological/ecological analysis of reference sites. Toxicological analyses usually are conducted using laboratory-based measurement endpoints, while an ecological perspective typically is gained through surveys/summaries of field-based endpoints describing benthic community structure. Irrespective of whether measurement endpoints are derived from laboratory or field data, because of the relatively undefined nature of what exactly constitutes an acceptable benthic community, endpoints used under different scenarios can differ quite markedly.

Although field-based measurement endpoints, such as organism abundance and community diversity, may be easier to relate to assessment endpoints than results obtained from laboratory tests, there are a number of reasons why laboratory assays are more commonly used in assessments of sediment toxicity than field-based analyses. From a practical/logistical standpoint, laboratory assays typically are less expensive and more rapid than surveys of benthic community structure. This enables, for example, more intensive sampling of sites, particularly at screening stages of an assessment. In addition, less expertise is generally required for interpretation of the measurement endpoints (*e.g.*, decreased survival, growth, or reproduction) typically used in laboratory tests as opposed, for example, to evaluating differences in diversity or similarity indices associated with extant benthic communities. This facilitates more ready application of results to regulatory/management scenarios in which the decision-maker may not be a trained biologist. In this context, measurement endpoints based on laboratory assays also can be very useful as a "real-time" tool for managers monitoring the effectiveness of remedial activities.

Another factor that contributes to the extensive use of laboratory assays for sediment assessments is related to the general concept of cause and effect. The structure and function of benthic communities is not necessarily dictated only by sediment contaminants; differences in species abundance or composition also may be influenced by factors such as transient toxicity events associated with water overlying the sediment, physical disturbance, lack of suitable habitat (*e.g.*, substrate, dissolved oxygen), resource limitations, and/or biotic interactions (*e.g.*, competition, predation). On the other hand, if a sediment exerts overt toxicity to one or more sensitive sentinel species in a laboratory setting, it is reasonable (in the absence of sampling artifacts) to believe that this adverse effect would be manifested in some manner in the field. Assessment of sediments in the laboratory also facilitates the development of dose-response relationships. The greater degree of control present in a laboratory setting affords the potential for more accurate exposure assessment than can be derived solely from a field analysis; this is particularly true for most contaminated sediments where one usually is dealing with complex mixtures of contaminants.

This is not to imply that field observations are not critical in assessing contaminated sediments. The incorporation of both field and laboratory data into ERAs of contaminated sediments is essential because the current state-of-science does not support exclusive reliance upon either type of measurement endpoint. As a simple example, laboratory assays may be used to screen for the presence/absence of toxicity, while subsequent field surveys could be performed to explore how the toxicity is manifested (*e.g.*, through absence of key species, reduced biomass). Without the field component, evaluation of the relevance of measurement endpoints as indicators of assessment endpoints is far more uncertain. However, to use combinations of laboratory and field data for probabilistic risk assessments, it is necessary to develop quantitative relationships between the two types of endpoints. Below are described some examples of how laboratory and field data concerning the biological effects of contaminated sediments have been related to one another; also briefly discussed are technical approaches relative to exposure assessment for complex mixtures. From the standpoint of laboratory assays, the focus is on those that are widely accepted and have had some degree of standardization, under the assumption that for the purposes of ERA some degree of standard "currency" is desirable. However, by focusing on these standard methods, it is not my intention to imply that other types of species/endpoints might not be useful for sediment assessments; this is a decision that should be made on a situation-specific basis.

8.3 Laboratory-to-field extrapolation

Organisms selected for use in assessing contaminated sediments in the laboratory have included bacteria, plants, clams, annelids, insects, micro- and macro-crustaceans, and fishes (Giesy and Hoke 1989; Burton 1991; Ingersoll 1995). Assays employing these organisms have been conducted with test fractions ranging from solid phase (whole) sediments to aqueous media such as pore water and elutriates to solvent extracts of sediments. However, the most commonly utilized laboratory assays for either freshwater or marine sediments consist of exposure of benthic macroinvertebrates to whole sediments. For marine sediments, standard methods have been developed for testing the amphipods *Rhepoxynius abronius, Ampelisca abdita, Leptocheirus plumulosus,* and *Eohaustorius estuarius,* while for freshwater sediments, standard assays exist for the amphipod *Hyalella azteca* and larvae of the midge *Chironomus tentans* (USEPA 1994a, 1994b). Tests with these species typically are 10 days in length, with the primary endpoint consisting of survival or survival and growth (*e.g., C. tentans*). These particular species are used for a number of reasons, including a) general technical acceptance by the scientific community, b) ease of culturing and/or handling in a laboratory setting, c) degree of contact with sediments, d) availability of data concerning relative sensitivity to contaminants, e) existing evaluations of within- and among-laboratory test variability, and f) evidence that assay results with the different species are predictive of impacts in the field.

One factor that limits the predictive power of current laboratory tests is that they are relatively short-term assessments of field situations in which chronic toxicity may be a dominant process. In recognition of this limitation, research to extend test duration and incorporate sublethal endpoints is ongoing with several species that are used in the 10-d tests (*L. plumulosus, H. azteca, C. tentans*). In any case, the discussion below is germane to the use of either short- or long-term laboratory assays.

The major conceptual problem in extrapolating laboratory measurement endpoints to population-level, and especially community-level measurement (or assessment) endpoints, is the lack of a mechanistic understanding of the relationship between the two different types of endpoints. That is, the laboratory test species serve only as sentinels to detect toxicity, which as a generic measure may not provide a basis for quantitative prediction of effects in variable field settings. This is due primarily to a lack of understanding of key processes controlling population and community dynamics, relative to the response of individual organisms. To address this problem, scientists involved in sediment assessments have sought to develop empirical relationships between the results of laboratory assays and field observations. As described below, this certainly is a viable approach, however, it must be recognized that definition of these types of relationships is by necessity often somewhat site-specific, *i.e.*, dependent both on the suite of contaminants present and the existing (or desired) benthic community.

There are at least two levels at which one may wish to extrapolate results of tests with laboratory organisms to population/community-level endpoints in the field. The first scenario is that in which responses of laboratory organisms are used to predict potential effects on populations of the same, or similar, species in the field. In these instances, simple demographic models and/or empirically derived relationships can be used to predict the effects of decreased survival and/or reproduction (in the laboratory) on population trends in the field, and confirm these predictions with limited field analyses. As one example of this, Wentsel et al. (Wentsel, McIntosh, Anderson 1977; Wentsel, McIntosh, Atchison 1977; Wentsel et al. 1978) correlated the presence/absence of C. tentans in metal-contaminated lake sediments with decreases in growth/emergence (and, hence, reproduction) of the midge in laboratory tests with the field sediments. Another example of this type of extrapolation is provided by Giesy et al. (1988), who reported that decreases in the growth of C. tentans of 30% or more in laboratory tests with sediments from the Detroit River corresponded with the absence of chironomids in the field. In a study from the marine environment, Swartz et al. (1985) noted that the toxicity of sediments along a contamination gradient on the Palos Verdes Shelf to R. abronius was predictive of the relative abundance of amphipods in the field (Table 8-1). Swartz et al. (1994) also established a relationship between the relative abundance of amphipods in DDT-contaminated sediments in the field and the results of laboratory toxicity assays with R. abronius, E. estuarius, and H. azteca.

The more demanding assessment scenario is that in which results of laboratory assays with one or more of the standard sentinel species must be extrapolated to the field to predict population-level effects in a dissimilar species, or more difficult yet, community-level impacts. Swartz et al. (1985) reported an example of the extrapolation of results of laboratory assays with R. abronius to population- and community-level impacts in contaminated sediments from the Palos Verdes Shelf. In this case, the presence of significant toxicity in the laboratory was correlated not only with marked decreases in amphipod abundance, but with species richness and the Infaunal Index (Table 8-1). Hoke et al. (1996) also provide an example of a correlation between toxicity of sediments in the laboratory (in this case to C. tentans and H. azteca) and community-level effects on benthic invertebrates at a site contaminated by DDT, DDD, and DDE. In these studies, increasing concentrations of total DDTR (DDT, DDD, and DDE) in sediments were related both to increased toxicity in laboratory toxicity tests and decreases in the total number of species present in the field (Figure 8-1). These data also serve to illustrate the variability that can exist in relationships between laboratory and field measurement endpoints; although there is an inverse relationship between species richness and toxicity, there clearly are instances where the two measurements appear to be disconnected. Unfortunately, in most situations it is impossible to ascertain whether a seeming lack of agreement between measurement endpoints is related to a lack of biological coherence or simply to sampling artifacts.

These examples of the extrapolation of laboratory results to field settings are not intended to be exhaustive but merely illustrative of how assay results have been used in this

Table 8-1   Toxicity to *Rhepoxynius abronius* of sediments from the Palos Verdes Shelf compared to measurements of benthic community structure

Parameter	Site						
	1	2	3	4	5	6	7
R.abronius survival %	77[a]	80[a]	80[a]	92	84	97	95
Species richness (S/0.1m^2)	15.8	18.6	28.2	41.2	96.6	72.4	73.4
Infaunal index	4.6	2.8	18.4	37.1	51.8	63.0	80.8
Amphipod density (N/0.1m^2)	0	0	2.4	1.6	46.0	52.2	65.4

[a]Significantly less than control.
Source: Swartz *et al.* 1985

context. It is important to reiterate that predictive relationships between most commonly utilized laboratory and field measurement endpoints (particularly at the community level) remain somewhat site-specific in nature. Clearly, these types of extrapolation efforts would be enhanced by more sensitive (chronic) assays as well as by a quantitative understanding of how effects in one or more laboratory species might translate into impacts at the population/community level.

8.4   Cause and effect

One important insight that greatly enhances the development of mechanistic models of relationships between laboratory and field measurement endpoints is the ability to identify contaminants responsible for sediment toxicity. As mentioned above, an important shortcoming of measurement endpoints based solely on field observations is the lack of definition of cause-and-effect relationships. However, because most contaminated sediments contain mixtures of potentially toxic chemicals, traditional laboratory toxicity assays, in and of themselves, also suffer from this shortcoming. But through the application of bioavailability predictions, structure activity, and interactive toxicity models, recent studies have shown that it is possible to identify those compounds responsible for sediment toxicity, and thereby develop plausible dose-response relationships. For example, Swartz *et al.* (1994) recently described the application of bioavailability and "toxic units" models to differentiate sediment toxicity due to DDT from contributions of PAHs, dieldrin, and metals. Swartz *et al.* (1995) also describe the use of bioavailability, quantitative structure activity, and interactive toxicity models to make quantitative predictions of the toxicity of sediments contaminated with complex mixtures of PAHs.

Another approach to the identification of contaminants responsible for sediment toxicity is via the use of toxicity-based fractionation schemes, commonly termed TIEs. Ankley and Schubauer-Berigan (1995) provide an overview of TIE methods, and describe the application of these procedures to contaminated sediments. It also is possible to use TIE

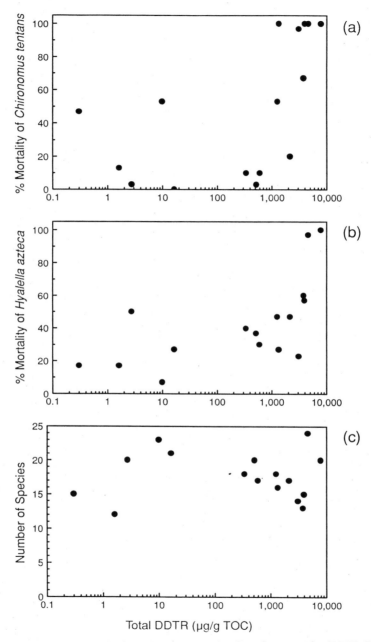

Figure 8-1 Laboratory toxicity of sediments contaminated with total DDT residue (DDTR: DDT, DDD and DDE) to *Chironomus tentans* and *Hyalella azteca* versus number of species in the field

procedures in conjunction with toxic units/mixture models and predictions of contaminant bioavailability. For example, Ankley *et al.* (1991) used basic TIE manipulations to help explore the utility of a proposed bioavailability/toxicity model for cationic metals in sediments.

Although diagnostic methods such as TIEs and bioavailability/interactive toxicity models thus far have been applied only infrequently to the interpretation of sediment toxicity, I feel that they are critical to ensuring the success of ERAs with contaminated sediments. The results of these types of analyses not only facilitate a mechanistic understanding of relationships between measurement and assessment endpoints, but they help provide a solid technical underpinning for the identification of appropriate remedial options.

8.5 Acknowledgments

Drs. R. Johnson, R. Swartz, R. Erickson, R. Hoke, C. Ingersoll, and P. Sibley participated in helpful discussions and provided a critical appraisal of an earlier version of this overview, which greatly improved the final version. This paper has been reviewed in accordance with official EPA policy.

8.6 References

Ankley GT, Schubauer-Berigan MK. 1995. Background and overview of current sediment toxicity identification evaluation procedures. *J Aquat Ecosystem Health* 4:133–149.

Ankley GT, Phipps GL, Leonard EN, Benoit DA, Mattson VR, Kosian PA, Cotter AM, Dierkes JR, Hansen DJ, Mahony JD. 1991. Acid-volatile sulfide as a factor mediating cadmium and nickel bioavailability in contaminated sediment. *Environ Toxicol Chem* 10:1299–1307.

Burton GA. 1991. Assessment of freshwater sediment toxicity. *Environ Toxicol Chem* 10:1585–1627.

Giesy JP, Graney RL, Newsted JL, Rosiu CJ, Benda A, Kreis Jr RG , Horvath FJ. 1988. Comparison of three sediment bioassay methods using Detroit River sediments. *Environ Toxicol Chem* 7:483–498.

Giesy JP, Hoke RA. 1989. Freshwater sediment toxicity bioassessment: rationale for species selection and test design. *J Great Lakes Res* 15:539–569.

Hoke RA, Ankley GT, Kosian PA, Cotter AM, VanderMeiden FM, Balcer M, Phipps GL, West C, Cox JS. 1996. Equilibrium partitioning as the basis for an integrated laboratory and field assessment of impacts of DDT, DDE and DDD in sediments. *Ecotoxicology:* In press.

Ingersoll CG. 1995. Sediment toxicity tests. In: Rand G, editor. Fundamentals of aquatic toxicology. 2nd ed. Washington DC: Taylor and Francis.

Swartz RC, Cole FA, Lamberson JO, Ferraro SP, Schults DW, DeBen WA, Lee II H, Ozretich JR. 1994. Sediment toxicity, contamination, and amphipod abundance at a DDT and dieldrin-contaminated site in San Francisco Bay. *Environ Toxicol Chem* 13:949–962.

Swartz RC, Schults DW, Ditsworth GR, DeBen WA, Cole FA. 1985. Sediment toxicity, contamination, and macrobenthic communities near a large sewage outfall. In: Boyle TP, editor. Validation and predictability of laboratory methods for assessing the fate and effects of contaminants in aquatic ecosystems. Philadelphia: American Soc for Testing and Materials (ASTM). STP 865. p 152–175.

Swartz RC, Schults DW, Ozretich RJ, Lamberson JO, Cole FA, DeWitt TH, Redmond MS, Ferraro SP. 1995. $\sum$PAH: a model to predict the toxicity of polynuclear aromatic hydrocarbon mixtures in field-collected sediments. *Environ Toxicol Chem* 14:1977–1987.

[USEPA] U.S. Environmental Protection Agency. 1994a. Methods for measuring the toxicity and bioaccumulation of sediment-associated contaminants with freshwater invertebrates. Duluth MN: USEPA. EPA-600/R-94/024.

[USEPA] U.S. Environmental Protection Agency. 1994b. Methods for assessing the toxicity of sediment-associated contaminants with estuarine and marine amphipods. Narragansett RI: USEPA. EPA-600/R-94/025.

Wentsel R, McIntosh A, Anderson V. 1977. Sediment contamination and benthic invertebrate distribution in a metal-impacted lake. *Environ Pollut* 14:187–193.

Wentsel R, McIntosh A, Atchison G. 1977. Sublethal effects of heavy metal contaminated sediment on midge larvae (Chironomus tentans). *Hydrobiologia* 56:53–156.

Wentsel R, McIntosh A, McCafferty PC. 1978. Emergence of the midge Chironomus tentans when exposed to heavy metal contaminated sediments. *Hydrobiologia* 57:195–196.

SESSION 5
CRITICAL ISSUES IN ECOLOGICAL RELEVANCE

_____

Chapter 9

Ecological significance of endpoints used to assess sediment quality

William H. Clements

9.1 Introduction

Criteria for selecting endpoints in ERAs include susceptibility to the stressor, ease of measurement, unambiguous definitions, and societal and ecological relevance (Suter 1993). To link sediment quality assessments with ERAs, similar criteria should be employed to select endpoints to assess sediment quality. Ecological relevance of population, community, and ecosystem responses to contaminants has received relatively little attention, although some scientists argue that this is probably the most important criterion (Cairns 1986). If the ultimate objective of an ERA is to protect ecological integrity, it follows that ecological relevance of endpoints should be given greater priority.

One of the challenges that we face when conducting any ERA is distinguishing ecologically important changes from change *per se*. In other words, simply because we are able to demonstrate statistically significant differences between reference and polluted sites does not necessarily mean that biological or ecological integrity (*sensu* Karr and Dudley 1981) is compromised. More importantly, of the myriad of endpoints that have been shown to respond to contaminants in sediments, which ones indicate significant ecological effects? Although this problem is most critical for sublethal effects measured at lower levels of organization (*e.g.*, physiological and biochemical responses), ecological relevance of changes at higher levels of organization (populations, communities, and ecosystems) should also be evaluated. Because it is unlikely that any single endpoint will be useful in all situations, a suite of measures that integrate responses across levels of organization is necessary (Karr 1991; Karr 1993; Clements and Kiffney 1994a). I suggest that linking responses among levels of organization and correlating sublethal responses with significant ecological effects is an important area of research for sediment risk assessments (Clements and Kiffney 1994b).

Ecologically relevant effects of contaminants on freshwater ecosystems may be divided into three general categories. First, direct effects of contaminants may include shifts in community composition, reduced species diversity, or changes in ecosystem function. These direct effects generally result from differences in sensitivity among taxa. Secondly, indirect effects of contaminants may include alterations in predator–prey or competitive interactions. For example, lower abundance of a particular species may result from reduced abundance of its prey or increased abundance of a competitor. Finally, direct and indirect effects of contaminants on benthic communities may influence higher trophic levels via loss of important prey resources or food chain transfer of contaminants. The primary objective of this presentation is to discuss the ecological significance of end-points used to evaluate contaminated sediments. I will briefly discuss the appropriate-ness of using benthic communities in sediment risk assessments and argue that structural and functional endpoints at several levels of organization are necessary to evaluate contaminated sediments. Using studies from the primary literature and results of my research conducted at a USEPA Superfund site, I will describe the importance of measuring direct, indirect, and food chain effects of sediment contaminants. Finally, I will provide recommendations for future research to evaluate ecological significance of endpoints used in sediment risk assessments.

9.2 Focus on benthic communities

Most assessments of sediment contamination have focused on responses of benthic macroinvertebrates. The use of benthic macroinvertebrate communities to assess effects of contaminants on aquatic ecosystems has a long history (Cairns and Pratt 1993), and changes in composition of benthic communities have been used as indicators of water quality since the early 1900s (Carpenter 1924; Richardson 1929). Because of the attention that benthic communities have received in water quality assessments, unambiguous operational definitions of community responses to contaminants have been developed (Plafkin et al. 1989). Because of their intimate association with sediments, benthic com-munities are highly susceptible to sediment contaminants. Benthic macroinvertebrates are important in the function of aquatic ecosystems and in the diet of higher trophic lev-els (e.g., sport fish). Therefore, endpoints associated with benthic macroinvertebrates have both ecological and societal relevance. Because benthic macroinvertebrates readily accumulate sediment contaminants, they represent an important link to these higher trophic levels. Finally, assessments of benthic community structure provide data that may be used to assess integrated responses at several levels of organization simulta-neously. For example, Kerans and Karr (1994) have recently developed an index of bio-logical integrity for benthic communities, which integrates structural and functional measures into a single value.

9.3 Criticisms of single-species toxicity tests

Criticisms of single-species laboratory toxicity tests for predicting effects of contaminants on natural populations, communities, and ecosystems are well known (Cairns 1986). The basis for most of this criticism is the generally untested assumption that laboratory re-sponses of surrogate species are protective of ecosystem structure and function. Valida-tion of single-species toxicity tests, using more sophisticated testing systems such as microcosms, mesocosms, and direct field experimentation, have been employed to test this assumption. Good correspondence between laboratory toxicity tests and field assess-ments has been reported; however, in many instances failure to account for indirect ef-fects has resulted in underestimates of contaminant effects (Schindler 1987; Gonzalez and Frost 1994).

Similar criticisms of single-species toxicity tests are also applicable to sediment toxicity tests, in which growth rates or survivorship of surrogate benthic species (*e.g.*, *Chironomus tentans, Hyalella azteca*) are employed to predict effects of contaminated sediments on natural communities in the field. Failure to account for indirect effects and the lack of research on effects of sediment contaminants on ecosystem processes limits the applicability of sediment toxicity tests. Development of approaches that integrate sediment toxicity tests with field measurements, such as the sediment quality triad (Chapman 1989; Chapman *et al.* 1991) will improve the predictive ability of simple laboratory procedures.

9.4 Structural and functional measures

Ecotoxicologists generally distinguish between two types of responses to contaminants: structural and functional. Structural responses are typically associated with community-level measures and include estimates of abundance, species richness, diversity, and community composition. These measures have received the most attention in field assessments of contaminated sediments. Functional measures generally include ecosystem processes, such as the rates of productivity, nutrient cycling, energy flow, and decomposition. Although there has been recent interest in using ecosystem-level endpoints in hazard assessments, functional measures have been largely ignored in assessments of contaminated sediments. Hill (1992) has developed a relatively simple procedure for measuring community respiration in sediments and has shown that changes in respiration rates are indicative of contaminated sediments. The effects of sediment contaminants on ecosystem processes will vary depending on the size of the system. Pelagic processes dominate ecosystem function in large lakes and rivers, whereas benthic processes are generally more important in smaller lakes and streams because of the greater ratio of sediment to water volume (Reice and Wohlenberg 1993).

There is debate in the literature concerning whether structural or functional measures are more appropriate for assessing ecological integrity (Cairns and Pratt 1986). Because of functional redundancy of ecosystems and variability of ecosystem responses, ecosystem processes may be relatively insensitive to stress (Cairns and Pratt 1986; Schindler 1987). Despite the criticism of single-species toxicity tests described above, population responses to contaminants in the field may be very sensitive to stress. Included in Odum's (1992) top 20 list of "great ideas in ecology" is the concept that "the first signs of environmental stress usually occur at the population level, affecting especially sensitive species."

Structural and functional responses are so intimately related that a distinction is somewhat artificial. Indeed, a holistic perspective of ecology requires that we consider communities and ecosystems as more than simply the sum of their component populations (Webster 1979). Because biological systems are arranged in a hierarchical fashion, sediment contaminants will probably affect structural and functional responses simultaneously. Differential sensitivity of benthic populations often results in structural changes in benthic communities. These changes in community composition may have important cascading effects on ecosystem function. For example, Wallace *et al.* (1982) reported that reduced abundance of shredders in streams treated with insecticides resulted in changes in organic matter processing. Stewart and Hill (1993) speculate that removal of grazing snails had indirect effects on higher trophic levels and increased the movement of toxicants through a lotic food web. Just as ecosystem-level assessments complement assessments at lower levels of organization (Suter 1993), understanding population responses to contaminants provides information necessary to elucidate mechanisms of changes at higher levels. I suggest that assessments of sediment quality should include both structural and functional measures at several levels of organization.

9.5 Direct effects of sediment contaminants on benthic communities

Most research on contaminant effects in freshwater ecosystems has focused on direct effects (Clements 1991). In particular, shifts in community composition and replacement of sensitive species by tolerant species have received considerable attention. For example, we know that certain groups of organisms, such as mayflies, are highly sensitive to heavy metals, whereas other groups, such as many caddisflies and most orthoclad chironomids are quite tolerant. Thus, shifts in community composition observed at metal-polluted locations relative to reference sites may be a result of the loss of sensitive populations (Clements et al. 1992). Because reduced species richness or diversity may result in reduced ecosystem stability or trophic complexity (Connell 1978; Pimm 1984), changes in these endpoints are ecologically relevant. Similarly, reduced abundance of keystone species may affect ecological integrity. Keystone species have been identified in several marine ecosystems, especially rocky intertidal habitats; however, considerably less research has been conducted in freshwater systems, and few studies have documented the role of keystone species in lakes and streams. The lack of research in freshwater ecosystems does not mean that keystone species are absent, but rather demonstrates the difficulty measuring the role of keystone species. Experimental manipulation is the most direct way to demonstrate the importance of species in an ecosystem. Experimental manipulation is inherently more difficult in lakes and streams than in the rocky intertidal zone where species are easily removed from an essentially two-dimensional habitat. Therefore, documenting the role of keystone species will be very difficult in freshwater ecosystems.

Direct effects of contaminants on ecosystem function are also ecologically relevant and should be considered in assessments of contaminated sediments. As noted above, there is debate among researchers regarding the sensitivity of ecosystem-level endpoints; however, protection of ecological integrity should also include protection of functional integrity. Several functional endpoints have been used to measure effects of contaminants on aquatic ecosystems, including changes in primary and secondary productivity, decomposition rates, and nutrient cycling (Rapport et al. 1985). Few studies have examined effects of contaminated sediments on ecosystem function in the field.

9.6 Influence of location and previous exposure on responses to contaminants

Because direct effects of contaminants on benthic communities may vary spatially, care must be taken to select endpoints that show similar responses among locations. Kiffney and Clements (1994) report that benthic communities from small, high elevation streams are more sensitive to metals than those from larger, low elevation streams. Previous exposure to contaminants may also affect population-, community-, and ecosystem-level responses. While acclimation or adaptation of benthic populations to contaminants has been measured (Bryan and Hummerstone 1971; Klerks and Weis 1987; Klerks and Levinton 1989, 1993), few studies have examined this process at higher levels of organization (Neiderlehner and Cairns 1992). Using stream microcosms, I have observed that benthic communities from sites polluted with low concentrations of heavy metals were more tolerant to metals than communities from unpolluted locations. If variation in susceptibility to contaminants among locations is a general phenomenon, results of these studies have important implications for developing SQC.

9.7 Indirect effects of sediment contaminants on benthic communities

Shifts in community composition may also result from indirect effects of contaminants on species interactions, such as competition and predation. There is theoretical and empirical support for the hypothesis that species interactions play a major role in structuring freshwater and marine benthic communities (Connell 1961; Dayton 1971, 1975; Menge 1976; Peckarsky and Dodson 1980; Walde and Davies 1984; Walde 1986; Menge and Sutherland 1987; Hart 1992). In particular, studies of predation and competition have received attention from benthic ecologists for many years, and there is general agreement among ecologists that these interactions are important in some habitats. Again, much of the experimental research demonstrating direct effects of predation or competition on community structure has been conducted in marine rocky intertidal habitats; however, ecologists have begun to examine the role of species interactions in freshwater benthic communities (Peckarsky and Dodson 1980; Walde and Davies 1984; Walde 1986; Hart 1992).

Ecologists have long recognized the importance of interactions between abiotic factors, such as environmental stress, and biotic factors (Park 1962; Tilman 1977; Dunson and Travis 1991). Surprisingly, the effect of contaminants on biotic interactions, such as predation and competition, has been largely ignored in ecotoxicology. Warner et al. (1993) found that interspecific competition between anurans was influenced by acidification. Dunson and Travis (1991) observed that competition for food between killifish was influenced by salinity. Clements et al. (1989) reported that net-spinning caddisflies (Trichoptera: Hydropsychidae) were more susceptible to stonefly (Plecoptera: Perlidae) predation in streams dosed with copper than in control streams. Finally, Wipfli and Merritt (1994) reported that reduced black fly density in a larvicide-treated stream altered species interactions. They concluded that community structure may be indirectly affected by applications of larvicides. These studies suggest that abiotic factors may influence the outcome of species interactions and indirectly alter community composition.

The effects of contaminants on species interactions will likely depend on the relative sensitivities of the interacting species (Clements et al. 1989; Arnott and Vanni 1993; Warner et al. 1993). Menge and Sutherland (1987) developed a model of community regulation that predicts changes in the relative importance of species interactions along environmental stress gradients. According to this model, disturbance is the most important factor structuring communities at extreme levels of environmental stress, whereas at intermediate and low levels of stress, the importance of competition and predation increases. If environmental contaminants may be considered a type of environmental stress, then similar responses may occur along contaminant stress gradients. With respect to the importance of predation, the specific outcome of environmental stress models differs depending on whether consumers or their prey are more susceptible to stress (Menge and Olson 1990). For example, if predators are more sensitive to stress than their prey, the consumer stress model predicts that predation rates will be lower in high-stress environments. This pattern was observed in rocky intertidal habitats subjected to extreme wave action (Menge and Olson 1990), streams where the hydraulic regime was unfavorable to predators (Peckarsky et al. 1990), and in acidified lakes (Locke and Sprules 1994). In contrast, if prey are more sensitive to stress, the prey stress model predicts that predation rates will be greater in stressful environments. This pattern was observed in the predator-prey experiments between stoneflies and caddisflies described above (Clements 1994). Regardless of the direction of effects, these studies demonstrate that indirect effects of contaminants on benthic communities are complex and may complicate assessments of direct effects.

One of the most important reasons for investigating indirect effects is that species interactions may actually be more sensitive to contaminants than direct effects. For example, field and laboratory experiments have shown that filter-feeding caddisflies (Trichoptera: Hydropsychidae) are relatively insensitive to Cu exposure (Clements et al. 1992). However, susceptibility of caddisflies to predation by stoneflies increased when both groups were exposed to metals (Clements et al. 1989). Similar results were reported by Kiffney (1995) for these same groups.

The lack of studies on indirect effects of contaminants on species interactions is surprising, given the prominent role that research on species interactions has played in aquatic ecology. The lack of research on indirect effects is probably related to the difficulty of designing and implementing field experiments. In the laboratory, relatively complex experimental designs are necessary to separate direct effects of contaminants from the indirect effects on species interactions. Separating these effects in the field would first require demonstrating that species interactions are important, and then demonstrating that these interactions are affected by contaminants. Both tasks require experimentally manipulating abundances of several species. For example, by conducting caging experiments in reference and polluted locations, one could evaluate the effects of predation on both communities. If predation effects were greater in the contaminated sites, this would support the hypothesis that contaminants increase the susceptibility of prey organisms to predation. Alternatively, microcosm experiments using communities obtained from reference and polluted locations could be employed to measure effects of contaminants on species interactions. In experimental streams I have measured effects of stonefly predation on benthic invertebrate communities obtained from reference and metal-polluted locations. Preliminary results show that communities from contaminated sites were more susceptible to stonefly predation than communities from reference sites.

9.8 Effects on higher trophic levels

Alterations in benthic macroinvertebrate communities resulting from contaminated sediment can have negative effects on fish predators. These impacts occur either through loss of preferred prey resources or by food chain transfer of contaminants. Although numerous studies have measured transfer of contaminants from benthic macroinvertebrates to fish (Reynoldson 1987; Dallinger *et al.* 1987; Hatakeyama and Yasuno 1987; Douben 1989; Clements *et al.* 1994; Woodward *et al.* 1994), relatively few studies have examined pollution-induced changes in prey availability and predator feeding habits (Jefree and Williams 1980; Clements and Livingston 1983; Stair *et al.* 1984; Rees 1994). Because many fish predators are opportunistic feeders, it is hypothesized that pollution-induced changes in prey abundance will alter feeding habits. For example, differences in macroinvertebrate abundance and community composition between reference and metal-polluted sites at the Arkansas River (a USEPA Superfund site) altered feeding habits of brown trout (*Salmo trutta*) collected from these two sites (Rees 1994). Jefree and Williams (1980) reported similar results for the purple-striped gudgeon (*Mogurnda mogurnda*) and speculated that selective predation by tolerant predators on abundant prey species in polluted habitats may dampen differences between reference and polluted sites. While these studies demonstrate that opportunistic predators are capable of switching to more abundant prey resources at polluted locations, changes in feeding habits may influence predator energy budgets. Optimal foraging theory predicts that if a predator's diet is influenced by natural selection, animals will have the greatest fitness if they maximize energy intake and minimize energy expenditures (Werner and Hall 1974). Although other factors, such as risk of predation and habitat availability, also influence predator fitness, these simple models suggest that any anthropogenic changes in feeding habits will affect a predator's energy budget. For example, shifts in brown trout feeding habits observed at polluted sites in the Arkansas River, from relatively large organisms such as mayflies and stoneflies to small orthoclad chironomids, may increase energy expenditures and have negative impacts on brown trout populations (Rees 1994). Similar responses are likely to occur in other situations where prey communities are altered by sediment contaminants.

Bioaccumulation of sediment contaminants by benthic macroinvertebrates is frequently measured in assessments of sediment quality. Because benthic macroinvertebrates may accumulate contaminants from sediments and transfer these contaminants to higher trophic levels, the concentration of contaminants in benthic organisms is ecologically relevant. Food chain transfer of sediment contaminants from benthic macroinvertebrates to fish and the influence of food chain structure on contaminant levels in top predators has been considered (Rasmussen *et al.* 1990; Stewart and Hill 1993; Clements *et al.* 1994; MacDonald *et al.* 1994). There is some controversy over the relative importance of dietary and aqueous routes of exposure to fish, particularly for heavy metals; however, several studies have shown that dietary exposure is significant (Hatakeyama and Yasuno 1987; Douben 1989; Clements *et al.* 1994). More importantly, researchers have demonstrated significant effects on growth and survivorship of fish feeding on contaminated prey (Woodward *et al.* 1994).

Because levels of contaminants in benthic organisms are often orders of magnitude higher than those in the water column, it follows that food chain transfer may be important. Even for contaminants such as heavy metals where dietary transfer from benthic invertebrates to fish is relatively inefficient, concentrations in fish can still reach harmful levels owing to selective consumption of contaminated prey (Dallinger and Kautzky 1985; Dallinger *et al.* 1987). This "food chain effect" hypothesized by Dallinger *et al.* (1987) is influenced by mechanisms of tolerance of prey populations. If tolerant prey compartmentalize or sequester contaminants, exposure to higher trophic levels may be enhanced. Finally, because concentrations of contaminants will most likely vary significantly among prey taxa (Kiffney and Clements 1993), feeding habits and predator preferences of fish will also influence contaminant accumulation (Rees 1994).

In summary, measurements of contaminant levels in benthic organisms are potentially useful measures of sediment quality if these measurements can be linked to direct effects on benthic organisms or to transfer to higher trophic levels. This latter step will require detailed information on prey abundance, contaminant levels in prey, laboratory studies of transfer efficiency, and feeding habits of predators.

9.9 Recommendations for future research

- Studies that integrate structural and functional responses at several levels of organization are necessary. Studies conducted at lower levels of organization will help elucidate mechanisms responsible for changes at higher levels.
- More studies documenting the correspondence or lack of correspondence between simple laboratory toxicity tests and field assessments are necessary.
- More consideration of the indirect effects of contaminants on species interactions is necessary. Assessing the importance of these indirect effects will be difficult and will require experimental manipulation. There is a rich body of literature on disturbance theory and the influence of disturbance on species interactions that should be considered when designing field studies.
- Measuring bioaccumulation of sediment contaminants and food chain transfer to higher trophic levels is an important area of research. However, in order to improve the ecological relevance of these endpoints, tissue concentrations of contaminants must be linked to some estimate of ecological effects.

9.10 References

Arnott SE, Vanni MJ. 1993. Zooplankton assemblages in fishless bog lakes: influence of biotic and abiotic factors. *Ecology* 74:2361–2380.

Bryan GW, Hummerstone LG. 1971. Adaptation of the polychaete *Nereis diversicolor* to estuarine sediments containing high concentrations of zinc and cadmium. *J Biol Assoc UK* 53:839–857.

Cairns Jr J. 1986. The myth of the most sensitive species. *BioScience* 36:670–672.

Cairns Jr J, Pratt JR. 1986. On the relation between structural and functional analyses of ecosystems. *Environ Toxicol Chem* 5:785–786.

Cairns J, Pratt JR. 1993. A history of biological monitoring using benthic macroinvertebrates. In: Resh DM, Rosenberg VH, editors. Freshwater biomonitoring and benthic macroinvertebrates. New York: Chapman and Hall. p 10–27.

Carpenter K E 1924. A study of the fauna of rivers polluted by lead mining in the Aberystwyth District of Cardiganshire. *Ann Appl Biol* 11:1–23.

Chapman PM. 1989. Current approaches to developing sediment quality criteria. *Environ Toxicol Chem* 8:589–599.

Chapman PM, Power EA, Dexter RN, Andersen HB. 1991. Evaluation of effects associated with an oil platform, using the sediment quality triad. *Environ Toxicol Chem* 10:407–424.

Clements WH. 1991. Community responses of stream organisms to heavy metals: a review of observational and experimental approaches. In: Newman MC, McIntosh AW, editors. Ecotoxicology of metals: current concepts and applications. Chelsea MI: Lewis. p 363–391.

Clements WH. 1994. Benthic community responses to heavy metals in the Upper Arkansas River Basin, Colorado. *J North Amer Benthol Soc* 13:30–44.

Clements WH, Cherry DS, Cairns Jr J. 1989. The influence of copper exposure on predator-prey interactions in aquatic insect communities. *Freshwater Biol* 21:483–488.

Clements WH, Cherry DS, Van Hassel JH. 1992. Annual variation of benthic communities exposed to heavy metals at the Clinch River (Virginia): evaluation of an index of metal impact. *Can J Fish Aquat Sci* 49:1686–1694.

Clements WH, Kiffney PM. 1994a. An integrated approach for assessing the impact of heavy metals at the Arkansas River, CO. *Environ Toxicol Chem* 13:397–404.

Clements WH, Kiffney PM. 1994b. Assessing contaminant impacts at higher levels of biological organization. *Environ Toxicol Chem* 13:357–359.

Clements WH, Livingston RJ. 1983. Overlap and pollution-induced variability in the feeding habits of filefish (Pisces: Monacanthidae) from Apalachee Bay, Florida. *Copeia* 1983:331–338.

Clements WH, Oris JT, Wissing TE. 1994. Accumulation and food chain transfer of benzo[a]pyrene and fluoranthene in aquatic organisms. *Arch Environ Contam Toxicol* 26:261–266.

Connell JH. 1961. The influence of interspecific competition and other factors on the distribution of the barnacle *Chthamalus stellatus*. *Ecology* 42:710–723.

Connell JH. 1978. Diversity in tropical rain forests and coral reefs. *Science* 199:1302–1309.

Dallinger R, Kautzky H. 1985. The importance of contaminated food uptake for the heavy metals by rainbow trout (*Salmo gairdneri*): a field study. *Oecologia* (Berlin) 67:82–89.

Dallinger R, Prosi F, Back H. 1987. Contaminated food and uptake of heavy metals by fish: a review and a proposal for further research. *Oecologia* (Berlin) 73:91–98.

Dayton PK. 1971. Competition, disturbance, and community organization: the provision and subsequent utilization of space in a rocky intertidal community. *Ecol Monogr* 41:351–389.

Dayton PK. 1975. Experimental evaluation of ecological dominance in a rocky intertidal algal community. *Ecol Monogr* 45:137–159.

Douben PET. 1989. Metabolic rate and uptake and loss of cadmium from food by the fish *Noemaccheilus barbatulus* L. (stone loach). *Environ Pollut* 59:177–202.

Dunson WA, Travis J. 1991. The role of abiotic factors in community organization. *Am Nat* 138:1067–1091.

Gonzalez MJ, Frost TM. 1994. Comparisons of laboratory bioassays and a whole-lake experiment: rotifer responses to experimental acidification. *Ecol Appl* 4:69–80.

Hart DD. 1992. Community organization in streams: the importance of species interactions, physical factors, and chance. *Oecologia* (Berlin) 91:220–228.

Hatakeyama S, Yasuno M. 1987. Chronic effects of Cd on the reproduction of the guppy (*Poecilia reticulata*) through Cd-accumulated midge larvae (*Chironomus yoshimatsui*). *Ecotox Environ Saf* 14:191–207.

Hill BH. 1992. Substrate-influenced respiration: a community-level measure of contaminant impacts in benthic microbial communities [abstract]. In: Society of Environmental and Chemistry (SETAC) Abstracts, 13th Annual Meeting. Pensacola FL: SETAC Pr. p 256.

Jefree RA, Williams NJ. 1980. Mining pollution and the diet of the purple-striped gudgeon *Mogurnda mogurnda* Richardson (Eleotridae) in the Finniss River, Northern Territory. *Australia Ecol Monogr* 50:457–485.

Karr JR. 1991. Biological integrity: a long-neglected aspect of water resource management. *Ecol Appl* 1:66–84.

Karr JR. 1993. Defining and assessing ecological integrity: beyond water quality. *Environ Toxicol Chem* 12:1521–1531.

Karr JR, Dudley DR. 1981. Ecological perspective on water quality goals. *Environ Manage* 5:55–68.

Kerans BL, Karr JR. 1994. The benthic index of biotic integrity (B-IBI) for rivers of the Tennessee Valley. *Ecol Appl* 4:768–785.

Kiffney PM. 1995. Influence of abiotic and biotic factors on the response of benthic macroinvertebrates to metals [dissertation]. Fort Collins: Colorado State Univ. 160 p.

Kiffney PM, Clements WH. 1993. Bioaccumulation of heavy metals by benthic organisms at the Arkansas River, Colorado. *Environ Toxicol Chem* 12:1507–1517.

Kiffney PM, Clements WH. 1994. Structural responses of benthic macroinvertebrate communities from different stream orders to zinc. *Environ Toxicol Chem* 13:389–395.

Klerks PL, Levinton JS. 1989. Effects of heavy metals in a polluted aquatic ecosystem. In: Levin SA, Harwell MA, Kelly JR, Kimball KD, editors. Ecotoxicology: problems and approaches. New York: Springer-Verlag. p 41–67.

Klerks PL, Levinton JS. 1993. Evolution of resistance and changes in community composition in metal-polluted environments: a case study on Foundry Cove. In: Dallinger R, Rainbow PS, editors. Ecotoxicology of metals in invertebrates. Ann Arbor MI: SETAC Special Publications Series. p 223–241.

Klerks PL, Weis JS. 1987. Genetic adaptation to heavy metals in aquatic organisms: a review. *Environ Pollut* 45:173–205.

Locke A, Sprules WG. 1994. Effects of lake acidification and recovery on the stability of zooplankton food webs. *Ecology* 75:498–506.

MacDonald CR, Metcalfe CD, Balch GC, Metcalfe TL. 1994. Distribution of PCB congeners in seven lake systems: interactions between sediment and food-web transport. *Environ Toxicol Chem* 12:1991–2003.

Menge BA. 1976. Organization of the New England rocky intertidal community: role of predation, competition and environmental heterogeneity. *Ecolog Monogr* 46:355–393.

Menge BA, Olson AM. 1990. Role of scale and environmental factors in regulation of community structure. *Trends Ecol Evol* 5:52–57.

Menge BA, Sutherland JP. 1987. Community regulation: variation in disturbance, competition, and predation in relation to environmental stress and recruitment. *Am Nat* 130:730–757.

Neiderlehner BR, Cairns Jr J. 1992. Community response to cumulative toxic impact: effects of acclimation on zinc tolerance of *aufwuchs*. *Can J Fish Aquat Sci* 49:2155–2163.

Odum EP. 1992. Great ideas in ecology for the 1990s. *BioScience* 42:542–545.

Park T. 1962. Beetles, competition, and populations. *Science* 138:1369–1375.

Peckarsky BL, Dodson SI. 1980. An experimental analysis of biological factors contributing to stream community structure. *Ecology* 61:1283–1290.

Peckarsky BL, Horn SC, Statzner B. 1990. Stonefly predation along a hydraulic gradient: a field test of the harsh-benign hypothesis. *Freshwater Biol* 24:181–191.

Plafkin JL, Barbour MT, Porter KD, Gross SK, Hughes RM. 1989. Rapid bioassessment protocols for use in streams and rivers: benthic macroinvertebrates and fish. Washington DC: U.S. Environmental Protection Agency (USEPA). EPA 440/4-89-001.

Pimm SL. 1984. The complexity and stability of ecosystems. *Nature* 350:321–326.

Rapport, DJ, Regier HA, Hutchinson TC. 1985. Ecosystem behavior under stress. *Am Nat* 125:617–640.

Rasmussen JB, Rowan DJ, Lean DRS, Carey JH. 1990. Food chain structure in Ontario lakes determines PCB levels in lake trout (*Salvelinus namaychush*) and other pelagic fish. *Can J Fish Aquat Sci* 47:2030–2038.

Rees DE. 1994. Indirect effects of heavy metals observed in macroinvertebrate availability, brown trout (Salmo trutta) diet composition, and bioaccumulation in the Arkansas River, Colorado [thesis]. Fort Collins: Colorado State Univ. 51 p.

Reice SR, Wohlenberg M. 1993. Monitoring freshwater benthic macroinvertebrates and benthic processes: measures for assessment of ecosystem health. In: Resh DM, Rosenberg VH, editors. Freshwater biomonitoring and benthic macroinvertebrates. New York: Chapman and Hall. p 287–305.

Reynoldson TB. 1987. Interactions between sediment contaminants and benthic organisms. *Hydrobiologia* 149:53–66

Richardson RE. 1929. The bottom fauna of the Middle Illinois River 1913–1925. Its distribution, abundance, valuation and index value in the study of stream pollution. *Bull Illinois Nat Hist Surv* 17:387–475.

Schindler DW. 1987. Detecting ecosystem responses to anthropogenic stress. *Can J Fish Aquat Sci* (Suppl) 1:6–25.

Stair DM, Tolbert VR, Vaughn GL. 1984. Comparison of growth, population structure, and food of the creek chub *Semotilus atromaculatus* in undisturbed and surface-mining-disturbed streams in Tennessee. *Environ Polut (Ser A)* 35:331–343.

Stewart AJ, Hill WR. 1993. Grazers, periphyton, and toxicant movement in streams. *Environ Toxicol Chem* 12:955–957.

Suter II GW. 1993. Ecological risk assessment. Chelsea MI: Lewis.

Tillman D. 1977. Resource competition between planktonic algae: an experimental and theoretical approach. *Ecology* 58:338–348.

Walde SJ. 1986. Effects of abiotic disturbance on a lotic predator-prey interaction. *Oecologia* (Berlin) 69:243–247.

Walde SJ, Davies RW. 1984. Invertebrate predation and lotic prey communities: evaluation of *in situ* enclosure/exclosure experiments. *Ecology* 65:1206–1213.

Wallace JB, Webster JR, Cuffney TF. 1982. Stream detritus dynamics: regulation by invertebrate consumers. *Oecologia* (Berlin) 53:197–200.

Warner SC, Travis J, Dunson WA. 1993. Effect of pH variation on interspecific competition between two species of hylid tadpoles. *Ecology* 74:183–194.

Webster JR. 1979. Hierarchical organization of ecosystems. In: Halfon E, editor. Theoretical systems ecology. New York: Academic Pr. p 119–129.

Werner EE, Hall DJ. 1974. Optimal foraging and the size selection of prey by the bluegill sunfish *Lepomis macrochirus*. *Ecology* 55:1042–1052.

Wipfli MS, Merritt RW. 1994. Disturbance to a stream food web by a bacterial larvicide specific to black flies: feeding responses of predatory macroinvertebrates. *Freshwater Biol* 32:91–103.

Woodward DF, Brumbaugh WG, DeLonay AJ, Little EE, Smith CE. 1994. Effects on rainbow trout fry of a metals-contaminated diet of benthic invertebrates from the Clark Fork River, Montana. *Trans Amer Fish Soc* 123:51–62.

SESSION 5
CRITICAL ISSUES IN ECOLOGICAL RELEVANCE
_____Chapter 10
Role of abiotic factors in structuring benthic invertebrate communities in freshwater ecosystems

David M. Rosenberg, Trefor B. Reynoldson, Kristin E. Day, Vince H. Resh

10.1 Introduction

Many contaminants in fresh waters are insoluble and may attach to suspended particulate material that eventually settles on the bottom of aquatic ecosystems. As a consequence, sediments can act as a sink for an array of organic and inorganic contaminants, which can affect the organisms that live in or near the sediment. These organisms may then mediate contaminant transfer back to the water column or to other trophic levels.

Ecological risk assessment is a process for determining the probability of adverse ecological effects in an ecosystem resulting from exposure to stressors (USEPA 1992; Parkhurst _et al._ 1994; Calow 1995). Ecological risk assessment is currently the focus of several intensive research programs, and methods are being developed for collecting chemical and biological data for inclusion in mathematical models used by environmental managers (_e.g._, Burns _et al._ 1994).

Ecological risk assessment of sediment contamination is a subdiscipline of ERA. Sediment ecological risk assessment has been applied to ponds, embayments, the bottom layers of lakes and estuaries, and the depositional areas of streams and rivers (_e.g._, Canfield _et al._ 1994; Pastorok _et al._ 1994). As with many other toxicological approaches used in fresh waters, SERAs frequently focus on benthic invertebrates (or _benthos_). These organisms are used for the following reasons (La Point and Fairchild 1992; Rosenberg and Resh 1993a; Metcalfe-Smith 1994; Davis 1995; Resh 1995): 1) they have been used historically in assessing environmental degradation; 2) they are widespread and can be affected by environmental perturbations in many different types of freshwater systems; 3) they live on and in the sediments and so are directly associated with chemical contaminants; 4) the presence of a large number of species offers a spectrum of responses to environmental stress; 5) their sedentary nature allows effective spatial analysis of pollutant or disturbance effects; 6) they have life cycles of months or years, and so can act as continuous monitors of environmental quality; and 7) their responses can be quantified in a manner that can be understood by managers, regulators, and the general public.

Most SERAs attempt to measure the abundance and richness of the benthos in specific areas of contamination, and then to compare the results to those obtained at control or reference sites (Canfield _et al._ 1994; Pastorok _et al._ 1994; Barbour _et al._ 1995). Much has been written to describe the collection, identification, and analysis of benthic data for biological monitoring of aquatic ecosystems (for reviews, see Hellawell 1986; Plafkin _et al._

1989; Rosenberg and Resh 1993b; Metcalfe-Smith 1994). However, despite numerous studies documenting alterations of benthic invertebrate communities as a result of environmental stress (*e.g.*, Rosenberg and Wiens 1976; Winner *et al.* 1980; Krieger 1984; La Point *et al.* 1984; Clements *et al.* 1988; Canfield *et al.* 1994), the success of correctly assessing the health or degradation of these communities depends on how well the responses caused by contamination can be discriminated from responses caused by other environmental factors (Dunson and Travis 1991; Hughes 1995).

Environmental assessments may confuse natural variability with environmental degradation because a thorough understanding of the many natural factors that can influence or regulate variability is lacking (Landis *et al.* 1994). Reynoldson (1984) and France (1990) have previously cautioned that the macrobenthos can respond to seemingly minor changes in substrate particle size, organic content, texture, and water quality as well as to the presence of contaminants. Furthermore, spatial heterogeneity in depositional areas can be high, which requires large numbers of sampling unit replicates to distinguish between natural variability and anthropogenic perturbation. For example, an SERA of the Upper Clark Fork River in Montana (Canfield *et al.* 1994) recognized that increased numbers of Chironomidae and Oligochaeta, and a predominance of metal-tolerant species in metal-contaminated sediments, suggested an imbalanced benthic community; however, they also noted that factors such as differences in habitat and, perhaps, intermittent physical disturbances could account for the observed community structure. Differences in habitat type are sources of spatial heterogeneity; intermittent disturbances are sources of temporal heterogeneity (Townsend and Hildrew 1994). An investigation of acidified rivers and lakes throughout the province of Ontario (Gibbons and Mackie 1991) revealed that reproductive output of the amphipod *Hyalella azteca* could be correlated with a number of environmental variables (*e.g.*, sulphate, calcium hardness, sediment particle size, seston, and organic matter of the fine sediment) in addition to the variable of interest, decreased pH. Thus, it is important in the SERA process to adequately describe the benthic communities that are being sampled and to understand the natural environmental factors affecting various habitats located within assessment and reference areas.

The objectives of this paper are 1) to describe environmental variables that are important in structuring natural benthic invertebrate communities in rivers and lakes, 2) to address the influence of spatial and temporal scales on habitat characteristics, 3) to discuss approaches that may be useful in differentiating between natural variability and anthropogenic stress in SERA, and 4) to identify implications for SERA and future research needs.

10.2 Rivers, lakes, and scales

Communities of organisms respond to and are structured by an array of abiotic and biotic variables. The significance of these variables to SERAs depends on the location of the variables (*i.e.*, in lakes or rivers) and on the spatial and temporal scales of interest. For example, rivers are oriented horizontally (*e.g.*, Vannote *et al.* 1980; Johnson *et al.* 1995), whereas lakes are oriented vertically (Ryder and Pesendorfer 1989). Rivers provide connections for transfer of materials, whereas lakes provide storage.

Morphology (*e.g.*, gradient, width, depth, substrate type), hydrodynamics, and tempera-ture together form the principal components of the abiotic milieu of streams. In contrast, morphology (*e.g.*, depth, area, volume) and climate determine the seasonal hydrodynam-ics of lakes (*e.g.*, spring and fall overturns, and wave dynamics of north-temperate lakes). Most of the energy flow in lakes begins with autochthonous production (photosynthesis), which is in contrast to allochthonous production that occurs in most parts of rivers. Other differences in abiotic and biotic variables of rivers and lakes are summarized in Table 10-1.

A close coupling exists between abiotic and biotic variables in fresh waters and their ef-fects on sediment contamination (Table 10-1). The sediments of lakes and rivers harbor the organisms of interest to SERA, so the different effects of abiotic and biotic variables on sediment contamination are of interest here. For example, mean particle size in rivers decreases in a downstream direction, whereas in lakes it decreases with depth. In both cases, mean particle size affects both processes within sediments, and the movement of sediments and contaminants into the water column. Floating plants are limited to slow-flowing reaches of rivers, whereas they are restricted to sheltered bays of lakes. Floating plants may directly affect the transport of contaminants to sediments when they die, settle, and decay. Many other examples of such interactions appear in Table 10-1.

The spatial-temporal organization of rivers and lakes has special relevance to SERA, and the variables being examined in an SERA must be appropriate to the spatial and tempo-ral scales chosen for analysis. It is important to note that those scales selected for exami-nation are artificial extracts from what is a natural continuum. Furthermore, factors that affect benthic invertebrates in rivers and lakes at different, arbitrarily chosen, spatial and temporal scales can act simultaneously (Resh and Rosenberg 1989).

10.2.1 Rivers

Physical factors operating at a variety of spatial and temporal scales ultimately create the plethora of riverine environments. Frissell *et al.* (1986) have identified a number of spa-tial scales at which river systems can be examined; these scales occur in an hierarchy ranging from the entire stream system (catchment) to the microhabitat level (Figure 10-1). Physical characteristics at any given scale generally are determined from the scales above in the hierarchy (Townsend and Hildrew 1994). That is, aquatic systems at progres-sively larger scales in the hierarchy are the result of progressively larger-scale and less-frequent geophysical events. However, it is the small-scale processes such as erosion and deposition that ultimately produce the habitats normally investigated (see below), and it is these processes that impinge directly on individual organisms. To explain this phenom-enon, we have selected several scales in an hypothetical river system.

The catchment of this hypothetical river covers about 250,000 km^2 and it has been in existence for a little more than 100,000 years. The physical forces responsible for its present form include the recent geology of the area, its glacial history, and long-term rain-fall patterns.

Table 10-1   Differences in some physical and biological attributes of lotic and lentic systems and their relationship to sediment contamination

Attributes	Lotic systems	Lentic systems	Activities in relation to sediment contamination
Erosion/deposition ratio	Decreases downstream	Decreases from nearshore littoral to greater depths	1, 3
Shoreline erosion	Extensive, induced by water currents	Localized, induced by wind-driven waves	1
Mean particle size	Decreases downstream	Decreases with depth	2, 3
Number of substrate types	Increases downstream	Determined by geomorphology and wave action	2
Distribution of substrate	Determined by wave currents; gravity driven	Determined by wind-induced currents, geology, and geography	1, 2, 3
Breadth/shoreline length ratio	Low	High	1
Watershed/surface area ratio	High	Low	1
Ice scouring effects	Robust, extensive	Localized to windward, near-shore littoral	3, 4
Flow characteristics	Unidirectional, horizontal	Three-dimensional	3, 4
Current	Gravitational movement, decreases vertically	Wind-induced, convectional	3, 4
Flow diversions	Common, habitat implications	Rare	1, 3
Water-level fluctuations	Flooding	Minor variations	1, 3, 4
Flooding effects on biota	Traumatic; reset event	Diminished	1, 3, 4
Groundwater/surface drainage ratio (summer)	High ratio decreases temperature; low ratio increases temperature	Significant only in seepage lakes; effect same as in rivers	2, 3
Oxygen content	Usually high	Variable	2, 3, 4?
Diurnal variation of dissolved O_2	Dependent upon photosynthesis and respiration rates, and decomposition	Not significant	2, 3
Phosphorus	Variable, headwaters to mouth	High in spring and at overturns	2
Total dissolved solids	Increase downstream	Temporal cycle	2, 3?
Turbidity	High; varies among rivers and spatially	Low; varies among lakes and temporally	3

Table 10-1   continued

Attributes	Lotic systems	Lentic systems	Activities in relation to sediment contamination
Rate of nutrient influx	Governed by terrestrial vegetation, flooding	Determined by internal cycling rates	1, 2, 3
Speed of nutrient transit	Rapid; determined by gradient and current	Slow; determined by wind and waves, and overturn	1, 2
Nutrient retention	Low	High	2, 3
Mineral uptake rates	Rapid	Low	2, 3
Floral distribution	Reliant upon nutrients, current	Dependent on depth, substrate	2
Phytoplankton abundance	Low; greatest in intermediate or high orders; limited by turbidity, turbulence	High	1, 2
Bryophytes/lichens	Prevalent; fast-flowing areas; shade tolerant; some limited by hard water	Scarce	1, 2, 4?
Macrophyte adaptations	Few; restricted to running waters	Many	1, 2, 4?
Floating plants	Limited to slow-flowing reaches; prevalent in tropical rivers	Restricted to sheltered bays	1?
Primary fish forage	Invertebrates	Forage fishes	2, 3
Major energy source	Detritus	Solar radiation	1, 2
Nutrient regime	Spiraling	Cycling	2
Reset events	Floods (severe); ice scouring	Overturn (moderate)	1, 2, 3, 4
Resilience	High	Low	2, 4
Community structure	Sequential (seasonal); successional (spatial); mainly stochastic	Harmonic; successional (temporal); mainly deterministic	1, 2, 3, 4
Community refugium	Boundary layer; hyporheic zone (*i.e.*, deep, inhabitable substrate)	Ecotone; shallow, inhabitable substrate	1, 2?

Source: Attributes and differences from tables 1-8 of Ryder and Pesendorfer (1989) *Can Spec Publ Fish Aquat Sci* 106:65–85. Reprinted by permission.

"Activities" column added: 1 = Direct effects on transport of contaminants to sediments; 2 = Effects on processes within sediments; 3 = Movement of sediments and contaminants into water column; 4 = Removal of contaminants from sediments; ? expresses uncertainty

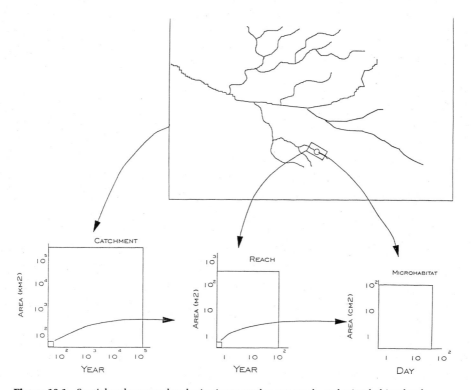

Figure 10-1 Spatial and temporal scales in rivers: catchment, reach, and microhabitat levels

If the scale is reduced several orders of magnitude to a stream reach, this consists of a riffle-pool sequence. The reach occupies an area of ≈500 m². Its present form was determined by the stream gradient in the area and the last 100-year flood.

Reducing the scale to the microhabitat level, a single rock sitting in the riffle of the stream reach is considered. The top of the rock covers an area of ≈100 cm². The rock has been there since the last major freshet produced enough tractive force to roll it from elsewhere to its present position (*e.g.*, within weeks). As long as this rock is stable, it serves as potential habitat for benthic invertebrates.

Next we describe the factors that affect benthic invertebrates at the three spatial-temporal riverine scales chosen.

    1) Catchment level. Several authors have used multivariate statistical approaches to examine factors that regulate lotic macroinvertebrates over large geographic scales. Ormerod and Edwards (1987) examined a small catchment (4200 km²) in Wales; Corkum and Currie (1987) and Corkum (1989) examined ≈100 rivers in Alberta, British Columbia, the Yukon, and Alaska, and Wright *et al.* (1984,

1988) studied rivers in all of Great Britain. Each of these studies discovered a number of variables that were correlated with benthic invertebrate community structure (Table 10-2). Little concordance is evident among the variables in the studies, partly because of the different geographic areas and partly because of differences in the variables measured.

Recently, we used a multivariate approach to design a biomonitoring program based on benthic invertebrates, for the Fraser River catchment in British Columbia (unpublished data). The program uses physical and chemical variables collected along with benthic invertebrates from 250 reference sites spread through the catchment. The aim of the program is to create a model, based on reference conditions, that uses conservative physical and chemical variables to predict the community structure of macroinvertebrates. Predicted community structure at impacted sites may then be compared to actual benthic communities, and the divergence would indicate the extent of remediation required. The method is an adaptation of the River InVertebrate Prediction And Classification System (RIVPACS) developed in Great Britain by Wright et al. (1984, 1988) and the BEnthic Assessment of SedimenT (BEAST) model developed for the Great Lakes by Reynoldson et al. (1995). These large-scale models, of necessity, do not identify what has caused the benthic invertebrate community to deviate from reference conditions; that information comes from smaller-scale experimental work.

2) Reach level. A number of factors influence benthic invertebrates at this level, which is the most-studied scale in rivers, including substrate stability, discharge, temperature, and food availability (Resh and Rosenberg 1984). For example, the stability of substrate at a given discharge will determine the numbers of invertebrates present in a stretch of stream (e.g., Cobb et al. 1992). Invertebrate numbers may be reduced during a spate in patches of small, unstable gravel or shale, whereas at the same discharge they are not affected in patches of larger, more stable substrate. Densities and diversities of invertebrates in affected reaches can recover quickly to pre-spate levels during periods of low flow (Cobb et al. 1992).

Changes in the above factors that result from anthropogenic activities at the reach level can also affect benthic invertebrates. For example, extensive clearcutting of forests may result in increased discharge that moves substrate at unexpected times of the year. If the natural capacity of the system to absorb disturbance is exceeded, the ability of the benthic invertebrate community to recover may be substantially altered. Another example may be dams that release water much colder than can be tolerated by downstream benthic fauna historically adapted to warmer water (e.g., Lehmkuhl 1972). Such types of abiotic factors must be considered alongside chemical stressors in SERA.

But what about the pool part of the riffle-pool sequence? In general, the combination of subhabitats collectively referred to as a pool has received less study

Table 10-2   Summary of catchment-level (and larger) variables that correlate with benthic invertebrate community structure in lotic systems

Variable	United Kingdom[1]	Pacific Northwest[2]	South Wales[3]
Distance from source	+	+++	+++
Slope	+++	+	+++
Latitude	—	+++	—
Altitude	+	+	+++
Discharge category	+++	—	+
Geology	—	+++	—
Vegetation cover	—	+++	—
Physiography	—	+	—
Land use	—	+++	—
Stream order	—	—	+
Mean channel width	+++	+	—
Depth category	+	—	—
Substrate heterogeneity	+	—	—
Date	+	+++	—
Water width	+	+++	—
Water depth	+	+	+++
Surface velocity	+++	+	—
Mean substrate	+++	+	+
Dominant particle size	+++	—	—
% macrophyte cover	+++	+	—
Overhanging vegetation	—	+++	—
pH	+++	+	+++
Oxygen	+	—	—
Nitrate	+	—	+++
Chloride	+++	—	—
Phosphate	+++	—	+
Alkalinity	+++	—	+++
Conductivity	—	+++	+++

[1] Source: Wright *et al.* 1984.
[2] Sources: Corkum and Currie 1987; Corkum 1989.
[3] Source: Ormerod and Edwards 1987.
+ = variable measured; +++ = variable correlated with invertebrate community structure

than riffles. However, pools are depositional habitats and fine sediments collect in them, so they can be places where contaminants concentrate in streams. Likely, pools and their invertebrate fauna behave more like lacustrine habitats of a similar spatial and temporal scale than do riffles.

3) Microhabitat level. Two major factors affect the fauna at this spatial and temporal scale: a) tractive force of water flow on a rock, which determines whether the rock is stable and provides suitable habitat or whether it is mobile and discourages colonization by stream invertebrates; and b) the hydraulic regime of water flowing over a rock, which determines the species present and

where they are located on the rock (*e.g.*, Newbury 1984; Statzner *et al.* 1988). For example, the filter-feeding caddisfly *Brachycentrus occidentalis* typically occurs on the top of a cobble as the flow accelerates to pass over the obstruction posed by the cobble (*e.g.*, Wetmore *et al.* 1990; Figure 10-2). Here, the streamlines contract as flow accelerates so that an organism extending its filtration apparatus up into the flow will intercept a high proportion of suspended material. A filter feeder in this position would have access to higher rates of food delivery than if it were to occur in the deeper, uniform flow area just approaching the cobble. Distribution of another filter feeder, such as larvae of the biting black fly *Simulium vittatum*, can be similar to *B. occidentalis* for the same reason. In conducting SERAs, this example demonstrates the importance of sampling similar microhabitats in reference and affected areas to avoid confusing microhabitat effects with contaminant effects.

Sediment ecological risk assessment is normally done at scales above the microhabitat level, but knowledge at the microhabitat level is required to provide a mechanistic understanding of the effects of disturbance at these larger scales. For example, consider a stream reach beside a cobble-bed stream that is cleared for farming, crops are grown, and pesticides are applied in the normal course of farming practice. During heavy rainfall, soil erosion occurs and organic particles coated with pesticide are washed into the stream. The filter-feeding community is then at risk from these contaminants. This result may not be considered detrimental if Simulium black fly larvae are killed (and biting adults are not produced), but if other non-pest species like *B. occidentalis* are affected, a significant portion of the energy processing capability of the stream could be impacted. Temporally, the threat to filter feeders would persist until the soil erosion event stops, the next spate rearranges the stream substrate, or the insect pupates and emerges as an adult.

The stream benthic community that is sampled in the normal course of an SERA on this hypothetical stream reach is different from the community occurring before farming was initiated and also differs from that occurring in a reference stream. For the above example, it is clear that a microhabitat-level understanding can explain the mechanisms behind a specific response; such elucidation then permits the development of a management plan to ameliorate these effects.

10.2.2 Lakes

No widely used hierarchical classification system exists for lentic waters that is similar to Frissell *et al.*'s (1986) system for lotic waters. For comparison, we propose a similar scheme, based upon a regional lake system. An hypothetical regional lake catchment covering about 500,000 km^2 is shown in Figure 10-3. The geological foundation for the catchment was set in the Precambrian period, some three billion years ago, but the present system of lakes (and streams) has existed for ≈10,000 years, since the last major glaciation. As with the hypothetical river system previously described, major physical

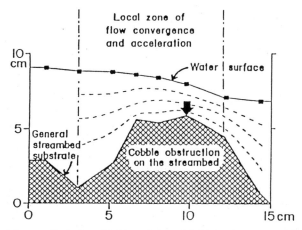

Figure 10-2 Local flow profile for larval habitat of the caddisfly *Brachycentrus occidentalis* in Wilson Creek, Manitoba. Measurements of water-surface elevation, bed elevation, and mean flow velocity taken at stations marked ■. Heavy arrow shows position of insect larva. Streamlines (dashed) are adjacent to larval location to illustrate local zone of flow convergence and acceleration. Source: Wetmore *et al.* 1990. Reprinted by permission of the Journal of the North American Benthological Society.

disturbances and climate change operating on a global scale and over long periods of time have structured the basic environment of this system.

Next, the scale is reduced several orders of magnitude to a bay on one of the lakes. It occupies an area of about 10 km² and was formerly a glacial meltwater channel whose outflow was blocked some 7000 years ago.

The scale can again be reduced several orders of magnitude to a patch of fine-grained sediment at the marshy end of the bay. The patch covers an area of 1 m² and it contains particulate organic material, which mostly originated from the last phytoplankton bloom (*e.g.*, 30 d before) and which provides food for a diverse assemblage of invertebrates such as chironomid larvae, mayfly nymphs, sphaeriid molluscs, and oligochaetes. The sediment patch may be altered by freshets in the creek that drains into the marshy end of the bay or by seasonal storm events that are strong enough to produce waves that rework the bottom sediments of the bay.

The factors affecting benthic invertebrates at the three spatial-temporal scales established are as follows:

1) Regional level. Johnson and Wiederholm (1989) and Reynoldson *et al.* (1995) used multivariate approaches to examine factors that affect benthic invertebrate community structure in lakes over large geographic areas (Table 10-3). Six groups of lakes were identified from the 68 oligo-mesohumic lakes in Sweden examined by Johnson and Wiederholm (1989), based on invertebrate species composition. Three of these groups were most influenced by low pH, high $SO_4^=$,

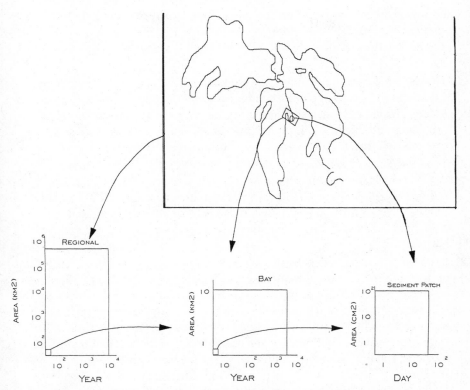

Figure 10-3   Spatial and temporal scales in lakes: regional, bay, and sediment-patch levels

high temperature, and high phytoplankton biovolumes. Two more were associated with high pH and low total phosphorus (TP) concentrations. The last group consisted of only one lake, from which only two taxa of macroinvertebrates were collected. The best predictors of community structure for all groups were depth, silica, bicarbonate (alkalinity), phytoplankton production, and pH.

Five species groups were identified from the 96 sites examined by Reynoldson *et al.* (1995) in the Laurentian Great Lakes. Of the 25 environmental variables they examined, the best water-column predictors of occurrence in a group were depth, NO_3^--nitrogen, alkalinity, and pH. The most relevant sediment variables were percent silt, organic content (measured as % loss on ignition), and concentrations of Al^{+++} and Na^+.

Other studies of benthic invertebrate communities in lakes have used a multiple regression approach to relate biomass of benthic invertebrates to environmental variables (Hanson and Peters 1984; Rasmussen and Kalff 1987; Table 10-3). Hanson and Peters (1984) found that TP was the best univariate predictor of biomass in the 38 lakes of their study; combining TP and lake area slightly improved the relationship. The best model from Rasmussen and Kalff's (1987) study of 131 lakes involved chlorophyll concentration, depth, slope, water color, and temperature. Both of the above studies included lakes from several continents.

The results of these four studies suggest that relatively simple and consistent variables can be used to predict benthic invertebrate community structure or biomass in lakes. Depth was an important predictor in all four studies, phytoplankton production (chlorophyll) was important in three, and pH, temperature, phosphorus, and nitrogen were important in two (Table 10-3).

The BEAST model developed by Reynoldson *et al.* (1995) for the Laurentian Great Lakes (described above) proposes a multivariate approach for assessing sediment quality. The model has already been used to measure the extent of sediment contamination (compared to reference conditions) in a metal-contaminated harbor and to recommend remediation in certain parts of the harbor.

2) Bay level. The hypothetical bay is long and narrow and is joined to the main body of the lake by a shallow inlet (*e.g.*, see Bodaly and Lesack 1984). Therefore, it receives most of its water and nutrients from the main lake only when water levels are high enough to flow over the inlet, as for example during spring high-water events. This situation has existed since the bay was formed, some 7000 years ago. A small creek flows into the opposite end of the bay; it also provides nutrients and fine sediments. The creek end of the bay is a marshy area, which supports the highest benthic invertebrate abundances in the bay. The shoreline around the rest of the bay is composed of a variety of substrate types, most of which are either bedrock or sand and silt. This variety of shoreline type produces a spatially disjunct distribution of habitats suitable for benthic invertebrates. Leaf litter inputs in the autumn are important sources of energy because the bay is long and narrow (*i.e.*, the ratio of terrestrial area:bay surface area is high), and it is oligotrophic.

A contaminant spill into the bay would have negative, long-term consequences for the benthic invertebrate community because water renewal and sediment retention times are so long. In a worst-case scenario, prevailing winds could move the contaminant to the marshy end of the bay where benthic populations would be most affected; perhaps the contaminant (*e.g.*, oil) would persist there over time. Although remediation and assessment will be undertaken at this spatial scale, we shall see below that biological responses occur at a smaller scale.

3) Sediment patch level. Three major factors determine the occurrence and abundance of benthic invertebrates in a fine-grained sediment patch at the marshy end of the bay: a) food of sufficient quality and quantity (*i.e.*, productivity), b) tolerable temperature extremes, and c) sufficient dissolved oxygen. The

Table 10-3 Summary of catchment-level (and larger) variables that correlate with benthic invertebrate community structure in lentic systems

Variable	Lentic system			
	Laurentian Great Lakes[1]	Sweden[2]	38 lakes on various continents[3]	131 lakes on various continents[4]
Latitude	+++	—	—	—
Geology	+	—	—	—
Vegetation cover	+	—	—	—
Physiography	+	—	—	—
Land use	+	—	—	—
Surface area	—	—	+++	+
Slope	—	—	—	+++
Substrate	+	—	—	—
Date	+	+	—	—
Water depth	+++	+++	+++	+++
Mean substrate	+	—	—	—
Dominant particle size	+++	—	—	—
Sediment chemistry	+++	—	—	—
pH	+++	+++	—	—
Oxygen	+	+	—	—
Nitrogen (TN, nitrate)	+++	+++	—	—
Chloride	+	—	—	—
Sulphate	—	+++	—	—
Phosphorus (TP, phosphate)	+	+++	+++	+
Alkalinity	+++	+	—	—
Conductivity	—	—	—	+
Temperature	+	+++	—	+++
Color	—	+	—	+++
Major ions	+	+++	—	—
Water clarity	—	+	—	+
Phytoplankton production	—	+++	+++	+++
Zooplankton production	—	+	—	+

[1] Source: Reynoldson *et al.* 1995.
[2] Source: Johnson and Wiederholm 1989.
[3] Source: Hanson and Peters 1984.
[4] Source: Rasmussen and Kalff 1987.

+ = variable measured; +++ = variable correlated with invertebrate community structure
TN = total nitrogen; TP = total phosphorus.

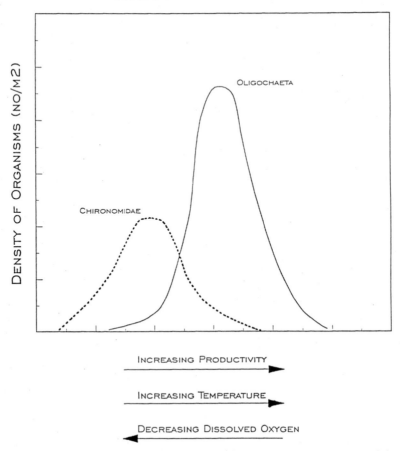

Figure 10-4 Effect of productivity, temperature, and dissolved oxygen on occurrence and abundance of Chironomidae and Oligochaeta in a fine-sediment patch at the bottom of a bay

effect of these factors on the occurrence and abundance of Chironomidae and Oligochaeta is illustrated in Figure 10-4.

As with the stream example, investigation at this smallest scale can provide a mechanistic understanding of events that occur at larger scales. For example, if a lumbering operation is started in the catchment of the bay and the bay is eventually clear cut, normal seasonal rainfall washes sediment and organic matter into the bay; nutrient levels then increase. Algal production is enhanced and the rise in organic matter deposited in the sediment increases bacterial oxygen demand in the sediment. By late summer, when maximum water temperatures occur, the waters over this sediment patch become anoxic. Thus, a sediment patch that was formerly dominated by a diverse array of oxygen-sensitive invertebrates shifts to one dominated by tubificid oligochaetes that tolerate anoxia. (In the event of a coincident oil spill, effects caused by anoxia must be separately discerned from those caused by chemical contamination). The altered community will probably persist until sufficient reforestation occurs to prevent any further soil erosion and organic matter inputs to the bay. The invertebrates that formerly inhabited the sediment patch may then recolonize it. Note, again, that processes affecting benthic communities at this spatial scale are a composite of events occurring at even smaller spatial scales (*e.g.*, particle ingestion, fluxes across biological membranes). As discussed for rivers and streams, an understanding of these smaller scale processes is critical to the development of remediation plans for lentic environments.

In summary, spatial and temporal variability are characteristic of aquatic ecosystems and must be accounted for in the design of SERAs, the elucidation of results obtained, and the extrapolation of study conclusions to management options. Such variability involves populations and communities, can occur in individual habitats or at ecosystem scales, and can operate at time scales ranging from hours to years. Such variability also should be expected by investigators trying to determine perturbation-induced changes in aquatic ecosystems.

The most likely scale of investigation for SERA will be the reach level in lotic ecosystems and a bay (or part of a lake) in lentic ecosystems. The environmental factors important in structuring these environments operate at the next highest scale, but the most appropriate sampling unit will be the next lowest scale (*e.g.*, the cobble substrate in a stream or the fine sediment patch in a lake). Although studies at the smallest spatial and temporal scales are usually inappropriate for SERA, it is critical to understand processes that operate at these scales because they form the underlying biological fabric for benthic communities. After all, it is at these scales that community structure begins.

10.3 Identifying anthropogenic impact

The success of any SERA lies in the ability of the methods used to distinguish between impacted and unimpacted conditions. Measures of both functional and structural change have been used to identify anthropogenic impacts. The sediment quality triad approach (Chapman *et al.* 1991, 1992) integrates data from physical and chemical analyses, laboratory exposure to whole sediments, and benthic invertebrate community structure to determine effects. The nature of the impact of concern should determine the measure used. If the concern is related to contaminant accumulation in vertebrates, then it is necessary to understand the uptake of contaminants from sediments and their movement through the food web (see Luoma and Fisher, Chapter 14). If the concern is one of general environmental degradation, then measurement of change in the structure of benthic communities is appropriate.

An array of methods is available for community structure assessment. Usually, the measured attributes are based on taxa richness, enumerations, or some integrative measure of the two (*e.g.*, diversity or similarity indices; Resh and Jackson 1993). Univariate approaches are most frequently used (Norris and Georges 1993) and can be quantitative (*e.g.*, hypothesis setting, rigorous statistical design, replication; Resh and McElravy 1993) or qualitative (*i.e.*, rapid assessment procedures; Resh and Jackson 1993). These methods usually compare control sites to impacted sites by inference.

The use of rapid assessment procedures as an alternative to traditional quantitative approaches has been embraced by regulatory agencies in the United States (Davis and Simon 1995). These methods (Plafkin *et al.* 1989; Resh *et al.* 1995) involve reduced sampling and identification costs compared to quantitative studies using benthic invertebrates. Reference community types are often based on an ecoregion approach (Omernik 1995). In most rapid assessment programs, several different measures of benthic communities (*i.e.*, the "multimetric" approach) are used in evaluations. The best measurements to use in rapid assessment programs are discussed in Resh and Jackson (1993) and Resh *et al.* (1995).

Multivariate approaches can be divided into two types: 1) multiple regression and 2) paired matrices. Multiple regression develops relationships between benthic invertebrates and environmental attributes but considers only univariate measures of community structure. The paired matrices approach tries to identify pattern and structure in the benthic invertebrate community and then relate that to environmental variables (*e.g.*, Wright *et al.* 1984; Corkum and Currie 1987; Johnson and Wiederholm 1989). The purpose of this approach is to establish predictive models of communities that can be used to assess environmental impairment. If the predictive models are based on reference conditions, then the models derived can be used to measure divergence from the reference state and the degree of risk associated with an environmental stressor. For example, the approach developed by Reynoldson *et al.* (1995) has allowed the setting of appropriate, site-specific and biologically based sediment-quality objectives for the Laurentian Great Lakes using easy-to-measure habitat characteristics. Future bioassessment programs will probably combine elements of rapid assessment and multivariate methods (Resh *et al.* 1995).

10.4 Implications for sediment ecological risk assessment

What are the critical needs for doing SERA in rivers and lakes? First, the appropriate spatial and temporal scales for the assessment must be determined. Three discrete spatial-temporal scales were presented above for a hypothetical river and lake system, but in reality spatial and temporal scales both comprise a continuum of conditions. The scales chosen ultimately will depend on the questions being asked and the requirements of environmental managers and decision makers. The factors being measured need to match the scales chosen; mismatches probably will not provide answers to the questions being asked. As the scales increase, the risk assessment process becomes more difficult; for example, increasing geographic size of the study area will result in changes in species composition and will require more habitats to be sampled. At the same time, better methods are needed to assess the influence of spatial and temporal variations in structure and function of benthic communities on both accumulation and effects of sediment-based contaminants.

Second, it is important to understand the environmental factors that affect benthic invertebrate communities at the spatial and temporal scales chosen and the natural variability of these factors. Natural variability may decrease in moving from larger to smaller scales, but certainly this is not always true. The impacts of anthropogenic stresses can only be understood in the context of deviations from natural variability. The establishment of regional reference bases for benthic invertebrate community structure, either through a biocriteria approach (*e.g.*, Davis and Simon 1995) or through multivariate approaches (*e.g.*, Wright *et al.* 1984, 1988; Reynoldson *et al.* 1995), allow the effects of anthropogenic impacts to be determined over wide geographic areas.

Third, bioassessment methods traditionally have used univariate approaches. This may have been suitable for the severe organic pollution problems characteristic of the first half of the twentieth century in developed and industrialized countries, but the many confounding factors associated with modern-day multiple inputs of toxicant mixtures may call for the increased use of multivariate approaches.

10.5 Future research needs

A key future need is that a constant proportion of funding in risk assessment research should be directed to questions involving basic systematic and ecological research. The perilous state of contemporary systematics and recommendations for its resurrection are outlined by Wheeler (1995). Systematics underlies our understanding of the natural world because species are the basic units in that world. Knowledge of species is fundamental to biomonitoring and SERA.

The need for applied freshwater ecology is obvious, but it is impossible to apply knowledge that does not exist (Johnson *et al.* 1993). An understanding of benthic invertebrate ecology is a prerequisite to implementing a biological approach to ecosystem management, as practiced in SERA.

Last, it is important to determine the degree to which experimental results derived at one spatial-temporal scale can be extrapolated to other scales (*e.g.*, see Fee and Hecky 1992; Fee *et al.* 1992; Turner *et al.* 1995). Continued support of long-term field research facilities that have access to lakes and rivers of different sizes is essential to this need.

10.6 Acknowledgments

We thank G. Biddinger and T. Dillon, co-chairs of the Pellston Workshop on Sediment Risk Assessment, for providing the impetus and a forum for this paper. Reviews by C.G. Ingersoll, D. Malley, and D. Morrissey helped improve the paper. D. Laroque did most of the word processing, and A. Wiens prepared initial drafts of some of the figures.

10.7 References

Barbour MT, Stribling JB, Karr JR. 1995. Multimetric approach for establishing biocriteria and measuring biological conditions. In: Davis WS, Simon TP, editors. Biological assessment and criteria. Tools for water resource planning and decision making. Boca Raton FL: Lewis. p 63–77.

Bodaly RA, Lesack LFW. 1984. Response of a boreal northern pike (*Esox lucius*) population to lake impoundment: Wupaw Bay, Southern Indian Lake, Manitoba. *Can J Fish Aquat Sci* 41:706–714.

Burns LA, Ingersoll CG, Pascoe GA. 1994. Ecological risk assessment: application of new approaches and uncertainty analysis. *Environ Toxicol Chem* 13:1873–1874.

Calow P. 1995. Risk assessment: principles and practice in Europe. *Aust J Ecotoxicol* 1:11–13.

Canfield TJ, Kemble NE, Brumbaugh WG, Dwyer FJ, Ingersoll CG, Fairchild JF. 1994. Use of benthic invertebrate community structure and the sediment quality triad to evaluate metal-contaminated sediment in the Upper Clark Fork River, Montana. *Environ Toxicol Chem* 13:1999–2012.

Chapman PM, Power EA, Burton Jr GA. 1992. Integrative assessments in aquatic ecosystems. In: Burton Jr GA, editor. Sediment toxicity assessment. Chelsea MI: Lewis. p 313–340.

Chapman PM, Power EA, Dexter RN, Andersen HB. 1991. Evaluation of effects associated with an oil platform, using the sediment quality triad. *Environ Toxicol Chem* 10:407–424.

Clements WH, Cherry DS, Cairns Jr J. 1988. Impact of heavy metals on insect communities in streams: a comparison of observational and experimental results. *Can J Fish Aquat Sci* 45:2017–2025.

Cobb DG, Galloway TD, Flannagan JF. 1992. Effects of discharge and substrate stability on density and species composition of stream insects. *Can J Fish Aquat Sci* 49:1788–1795.

Corkum LD. 1989. Patterns of benthic invertebrate assemblages in rivers of northwestern North America. *Freshwater Biol* 21:195–205.

Corkum LD, Currie DC. 1987. Distributional patterns of immature Simuliidae (Diptera) in northwestern North America. *Freshwater Biol* 17:201–221.

Davis WS. 1995. Biological assessment and criteria: building on the past. In: Davis WS, Simon TP, editors. Biological assessment and criteria. Tools for water resource planning and decision making. Boca Raton FL: Lewis. p 15–29.

Davis WS, Simon TP, editors. 1995. Biological assessment and criteria. Tools for water resource planning and decision making. Boca Raton FL: Lewis.

Dunson WA, Travis J. 1991. The role of abiotic factors in community organization. *Am Nat* 138:1067–1091.

Fee EJ, Hecky RE. 1992. Introduction to the Northwest Ontario Lake Size Series (NOLSS). *Can J Fish Aquat Sci* 49:2434–2444.

Fee EJ, Shearer JA, DeBruyn ER, Schindler EU. 1992. Effects of lake size on phytoplankton photosynthesis. *Can J Fish Aquat Sci* 49:2445–2459.

France RL. 1990. Theoretical framework for developing and operationalizing an index of zoobenthos community integrity: application to biomonitoring with zoobenthos communities in the Great Lakes. In: Edwards CJ, Regier HA, editors. An ecosystem approach to the integrity of the Great Lakes in turbulent times. Ann Arbor MI: Great Lakes Fishery Commission. Special Publication 90-4. p 169–193.

Frissell CA, Liss WJ, Warren CE, Hurley MD. 1986. A hierarchical framework for stream habitat classification: viewing streams in a watershed context. *Environ Manage* 10:199–214.

Gibbons WN, Mackie GL. 1991. The relationship between environmental variables and demographic patterns of *Hyalella azteca* (Crustacea:Amphipoda). *J North Am Benthol Soc* 10:444–454.

Hanson JM, Peters RH. 1984. Empirical prediction of crustacean zooplankton biomass and profundal macrobenthos biomass in lakes. *Can J Fish Aquat Sci* 41:439–445.

Hellawell JM. 1986. Biological indicators of freshwater pollution and environmental management. London: Elsevier.

Hughes RM. 1995. Defining acceptable biological status by comparing with reference conditions. In: WS Davis, TP Simon, editors. Biological assessment and criteria: tools for water resource planning and decision making. Boca Raton FL: Lewis. p 31–47.

Johnson BL, Richardson WB, Naimo TJ. 1995. Past, present, and future concepts in large river ecology. *BioScience* 45:134–141.

Johnson RK, Wiederholm T. 1989. Classification and ordination of profundal macroinvertebrate communities in nutrient poor, oligo-mesohumic lakes in relation to environmental data. *Freshwater Biol* 21:375–386.

Johnson RK, Wiederholm T, Rosenberg DM. 1993. Freshwater biomonitoring using individual organisms, populations, and species assemblages of benthic macroinvertebrates. In: Rosenberg DM, Resh VH, editors. Freshwater biomonitoring and benthic macroinvertebrates. New York: Chapman and Hall. p 40–158.

Krieger KA. 1984. Benthic macroinvertebrates as indicators of environmental degradation in the southern nearshore zone of the central basin of Lake Erie. *J Gt Lakes Res* 10:197–209.

Landis WG, Matthews GB, Matthews RA, Sergeant A. 1994. Application of multivariate techniques to endpoint determination, selection and evaluation in ecological risk assessment. *Environ Toxicol Chem* 13:1917–1927.

La Point TW, Fairchild JF. 1992. Evaluation of sediment contaminant toxicity: the use of freshwater community structure. In: Burton Jr GA, editor. Sediment toxicity assessment. Boca Raton FL: Lewis. p 87–110.

La Point TW, Melancon SM, Morris MK. 1984. Relationships among observed metal concentrations, criteria, and benthic community structural responses in 15 streams. *J Water Pollut Control Fed* 56:1030–1038.

Lehmkuhl DM. 1972. Change in thermal regime as a cause of reduction of benthic fauna downstream of a reservoir. *J Fish Res Board Can* 29:1329–1332.

Metcalfe-Smith JL. 1994. Biological water-quality assessment of rivers: use of macroinvertebrate communities. In: Calow P, Petts GE, editors. Volume 2, The rivers handbook. London: Blackwell Scientific. p 144–172.

Newbury RW. 1984. Hydrologic determinants of aquatic insect habitats. In: Resh VH, Rosenberg DM, editors. The ecology of aquatic insects. New York: Praeger. p 323–357.

Norris RH, Georges A. 1993. Analysis and interpretation of benthic macroinvertebrate surveys. In: Rosenberg DM, Resh VH, editors. Freshwater biomonitoring and benthic macroinvertebrates. New York: Chapman and Hall. p 234–286.

Omernik JM. 1995. Ecoregions: a spatial framework for environmental management. In: Davis WS, Simon TP, editors. Biological assessment and criteria: tools for water resource planning and decision making. Boca Raton FL: Lewis. p 49–62.

Ormerod SJ, Edwards RW. 1987. The ordination and classification of macroinvertebrate assemblages in the catchment of the River Wye in relation to environmental factors. *Freshwater Biol* 17:533–546.

Parkhurst BR, Warren-Hicks W, Cardwell RD, Volison J, Etchison T, Butcher JB, Covington SM. 1994. Methodology for aquatic ecological risk assessment. Seattle WA: The Cadmus Group, Inc., Laramie and Durham, Parametrix, Inc. Contract No. RP91-AER-1.

Pastorok RA, Peek DC, Sampson JR, Jacobson MA. 1994. Ecological risk assessment for river sediments contaminated by creosote. *Environ Toxicol Chem* 13:1929–1941.

Plafkin JL, Barbour MT, Porter KD, Gross SK, Hughes RM. 1989. Rapid bioassessment protocols for use in streams and rivers: benthic macroinvertebrates and fish. Washington DC: U.S. Environmental Protection Agency (USEPA). EPA/444/4-89-001.

Rasmussen JB, Kalff J. 1987. Empirical models for zoobenthic biomass in lakes. *Can J Fish Aquat Sci* 44:990–1001.

Resh VH. 1995. Freshwater benthic macroinvertebrates and rapid assessment procedures for water quality monitoring in developing and newly industrialized countries. In: Davis WS, Simon TP, editors. Biological assessment and criteria: tools for water resource planning and decision making. Boca Raton FL: Lewis. p 167–177.

Resh VH, Jackson JK. 1993. Rapid assessment approaches to biomonitoring using benthic macroinvertebrates. In: Rosenberg DM, Resh VH, editors. Freshwater biomonitoring and benthic macroinvertebrates. New York: Chapman and Hall. p 195–233.

Resh VH, McElravy EP. 1993. Contemporary quantitative approaches to biomonitoring using benthic macroinvertebrates. In: Rosenberg DM, Resh VH, editors. Freshwater biomonitoring and benthic macroinvertebrates. New York: Chapman and Hall. p 159–194.

Resh VH, Norris RH, Barbour MT. 1995. Design and implementation of rapid assessment approaches for water resource monitoring using benthic macroinvertebrates. *Aust J Ecol* 20:108–121.

Resh VH, Rosenberg DM, editors. 1984. The ecology of aquatic insects. New York: Praeger.

Resh VH, Rosenberg DM. 1989. Spatial-temporal variability and the study of aquatic insects. *Can Entomol* 121:941–963.

Reynoldson TB. 1984. The utility of benthic invertebrates in water quality monitoring. *Water Qual Bull* 10:21–28.

Reynoldson TB, Bailey RC, Day KE, Norris RH. 1995. Biological guidelines for freshwater sediment based on BEnthic Assessment of SedimenT (the BEAST) using a multivariate approach for predicting biological state. *Aust J Ecol* 20:198–219.

Rosenberg DM, Resh VH. 1993a. Introduction to freshwater biomonitoring and benthic macroinvertebrates. In: Rosenberg DM, Resh VH, editors. Freshwater biomonitoring and benthic macroinvertebrates. New York: Chapman and Hall. p 1–9.

Rosenberg DM, Resh VH, editors. 1993b. Freshwater biomonitoring and benthic macroinvertebrates. New York: Chapman and Hall.

Rosenberg DM, Wiens AP. 1976. Community and species responses of Chironomidae (Diptera) to contamination of fresh waters by crude oil and petroleum products, with special reference to the Trail River, Northwest Territories. *J Fish Res Board Can* 33:1955–1963.

Ryder RA, Pesendorfer J. 1989. Large rivers are more than flowing lakes: a comparative review. In: Dodge DP, editor. Proceedings of the International Large River Symposium. *Can Spec Publ Fish Aquat Sci* 106:65–85.

Statzner B, Gore JA, Resh VH. 1988. Hydraulic stream ecology: observed patterns and potential applications. *J North Am Benthol Soc* 7:307–360.

Townsend CR, Hildrew AG. 1994. Species traits in relation to a habitat templet for river systems. *Freshwater Biol* 31:265–275.

Turner MG, Gardner RH, O'Neill RV. 1995. Ecological dynamics at broad scales. Ecosystems and landscapes. *BioScience Suppl (Science and Biodiversity Policy)*:S29–S35.

[USEPA] U.S. Environmental Protection Agency. 1992. Framework for ecological risk assessment. Washington DC: USEPA. EPA/630/R-92/001.

Vannote RL, Minshall GW, Cummins KW, Sedell JR, Cushing CE. 1980. The river continuum concept. *Can J Fish Aquat Sci* 37:130–137.

Wetmore SH, Mackay RJ, Newbury RW. 1990. Characterization of the hydraulic habitat of *Brachycentrus occidentalis*, a filter-feeding caddisfly. *J North Am Benthol Soc* 9:157–169.

Wheeler QD. 1995. Systematics and biodiversity. Policies at higher levels. *BioScience Suppl (Science and Biodiversity Policy)*:S21–S28.

Winner RW, Boesel MW, Farrell MP. 1980. Insect community structure as an index of heavy-metal pollution in lotic ecosystems. *Can J Fish Aquat Sci* 37:647–655.

Wright JF, Armitage PD, Furse MT, Moss D. 1988. A new approach to the biological surveillance of river quality using macroinvertebrates. *Int Ver Theor Angew Limnol Verh* 23:1548–1552.

Wright JF, Moss D, Armitage PD, Furse MT. 1984. A preliminary classification of running-water sites in Great Britain based on macro-invertebrate species and the prediction of community type using environmental data. *Freshwater Biol* 14:221–256.

SESSION 5
CRITICAL ISSUES IN ECOLOGICAL RELEVANCE

Chapter 11

Nonequilibrium dynamics and alternatives to the recovery model

Wayne G. Landis, Robin A. Matthews, Geoffrey B. Matthews

11.1 Introduction

Equilibrium models for the dynamics of ecological systems are often used as a framework in which to place the goals of ERA. There is a growing body of evidence that equilibrium models are not good descriptions of ecological events and that alternatives do exist. Nonequilibrium models are powerful tools in explaining and predicting events across both aquatic and terrestrial systems (Reice 1994). Given tools derived from conventional statistics (Johnson 1988; Kersting 1988) and machine learning (Matthews, Matthews, and Hachmoller 1991; Matthews, Matthews, and Ehinger 1991; Matthews *et al.* 1995a, 1995b; Landis *et al.* 1994), it is now possible to detect differences in noisy ecological systems. Are these differences significant? If the detectable differences are representations of the potential for ecological systems to react differently to a subsequent stressor event, then the differences are biologically and ecologically significant. The sections below present current definitions of recovery, data that indicate the persistence of information concerning stressor events and the heterogeneity of sediment, and a nonequilibrium hypothesis for the reaction of a sediment system to a stressor event. We also propose the replacement of the term *recover* in environmental toxicology with *restructure* or *reconstruct* to more accurately reflect the process undergoing investigation or prediction.

11.2 Definitions of recovery

Recovery is often stated as a goal or desired state in ERA. As often defined, *recovery* is the return to a state so that the assessment endpoints are of values not statistically different compared to those before the stressor event. Another common definition is that the impacted system has moved to a state not statistically distinguishable from the surrounding environmental mosaic as exemplified by a reference site. Both of these types of definitions are dependent on the finding of no statistically significant or no projected significant difference between the site of interest and the reference site or state. Therefore, these definitions depend upon opinion or upon the available statistical tools and their power. A more precise definition is as follows: Return after a disturbance to a state that reacts to subsequent events as if the initial disturbance had not occurred.

This type of definition recognizes that for recovery to occur, the information about the prior stressor event needs to be erased from the system. In effect, the system is ahistorical (see Lewontin 1969). However, many lines of evidence suggest that ecological systems are historical and are by definition complex systems. If ecological systems are historical, then recovery as defined above cannot occur.

11.3 Persistence of information in ecological systems

Using tools derived from machine learning, detectable differences in ecological structures subjected to different stressor regimens have been found to be persistent. Streams (Matthews, Matthews, and Hachmoller 1991), microcosm experiments (Landis, Matthews, Markiewicz, Shough, and Matthews 1993; Landis, Matthews, Markiewicz, and Matthews 1993; Landis *et al.* 1994; Landis, Matthews, Markiewicz, and Matthews 1995), and biomarkers in vole populations (Fairbrother *et al.*, in press), all have demonstrated that detectable differences are the rule. Even as differences disappear at certain times, they can reappear in a treatment related fashion. Other laboratory research on the structure and dynamics of ecological systems (Drake 1991; Drake *et al.* 1993) have recognized the persistence of information. Patterns abound in systems, and historical events are written into the structure and dynamics of ecological structures. Ecological systems are complex by definition in the terminology of Nicolis and Prigogine (1989). Complex systems and ecological systems (Brooks *et al.* 1989) have a common property of being irreversible.

Our research group (Landis, Matthews, Markiewicz, Shough, and Matthews 1993; Landis, Matthews, Markiewicz, and Matthews 1993; Landis *et al.* 1994; Landis, Matthews, Markiewicz, and Matthews 1995) has also described the persistence of information within even simple model ecological systems, *i.e.*, the standardized aquatic microcosm (SAM) (ASTM 1991) and the mixed flask culture (MFC) (Leffler 1984; Stay *et al.* 1988, 1989) when dosed with the water-soluble fraction of jet fuels. The 64-d SAM protocol is comprised of 10 algal, 3 invertebrate, and 1 bacterial species introduced into 3 L of sterile, chemically defined medium. Four treatment groups are comprised of six replicates. The MFC uses naturally occurring assemblages of aquatic organisms that are collected from local streams and lakes, brought back to the laboratory and allowed to reassemble and restructure during a 3-month equilibration period. Subsamples are then used to inoculate 1 L experimental vessels, and a cross- and re-inoculation procedure is used to attempt homogeneity among the experimental replicates. As with the SAM experiments, 4 treatments with 6 replicates are used.

We use tools derived from artificial intelligence (AI) and machine learning research (Matthews *et al.* 1995a, 1995b) to look for patterns within the diverse dataset typical of ecological experiments. Typical analytical methods, such as analysis of variance (ANOVA), are generally not able to effectively combine data as disparate as dichotomous sequence presence/absence with continuous data such as pH, algal counts, and soil respiration. These kinds of data are characteristic of problems dealt with traditionally by AI. Handling the visual system of a robot, for instance, in a room full of tools, debris, light and shadows, is a classic problem of analyzing large quantities of "dirty" data into meaningful categories. In previous work, we have shown that environmental datasets are amenable to AI techniques (Matthews, Matthews, and Hachmoller 1991; Matthews, Matthews, and Ehinger 1991; Landis, Matthews, Markiewicz, Shough, and Matthews 1993; Landis, Matthews, Markiewicz, and Matthews 1993; Landis *et al.* 1994).

In these sets of experiments (Landis, Matthews, Markiewicz, Shough, and Matthews 1993; Landis, Matthews, Markiewicz, and Matthews 1993; Landis, Matthews, Markiewicz, and Matthews 1995), our AI and other analysis tools filtered a complex dataset to reveal patterns that went unnoticed by unaided human ecologists. We discovered a persistent relationship between the treatment group and the clusters within the data over the course of these experiments. The effects were persistent beyond the detection of toxicant within the experimental systems. These results have led to the development of the community conditioning hypothesis by Matthews *et al.* (1996). These model ecological systems and analysis techniques have been adapted to sediment microcosms.

11.4 Sediment microcosms

The experiments described above were carried out with the jet fuel delivered to the microcosm system as a percentage of a water soluble fraction. However, Sandberg (1993) performed MFC microcosm experiments that injected jet fuel toxicant into the sediment component of a mixed-flask-culture–type microcosm. Four treatment groups were used. Unlike the conventional jet fuel experiments, the toxicant would reappear within the water column after stirring was conducted to sample the various organisms. Cosine and vector distances and nonmetric clustering coupled with association analysis were used to examine the dataset (Landis *et al.* 1994). It was found that the effects of the introduction of the fuel were observable during the first 21 days after dosing, were not seen during days 21 through 35, then reappeared and persisted during the course of the two-month-long experiment (Figure 11-1). In other words, the information related to treatment type remained within the model ecological systems.

11.5 Heterogeneity of sediments in time and space

Unlike laboratory experiments that strive to enhance the uniformity of the environment, sediments are heterogeneous in space and time. A study by Wiegers (1994) sampled a typical freshwater pond, Claypit Pond, that was known to be contaminated with heavy metals. Fifteen cores were taken, and each 5 cm was analyzed for metals using several extraction techniques. Presented in Table 11-1 are the variance-to-mean ratios of the 15 sample sites presented by depth. A variance-to-mean ratio of 1 is indicative of a normal distribution. A variance-to-mean ratio of greater than 1 indicates a contagious or clumped distribution. A ratio of less than 1 indicates an even distribution (Elliott 1971). As can be seen in Table 11-1, the variance-to-mean ratio for the metals studied is generally greater than 1 for chromium, lead, and zinc. Assuming that the greater depths were indicative of past conditions, it can also be seen that this ratio has varied greatly. A variance-to-mean ratio as high as 53 for chromium was observed in these sediment cores.

Such a degree of heterogeneity indicates that Claypit Pond, and other sediment environments, are patchy environments. The heterogeneity in the environment and the disturbance regimen is likely to be responsible for the biotic structure of the ecological system (Reice 1994). Nonequilibrium factors are then critical for the maintenance of the structure, a property of ecological systems often used as an assessment endpoint in ERA.

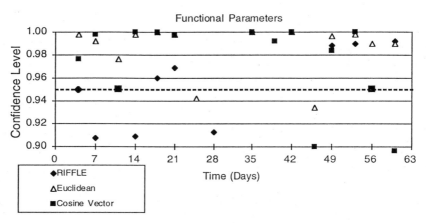

Figure 11-1 Clustering and association analysis of the sediment mixed-flask culture functional data. For
most of the period of the experiment, clustering demonstrated a statistically significant
association with treatment. Horizontal dashed line denotes 95% confidence level. Analysis
variables include pH, measurements of dissolved oxygen, and absorbence of light. A
parallel analysis with biotic (structural parameters) showed a similar pattern. During days
21 through 35, none of the clustering methods identified clusters associated with
treatment; however, clusters associated with treatment reappeared after day 35.

Not only is the chemical composition of a sediment heterogeneous, so is the distribution
of organisms. The protistan species of the *Paramecium aurelia* and *P. bursaria* complexes
usually exhibit highly contagious distributions (Landis 1981, 1982), as do many other
benthic organisms.

11.6 A hypothesis called "community conditioning"

Our results and the repeated confirmation of the heterogeneity of ecological systems have
led to our proposing the "community conditioning hypothesis": ecological systems tend
to preserve information about every event in their etiology (Matthews *et al.* 1996). The
information can be held in the interaction among the organisms within a community, in
the frequency of genetic markers within a population, or in the pattern of biomarkers
(Landis *et al.* 1996). The corollaries to this hypothesis are these:

1) Communities are a product of their unique etiology, which is the historical
   collection of physical, chemical, and biological events leading up to a point in
   time. No two ecological systems will ever be identical.

2) Events that alter the structure or function of populations within the community
   become a part of the history of the community and are difficult to erase. The
   influence of the event may increase or decrease over time, but it is not lost from
   the history of the community.

3) Information can be stored in an uncataloged array of biotic and abiotic forms,
   such as varieties of detritus, phenotypic fitness, sediment composition, or the
   genetic structure of constituent populations. Any subset of community mea-

Table 11-1 Heterogeneity of sediments: variance divided by mean sediment concentration in Claypit Pond

Digestion	Pond depth (cm)[b]	Variance-to-mean sediment concentration ratio (µg/g)[a]			
		Chromium	Copper[c]	Lead	Zinc
Strong acid	0–5	5.6	1.1	7.4	4.8
	5–10	16	1.5	9.3	6.1
	10–15	53.6	1.6	8.2	8.7
	15–20	22.1	0.7	13.6	5.4
	20–25	1.5	1.2	3.3	2.4
Acetic acid	0–5	1.2	0.3	0.06	6.1
	5–10	4.5	0.2	0.3	13.6
	10–15	6.8	0.2	0.02	13.5
	15–20	6.7	0.6	1.01	7.6
	20–25	0.7	0.7	0	1.2

[a] Often variance-to-mean ratio is greater than 1, indicating that contaminant distribution is highly clumped . Ratios also vary with depth of sediment, indicating that distribution of metal contaminant varies over time.
[b] Cores taken from 15 sites within the pond.
[c] Note that variance-to-mean ratio for copper is less than 1 for both digestion techniques, indicating even distribution of this metal.

surements, such as single-species population counts or reproduction dynamics, cannot be assumed to be representative of the entire community.

4) Information may be retained by properties of the community that remain hidden, unmeasured or unmeasurable, for indefinite time periods. The potential of this conditioning to alter the future trajectory of the community may remain undiminished.

5) Almost all environmental events can leave lasting effects. "No observed effect" does not imply "no effect."

The characteristic of ecological systems to preserve information about their etiology poses an interesting set of considerations. Given that no two ecological systems will have identical etiologies, similarities instead of differences should perhaps be the emphasis. A general assumption is made that if two systems are different, the cause is proximate. Given different etiologies, the cause may be either proximate or due to historical events widely separated in time. It is difficult to know which.

11.7 Differences and similarities

One of the implications of the improvement in our ability to recognize differences in ecological systems is that differences are the rule rather than the exception. The recognition of the nonequilibrium nature of ecological systems and the incorporation of information about etiologies state that no two systems will be the same. Turning the question around, it is perhaps more sensible to measure similarities of the systems. We are currently exploring the application of AI to the measurement of similarity with both static and dynamic data. Initial results are encouraging. Initial analysis, using the similarity measurement program RIGGLE, confirms the finding that Treatment 1 (non-dosed) microcosms are most similar to the treated replicates within an experiment compared to Treatment 1 replicates of other experiments. Analyses using generated data also illuminate the ability of the technique to measure similarities statistically and dynamically. Comparisons of static and dynamic similarity can demonstrate out-of-step dynamics and other characteristics of the dynamics of populations.

11.8 Implications for sediment risk assessment

Future ecological assessments can be set so that similarity scores are the determining factor. Underlying assumptions, which mirror the reality of the complex nature of ecological structures, are that no two systems are identical and that responses to a future stressor may be widely divergent. If long-term management of ecological structures is to be a national goal, and a mix of multiple stressors a long-term reality, then an understanding of these basic tenets will be essential to accurate and useful assessments.

As ecological systems incorporate historical information, regulatory decisions and policies must be based on the knowledge that toxic insults, as well as other disturbances, will have effects. At a particular instance in time, we may not be able to obtain evidence of the effects by measuring a subset of variables within the community. Detection of effects will depend upon the nature of the effect and our ability to predict its time and form of appearance. What is required is the regulatory courage to say which effects we will choose to look for in the community, which effects will be allowed, and which will not. Relying upon the fallacies of "no effect" or "no observable effect" is unsound and potentially disastrous if implemented as policy. The realization must come about that unacceptable effects will occur, even without anthropogenic inputs. Systems change, all of the time, and our actions are an inevitable part of that change.

11.9 Terminology

The term *recovery* has been used to mean a variety of dynamics and outcomes, both in ecology and environmental toxicology. Given that ecological systems are nonequilibrium structures, it makes no sense to use a term that has its basis in classical stability-recovery dynamics. After all, to recover a typical sediment-based ecological system, it will be necessary to recreate the layers of heterogeneity and dynamic diversity that characterize a complex, irreversible system (Landis, Matthews, and Matthews 1995). *Restructure* or *reconstruct* may be more accurate descriptions of the process and the predictions of ecological risk assessment and have been suggested as possible replacements for *recovery* (Reice 1994).

Why all the bother over semantics and terminology? The answer is straightforward. The use of terminology such as *recovery* also brings with it the recovery-stability model of the ecology of the 1920s to 1970s (with notable exceptions). No matter how it may be operationally defined, most investigators and regulators bring with the term *recovery* the recovery-stability metaphor with all of its misconceptions. A similar problem exists with the use of *ecosystem health* as a metaphor for the status of an ecological system (Suter 1993). The use of new terminology to more accurately reflect the nonequilibrium dynamics of ecological systems should better define the processes and goals of ecological and sediment risk assessment.

11.10 Acknowledgments

We would like to thank our fellow researchers involved in the programs of the Institute of Environmental Toxicology and Chemistry, notably A. Markiewicz, S. Kelly, R. Sandberg, M. Roze, C. Pickreign, and J. Wiegers. This research has been supported by USAFOSR Grant No. F49620-94-1-0285 and NOAA Grant No. NA470A0141.

11.11 References

[ASTM] American Society for Testing and Materials. 1991. Standard practice for the standardized aquatic microcosm: fresh water. In: Volume 11.04, Annual book of ASTM standards. Philadelphia: ASTM. E1366-91. p 1017–1051.

Brooks DR, Collier J, Maurer BA, Smith JDH, Wiley EO. 1989. Entropy and information in evolving biological systems. *Biol Philos* 4:407–432.

Drake JA. 1991. Community-assembly mechanics and the structure of an experimental species ensemble. *Amer Nat* 137:1–26.

Drake JA, Flum TE, Witteman GJ, Voskuil T, Hoffman AM, Creson C, Kenny DA, Huxel GR, Larue CS, Duncan JR. 1993. The construction and assembly of an ecological landscape. *J Anim Ecol* 62:117–130.

Elliott JM. 1971. Some methods for the statistical analysis of samples of benthic invertebrates. Freshwater Biological Association Scientific Publication No. 25.

Fairbrother A, Landis WG, Dominguez S, Shiroyama T, Buchholz P, Roze MJ, Matthews GB. In press. A novel nonmetric multivariate approach to the evaluation of biomarkers in terrestrial field studies. *Ecotoxicology*.

Johnson AR. 1988. Evaluating ecosystem response to toxicant stress: a state space approach. In: Adams WJ, Chapman GA, Landis WG, editors. Volume 10, Aquatic toxicology and hazard assessment. Philadelphia: American Soc for Testing and Materials (ASTM). STP 971. p 275–285.

Kersting K. 1988. Normalized ecosystem strain in micro-ecosystems using different sets of state variables. *Verhandlungen Internationale Vereinigung fur Theoretische und Angewandte Limnologie* 23:1641–1646.

Landis WG. 1981. The ecology, interactions, and the role of the killer trait in five species of the Paramecium aurelia complex inhabiting the littoral zone. *Can J Zool* 9:1734–1743.

Landis WG. 1982. The spatial and temporal distribution of *Paramecium bursaria* in the littoral zone. *J Protozool* 29:159–161.

Landis WG, Matthews GB, Matthews RA, Sergeant A. 1994. Application of multivariate techniques to endpoint determination, selection and evaluation in ecological risk assessment. *Environ Toxicol Chem* 13(12):1917–1927.

Landis WG, Matthews RA, Markiewicz AJ, Matthews GB. 1993. Multivariate analysis of the impacts of the turbine fuel JP-4 in a microcosm toxicity test with implications for the evaluation of ecosystem dynamics and risk assessment. *Ecotoxicology* 2:271–300.

Landis WG, Matthews RA, Markiewicz AJ, Matthews GB. 1995. Non-linear oscillations detected by multivariate analysis in microcosm toxicity tests with complex toxicants: implications for biomonitoring and risk assessment. In: Hughes S, Biddinger GR, Mones E, editors. Volume 3, Environmental toxicology and risk assessment. Philadelphia: American Soc for Testing and Materials. ASTM 1218. p 133–156.

Landis WG, Matthews RA, Markiewicz AJ, Shough NA, Matthews GB. 1993. Multivariate analyses of the impacts of the turbine fuel Jet-A using a microcosm toxicity test. *J Environ Sci* 2:113–130.

Landis WG, Matthews RA, Matthews GB. 1995. A contrast of human health risk and ecological risk assessment: risk assessment for an organism versus a complex non-organismal structure. *Hum Ecol Risk Assess* 1:485–488.

Landis WG, Matthews RA, Matthews GB. 1996. The layered and historical nature of ecological systems and the risk assessment of pesticides. *Environ Toxicol Chem* 15: 432–440.

Leffler JW. 1984. The use of self-selected, generic aquatic microcosms for pollution effects assessment. In: Harris HH, editor. Concepts in marine pollution measurements. College Park: Maryland Sea Grant College, Univ of Maryland.

Lewontin RC. 1969. The meaning of stability. In: Diversity and stability in ecological systems. Brookhaven NY: Brookhaven National Laboratory. Brookhaven Symposia in Biology No. 22, BNL50175(C-56). p 13–24.

Matthews GB, Matthews RA, Hachmoller B. 1991. Mathematical analysis of temporal and spatial trends in the benthic macroinvertebrate communities of a small stream. *Can J Fish Aquat Sci* 48:2184–2190.

Matthews GB, Matthews RA, Landis WG. 1995a. Nonmetric clustering and association analysis: Implications for the evaluation of multispecies toxicity tests and field monitoring. In: Hughes JS, Biddinger GR, Mones E, editors. Volume 3, Environmental toxicology and risk assessment. Philadelphia: American Soc for Testing and Materials (ASTM). STP 1218. p 79–93.

Matthews GB, Matthews RA, Landis WG. 1995b. Nonmetric conceptual clustering in ecology and ecotoxicology. *AI Applications* 9:41–48.

Matthews RA, Landis WG, Matthews GB. 1996. Community conditioning: an ecological approach to environmental toxicology. *Environ Toxicol Chem* 15: 597–603.

Matthews RA, Matthews GB, Ehinger WJ. 1991. Classification and ordination of limnological data: a comparison of analytical tools. *Ecol Model* 53:167–187.

Nicolis G, Prigogine I. 1989. Exploring complexity. New York: WH Freeman.

Reice SR. 1994. Nonequilibrium determinants of biological community structure. *Am Scientist* 82:424–435.

Sandberg RS. 1993. Investigation of the effects of a pulsed release of Jet-A turbine fuel from sediments using a modified mixed flask culture (MFC) microcosm [thesis]. Bellingham WA: Western Washington Univ.

Stay FS, Katko A, Rohm CM, Fix MA, Larsen DP. 1988. Effects of fluorine on microcosms developed from four natural communities. *Environ Toxicol Chem* 7:635–644.

Stay FS, Katko A, Rohm CM, Fix MA, Larsen DP. 1989. The effects of atrazine on microcosms developed from four natural plankton communities. *Arch Environ Contam Toxicol* 18:866–875.

Suter II GW. 1993. A critique of ecosystem health concepts and indexes. *Environ Toxicol Chem* 12:1533–1539.

Wiegers J. 1994. Distribution of chromium, copper, lead and zinc in the sediments and macrophytes of Claypit Pond [thesis]. Bellingham WA: Western Washington Univ.

SESSION 5
CRITICAL ISSUES IN ECOLOGICAL RELEVANCE

Chapter 12

Workgroup summary report on critical issues of ecological relevance in sediment risk assessment

Kristin E. Day, William H. Clements, Ted DeWitt,
Wayne G. Landis, Peter Landrum, Donald J. Morrisey,
Mary Reiley, David M. Rosenberg, Glenn W. Suter II

12.1 Introduction

Ecological relevance, in ERA, can take on a number of definitions depending upon scientific, societal, or administrative perspectives.

The *scientific perspective* in ERA involves the study of basic ecological principles to provide an understanding of how ecosystems function, the development of methods to detect the effects of contaminants on the integrity of these ecosystems, and determination of the adequacy of predictive powers to determine how ecosystems will function when stressors are added or removed. In the strictest scientific sense, detectable changes in structure or function are ecologically relevant and may indicate environmental risks. These changes may occur within the bounds of natural variability and be benign, or they could be more extreme and result in collapse of the ecosystem. The detection, measurement, and prediction of ecological change falls into the realm of science, whereas the degree of allowable change is not only a scientific problem but also a societal and administrative problem.

Societal perspectives involve aesthetic, recreational, and economic valuations of a resource that is important to a stakeholder. It is difficult to define ecological relevance from a societal perspective because many stakeholders have their own individual valuation system, so items of importance may vary from conservation of a backyard stream to conservation of a National Forest Reserve, and from maintenance of a viable fishery to protection of all species within an ecosystem.

The *administrative perspective* involves enforcement of regulations and statutory mandates such as the CWA in the U.S. or the Fisheries Act in Canada. In a managerial context, ERA is necessary for new product assessment, navigation dredging, or site cleanup. Ecological relevance in this context again depends upon the various concerns of the stakeholders and what legislative mandates exist to protect these concerns.

Ideally, assessment endpoints in the ERA framework (USEPA 1992) are created for a given problem only after the three above contexts have been evaluated and integrated. A variety of measurement activities can then be assigned based on a common assessment endpoint (see Chapter 7). For example, it may be desirable for an industry to determine the potential for contamination of sediments downstream from their outfall by any product or by-product that results from the manufacturing process. An administrative (regulatory) context may require that the sediment in the stream be supportive of "fishable, swimmable water." A societal context may value the maintenance of clean sediment to enable continued clam and flatfish harvest. A scientific context may want to mechanistically understand bioaccumulation of the by-product from contaminated sediment and its effects on the clam or flatfish fishery. A mutually acceptable assessment endpoint may be stated as follows: "not more than a 50% reduction in 5% or less of the benthic species with no significant bioaccumulation of the compound of concern in fishery organisms."

Several potential measurement activities will flow from this assessment endpoint: 1) determination of exposure, 2) performance of acute and chronic toxicity tests (*e.g.*, effects on survival, growth, reproduction) using actual or surrogate species that live on or near the sediment, 3) TIE which establishes causation, and 4) comparison of potentially impacted benthic communities to reference communities in field studies. Such measurements may be comprehensive and valuable in a weight-of-evidence approach, but it is difficult to determine their ecological relevance and whether achieving an endpoint will prevent future permanent ecological damage. More research effort is required in determining the ecological relevance of individual measurements before we can reach a full understanding of risk.

Four areas of knowledge are required in the understanding of ecological relevance as they apply to an SERA. These areas depend on the convergence of administrative aims, societal values, and scientific perspectives, and are as follows: 1) ability to predict and detect change in response variables; 2) understanding change in response variables; 3) causation (*i.e.*, linking cause and effect); and 4) inference. How and why these knowledge areas influence SERAs will be discussed in the following subsections.

Finally, it should be noted that two issues — spatial and temporal scales and information on the extent of exposure — are germane to each of the knowledge areas. These issues are integrated into each of the four knowledge areas discussed below.

12.2 Predicting and detecting change

The assessment of ecological risk to aquatic biota from sediments contaminated with toxicants requires the ability to predict and detect toxicity. Such predictions may be based on simple inferences or increasingly more complex correlations using data from a variety of sources such as 1) biomarkers of exposure, 2) laboratory-based toxicity tests with single or multiple species, 3) QSARs between chemical structure and toxicity, 4) artificially enclosed multispecies experiments in the laboratory or under semi-realistic field conditions (mesocosms), 5) biological surveys conducted at field sites, or 6) complex mathematical models of ecological structure based on field-collected data from clean and contaminated sites. Each of these methods has utility under selected circumstances; each is an abstraction of reality and has advantages and disadvantages in terms of ecological relevance.

12.2.1 Biological models

In each of the model systems presented below, biological models (cellular or subcellular biomarkers, responses to toxicity by whole organisms, responses to toxicity by populations and communities, *etc.*) are used to predict toxicity and its effects on the structure and function of ecological systems. Some models are more successful than others in determining causality, especially at the mechanistic level, but may be less ecologically relevant. Other biological models can provide information about toxicity without providing an understanding of the actual stressor–target relationship. For example, mortality can be observed and impacts upon ecological systems can be evaluated without knowing the specifics of the interactions. Models that incorporate more than one species or attempt to model food webs may be quite ecologically relevant because both direct and indirect effects on communities and their interactions may be observed. In addition, biological models can serve to test the physical and conceptual models discussed later in this section.

Molecular biomarkers or bioindicators (*i.e.*, biochemical, histological, and physiological indicators of anthropogenic exposure) generally examine the mechanistic aspects of a toxicant's effect on an organism. Several biomarkers have been developed for aquatic biota. Examples include measures of the induction of hepatic P450 microsomal cytochromes, inhibition of brain or blood cholinesterase by anticholinesterase pesticides, aberrations of hemoglobin synthesis, chromosomal damage, *etc.* (Huggett *et al.* 1992). Molecular predictors may be as specific as CYP1A1, a P450 specifically induced by chlorinated organics such as PCBs, or as nonspecific as heat shock proteins. These types of predictors are relevant to measuring the status of individual organisms and to indicating exposure to specific or nonspecific contaminants. Extrapolation to the subsequent effects of contaminants on survival, growth, and reproduction within a population based on cellular or subcellular changes within individuals is often lacking. As a result, the ability of biomarkers and bioindicators to predict alterations in the structure or function of ecosystems is unknown. Few such indicators have been extensively field-validated; nor have many been developed for organisms that inhabit the benthos (Johnson *et al.* 1993).

Single-species toxicity tests have long been used to evaluate potential toxicological effects of substances found contaminating water (Adams 1995) or sediment (Burton and MacPherson 1995). These tests may range from rapid screening procedures such as the 5-min solid-phase microbial bioluminescence test (Microtox) to the 10-d acute amphipod test for survival using *Rhepoxynius abronius* (marine) or *Hyalella azteca* (freshwater) to more extensive partial or complete life cycle sublethal tests that use mortality, growth, or reproduction as endpoints. Toxicity tests with growth and reproduction as endpoints are sublethal estimates of the toxicological impacts and can measure effects that occur at lower concentrations than acute mortality. These tests are also available for a wide range of organisms from different trophic levels including macrophytes (rarely), benthic invertebrates (often) and fish (often). Standard methods are generally available that allow some comparability between laboratories and acceptable QA/QC (DeWitt *et al.* 1992; USEPA 1994; ASTM 1995) but do not reflect the wide range of conditions found in the environment. Most of these tests are based upon the calculation of a medium effect or lethal concentration or dose (LD50, EC50, IC50, *etc.*) when sediments are spiked with individual contaminants, or upon the magnitude of response for field-collected sediments. However, the most useful information for predictive purposes is an estimation of the dose- (concentration-) response curve and the subsequent ranking of toxicity. Often several endpoints are measured in one test that may yield several dose-response relationships. A variety of techniques are used to calculate these endpoints: probit, moving angle average, and Spearman-Karber, as examples. Most of these methods allow calculation of a confidence interval.

An NOEC can also be calculated to estimate portions of the dose-response curve that are not expected to result in a toxicological response. These NOEC calculations depend upon the number of replicates used in the toxicity test and the power of the ANOVA and other multiple comparison techniques used. In other words, a particular calculation is largely a statistical artifact rather than an absolute measure. A more realistic computation for predictive use is the EC5 or EC10, which can be estimated from a regression of the doses against responses.

Predictions of toxicity to biota living in contaminated sediments, based on single-species toxicity tests conducted under laboratory conditions, suffers from a number of disadvantages that have been reviewed extensively by Cairns (1983), Cairns *et al.* (1992), and Landis and Ho-Yu (1994). These include 1) homogeneity of test organisms in comparison to natural populations, 2) health of test species, 3) exposure conditions that may not mimic what occurs in nature, and 4) inability to evaluate indirect or secondary impacts (Hurlbert 1975).

There are also difficulties in determining just what the ecological relevance of any calculated dose-response endpoint really means to the integrity of the population under field conditions. For example, how does a 50% mortality rate for organisms exposed to sediment under controlled laboratory conditions extrapolate to detrimental effects on benthic populations exposed to predators and competitors under field conditions and suffering from natural mortality? What level of mortality actually results in the elimination or extinction of a population of organisms, and how long will it take at that level? Of greater difficulty is the interpretation of sublethal endpoints such as a reduction in growth or in reproductive output. Does a 10% or 20% reduction in biomass or growth of a cohort of organisms exposed under laboratory conditions allow a prediction of effects on a population that can no longer sustain itself?

Another difficulty of the traditional toxicity test is that the concentration of any given toxicant in the environmental matrix is taken as a representation of dose. However, the route as well as the rate of uptake of many compounds may differ among organisms, and many aspects of metabolism may alter the effective or target dose before it reaches the site of action. McCarty (1991), McCarty *et al.* (1992), and McCarty and Mackay (1993) have suggested that a measurement of the toxic potential of a compound through improvements in 1) our ability to model and predict the fate of chemicals in aquatic sediments, 2) the use of these data to estimate the accumulation of chemical residues in organisms and in assemblies of organisms in a food chain or web, and 3) the ability to relate these body or tissue residues to various acute and chronic effects as determined in toxicity tests and bioassays have enhanced the assessment of ecological risk from chemicals. Such links between critical body residue (CBR) and adverse biological responses would also help in the evaluation of toxic mixtures and the relative contributions of a chemical to cumulative toxicity.

A number of other major problems specific to whole-sediment toxicity tests in addition to those discussed above are 1) sediments used as reference or control sediments may differ in their physical/chemical characteristics from those used in treatments, 2) organisms are exposed to sediments that have been altered from their field condition during collection, and 3) organisms are usually only exposed to sediment with one grain size distribution and organic matter content, especially in spiked-sediment toxicity tests with single contaminants. When such toxicity tests are used to predict effects under natural conditions, ecological relevance may be questionable. Having stated this, however, there have been some validations which correlate community responses of the benthos with toxicity demonstrated in whole-sediment toxicity tests (Canfield *et al.* 1994; Kemble *et al.* 1994; Swartz *et al.* 1994; Day *et al.* 1995; Section 12.3.3; Chapter 18).

A wide variety of multispecies toxicity tests can be used to predict toxicity (reviewed by Kennedy *et al.* 1995). Systems that contain a benthic component are particularly relevant to sediment work. Artificial streams of various sizes have been used to predict toxicity in the field, but they usually require a riffle-pool structure where fine particulate material can be deposited to be relevant to the issues of contaminated sediments. Multispecies toxicity tests have some advantages as models of natural ecological structures, but like natural systems, these types of systems exhibit complex dynamics and are largely irreversible. A variety of endpoints as well as biodegradation, bioaccumulation, and the direct and indirect effects of a toxicant upon the model ecological structure can be observed, which helps in the predictive process.

Extrapolation from these model systems to the ecological structure of concern has been an area of some difficulty. Multispecies toxicity tests are unlike natural ecological structures in several important ways. First, the temporal and spatial heterogeneity of the experimental system is minimized to allow easier statistical evaluation of the dataset. Temperature, chemical composition, and the composition of the sediment are usually narrowly defined. Second, the species composition is strictly defined; even systems that rely upon natural innocula often specify particular species or functional groups as desirable components. Third, migration into the model systems is usually tightly controlled, although exceptions exist in model ecological systems that have been used to study metapopulation dynamics (Drake *et al.* 1993).

Evaluation of the data is a common technical difficulty with multispecies methods. Several techniques have been used, especially conventional univariate ANOVA. Regression methods to obtain a dose-response relationship have also been used (Liber *et al.* 1992).

Derived endpoints are often used such as species diversity, species number, species richness, and integrity measurements (Davis 1995). These derived variables are supposedly simple to understand and have low variance, but they are simplifications that lose the true variability inherent within the system. Many of these derived variables are also based upon assumptions from information theory or other theoretical mathematical models so the resultant measure is an information-losing projection based upon a transformation, which in turn is based upon a hypothesis about the workings of ecological systems. A variety of multivariate methods may provide a more accurate representation of ecological structure.

12.2.2 Quantitative structure-activity relationships

Quantitative structure-activity relationship models are mathematical equations derived to estimate the toxicity or other property of a chemical from its structure. Quantitative structure-activity relationships are usually applied to predict the toxicity of new chemicals based on similarities in structure or substructure of a molecule to those of other chemicals with known toxicity. Two basic sets of models exist. In cases when the structure of the receptor is well known, it is possible to use molecular modeling to examine the interactions such as hydrophobicity or hydrogen bonding between a receptor and the ligand. Although data of such detail exist only in a few instances, these models have proven especially useful in examining the effects of PCBs and hormone mimics (McKinney and Waller 1994).

Generally, QSAR models are statistical, using substructural keys and correspondence to toxicity data to generate predictions; multivariate linear regression and discriminate analysis are commonly used. One of the primary difficulties in generating QSARs is the lack of a detailed set of toxicity tests using a wide variety of chemical types. Often only a few chemical structures are tested using acute or chronic tests, producing large gaps in the map of chemical structure versus toxicological response. QSARs rely on single-species toxicity tests to provide the database; therefore, they suffer from the limitations to predicting ecological responses discussed above. Many of the acute or chronic data used in QSARs suffer from lack of repeatability, and two chemicals may have a very different toxicity value. Another limitation is the difficulty of accurately representing the structural aspects of the chemical that contribute to the toxicological response. Another long standing difficulty has been the inability of many programs to recognize sufficient substructural keys, relying instead upon other characteristics such as the log of the octanol-water partition coefficient. In spite of these difficulties, regression-type QSAR models have been developed for a number of endpoints. In environmental toxicology, include daphnid and fish toxicity as well as interspecies models (Enslein *et al.* 1987, 1989).

12.2.3 Stochastic and deterministic models

These types of predictive tools are generally derived from empirical relationships or from first principles. Taylor's Power Law (Taylor 1961; Elliott 1971) is an empirical relationship predicting a measure of contagion from the number of organisms within a collection. The relationship was derived from field data and seems to be consistent for a species. In contrast, the interactions and dynamics predicted by the classical Lotka-Volterra models for population competition are not directly related to responses found in the field. These models are simple constructs with terms such as K (carrying capacity) and a (competition coefficient). Other types of population models such as the island biogeography derivations of MacArthur and Wilson (1967) are also of this general type. In the early and mid-1970s, discrete models of population dynamics were generated using difference instead of differential equations. The use of these models, even simple ones, led to the co-discovery of strange attractors and chaotic dynamics.

Barnthouse (1992) and Suter (1993) describe the uses of common models in the field of risk assessment. The models used in fisheries and wildlife management to predict the risk of extinction as a result of toxicant addition are especially useful. These are typically demographic models based on a matrix representation of growth and mortality rates for each age or size class of fish species. Other models include probabilistic terms as part of the equations. Stochastic differential or stochastic difference models can provide as output a probability distribution from a number of iterations. Stochastic difference equations are similar except that they are discrete; instead of a smooth distribution, they can produce a rather dramatic range of possible outcomes.

Understanding the intrinsic dynamics of each specific case is a difficulty in the use of models. Nonlinear difference equations are perhaps an extreme case, having realms of stability, bifurcations, and eventually chaotic dynamics (May and Oster 1978). The transition to each of these states is abrupt but can be easily mapped.

Oreskes *et al.* (1994) discuss the use of models in the geophysical sciences. Their review is relevant to much of the modeling conducted as part of an SERA because of the open nature of geophysical systems. Oreskes *et al.* (1994) propose that models can only be confirmed by the use of experimental data but not be verified or validated. Models assist in the creation and falsification of hypotheses.

12.2.4 Univariate and multivariate methods for detecting anthropogenic impact

The success of any ERA, whether for sediments or any other environmental matrix, lies in the ability of the methods used to predict effects in impacted areas versus unimpacted areas. The inherent variability of natural ecosystems and the complications that natural stressors can have on the benthos make this a difficult task.

A number of approaches have been used to identify "normal" communities at a given site (Chapter 10). The most common approach tries to develop relationships between environmental and community attributes based on the structure of the biotic community. Community attributes are usually based on taxa richness, enumerations, or some integrative measure of the two (*e.g.*, biotic, diversity, or similarity indices). Empirical approaches generally are either univariate or multivariate in nature.

Univariate approaches are the most frequently used (Norris and Georges 1993) and can be quantitative (*i.e.*, hypothesis setting, rigorous statistical design, replication; Resh and

McElravy 1993) or qualitative (*i.e.*, rapid assessment procedures; Resh and Jackson 1993). These approaches usually compare reference sites to impacted sites, but only incorporate environmental attributes as determinants of community structure through inference. In addition, the majority of these approaches have been applied only in riffle areas, not in depositional zones.

Conventional multivariate approaches can be divided into two types: 1) multiple regression and 2) paired matrices. Multiple regression develops relationships between benthic invertebrates and environmental attributes, but it only considers univariate measures of community structure. The paired matrices approach tries to identify pattern and structure in the benthic invertebrate community and relate that to environmental variables (*e.g.*, Wright *et al.* 1984, 1988; Corkum and Currie 1987; Corkum 1989; Johnson and Wiederholm 1989; Reynoldson *et al.* 1995). The purpose is to establish predictive models of communities that can be used to assess environmental impairment. If the predictive models are based on reference conditions, then the models derived can be used to measure divergence from the reference state and the degree of risk associated with an environmental stressor (for an example in SERA, see Reynoldson *et al.* 1995).

In a different multivariate approach, data clusters are optimized based on their intended function (*i.e.*, the accurate prediction of properties of the data) rather than using a distance metric or similarity function. The clustering method is applicable, without further *ad hoc* assumptions or transformations of the data, as follows: 1) when features are heterogeneous (both discrete and continuous) and not combinable, 2) where some data points have missing feature values, and 3) where some features are irrelevant (*i.e.*, have large variance but little correlation with other features). Further, it provides an integral measure of the quality of the resulting clustering. In order to compute these relationships, the RIFFLE clustering program is used (Matthews and Hearne 1991). A variety of field (Matthews, Matthews, and Ehinger 1991; Matthews, Matthews, and Hachmoller 1991) and microcosm experiments (Landis *et al.* 1989, 1994; Landis, Matthews, Markiewicz, and Matthews 1993; Landis, Matthews, Markiewicz, Shough, and Matthews 1993; Matthews *et al.* 1995) show that the conceptual clustering is, in many respects, superior to traditional methods.

Identification of patterns and their relationship to treatments or other environmental descriptors is only the initial step. Visualization of the relationships and the resultant dynamics can aid in the interpretation of the data. WORM is a visualization tool used for investigating high-dimensional time series from ecotoxicological dose-response experiments involving treatment groups. It allows the considerable pattern recognition abilities of humans to focus on the task of investigating complex data. Visualization does not replace statistical methods but rather helps researchers to gain qualitative insights into the data and to form hypotheses that can then be investigated with formal quantitative methods. The tool uses two- and three-dimensional plots and animation to visually compare variance between treatment groups and between the replicates within each individual treatment group over the course of the experiment.

An AI program called RIGGLE has also been designed to perform a conceptual temporal analysis of hypervariate datasets by investigating the intervals between the data points rather than the data points themselves. A set theory based on similarity function is then employed to compare replicates of an experiment or of different experiments and to identify population dynamics that are similar but may not be occurring in exactly the same time frame. RIGGLE has been applied to SAM data with promising results (Landis, Matthews, Markiewicz, and Matthews 1993; Landis, Matthews, Markiewicz, Shough, and Matthews 1993; Landis *et al.* 1994).

12.3 Understanding change in sediment ecological risk assessments

12.3.1 Introduction

Any anthropogenic impact added to an aquatic ecosystem is layered onto the natural patterns of community structure and function that result from biotic and abiotic factors acting at several spatial and temporal scales. Therefore, natural variation in the structure and function of the benthos in aquatic ecosystems must be separated from change initiated and sustained by toxic sediments before the interpretation and ecological relevance of the impact can be understood. This requires an understanding of the ecosystem and its trajectory (change over time from natural causes).

12.3.2 Spatial and temporal variability of sediment ecosystems

Biotic communities in sediments appear to consist of mosaics of patches characterized by different species compositions. These patches are believed to represent different stages of recovery from a range of external physical and biological disturbances (reviewed by Hall *et al.* 1994) although the relative importance of the various forces that structure the benthic environment remain poorly known (Power *et al.* 1988; Posey 1990; Hall *et al.* 1994). Patterns of recovery from these kinds of disturbances are generally unpredictable because of variability in recolonization caused by immigration and larval recruitment. These disturbances may affect a variety of spatial scales, making the interpretation of anthropogenic insult difficult.

Recolonization and succession (*i.e.*, changes in assemblages following disturbance), depend on the supply of propagules of different species. This supply is subject to spatial and temporal variability such as the timing of the reproductive cycle (Whitlatch and Zajac 1985) and patterns of drift. For example, many muddy-sandy intertidal areas of Manukau Harbor, New Zealand, are dominated by the bivalves *Macomona liliana* and *Chione stutchburyi*. These assemblages are subject to frequent disturbance by wind-generated waves (Turner *et al.* 1995). Recolonization of the disturbed areas is at least partly dependent on the settlement of juvenile bivalves. This natural disturbance and recolonization provides heterogeneity in the spatial and temporal scales that must be recognized before the impact of anthropogenic insult can be properly interpreted.

Estimation of normal patterns of spatial and temporal variation is a function of the predicted scale and organizational level of the insult. Structural responses to insult may appear at levels of organization from suborganismal to ecosystem, including the organism, population, and community. Functional responses are likely to show similar hierarchical organization from physiological to population (reproduction, turnover) to community (primary production, decomposition, respiration, species interactions) levels. Each of these levels of organization is characterized by temporal magnitude ranging from physiological cycles (reproduction, respiration, lifetime), diurnal cycles, lunar cycles, to seasonal cycles. These normal functional interactions of levels of organization and time may be overshadowed by the presence of stochastic events. In fact, the major structuring pressures in natural systems may be dictated by rare stochastic events (Underwood and Denley 1984).

As an example of the potential interactions between natural spatial variation and the effects of contaminants, macrofaunal populations, and communities in sandy marine sediment in Botany Bay, Australia, were exposed to experimental addition of copper to the sediment (Morrisey *et al.* 1995 and in review). The resulting responses were not spatially consistent; they varied at scales of meters or hundreds of meters. This variation in responses corresponded to patchiness in the distribution of the animals, which existed at the start of the experiment. The patchiness was, presumably, a consequence of previous physical and biological interactions.

Bottom-up (*e.g.*, nutrient availability; Robbins *et al.* 1989) or top-down (*e.g.*, predation and competition) controls on populations may determine the quantity and species composition of benthic communities in the absence of other perturbations (Hunter and Price 1992). The relative importance of all of these structuring forces varies through time and can be tied to normal temporal cycles. For example, blooms of phytoplankton may reduce the intensity of competitive interactions among consumers.

Ecosystems are constantly moving from one state to another as a result of the continual interaction of natural perturbations and temporal cycles within a system. These successional changes can be characterized as a trajectory of change for a given system, (*e.g.*, change toward reconstruction after acute perturbation). The trajectory of system change will be a strong indicator of overall system function. In the case of Manukau Harbour and the bivalves, the ability to demonstrate the trajectory of population recolonization and succession would provide an assessment endpoint against which ecological relevance could be interpreted. Determination of such trajectories for populations requires large studies and measurement of many variables for a large number of reference sites against which to compare the test site. These trajectories may be visualized and measured in several ways, including time series and trend analyses (*e.g.*, Gottman 1981). Two recently developed conceptual clustering techniques, WORM and RIGGLE, have been described above.

The "Beyond BACI" (before-after/control-impact) designs for detecting environmental impacts (Underwood 1993) recognize that temporal trajectories in populations of organisms at different places need not be the same even in the absence of an impact. This is achieved by comparing differences in changes in trajectories before and after the putative impact between the impacted site and several control sites with differences in changes among the control sites. The differences in changes among control sites provides an estimate of natural variation in trajectories among places in the absence of an impact.

12.3.3 Effects of contaminants on sediment ecosystems

Assessment of the effects of contaminants presumes exposure; thus, to develop understanding of the ecological relevance of the measurement endpoint, the exposure condition producing the observed change needs to be determined. Direct contaminant effects are a combination of the exposure environment and duration and the sensitivity of the organism. Exposure of biota to contaminants in sediment is complicated by chemical interactions among the particles of sediment, the pore water, the overlying water, and the contaminant as well as the biological behavior of the organism. Benthic organisms may be exposed to the contaminant either by passive diffusion of the compound through the integument from the pore or overlying water and ingestion of particles of sediment coated with the contaminant. Any determination of exposure, as well as the interpretation of effects, is facilitated by knowledge of more than the external concentration as dose. Significant examples are available that demonstrate that the external dose may not be sufficient for describing exposure to benthic organisms (see below). Thus, interpretation of effects at varying hierarchies may be more accurately described by assessing concentrations within the community rather than individual organisms. This measurement of accumulation represents an integration of the toxicokinetics of exposure for a species (but see Chapter 18). Assessment of the exposure that results in a direct effect (the ultimate cause of any observed ecological effects) needs to be delineated and the causative interaction defined; interpretation of indirect effects will be more complicated.

Direct effects are those where the stressor acts on the ecological component of interest. *Indirect effects* are those where the stressor acts upon the component of interest through supporting components of the ecosystem. Direct effects at the individual level include mortality and changes in growth and behavior; at the population level, they might involve reproduction or demographic structure; and at the community level, they could involve changes in diversity and species dominance. Indirect effects at the individual level might be changes in growth of a predator because of food limitation manifested when a species of prey is removed through direct mortality. For example, Pratt *et al.* (1993) observed a reduction in the availability of food for deposit- and detritus-feeders when microbial populations in sediments were affected by contaminants.

The importance of species interactions (*e.g.*, predation and competition) in structuring benthic communities has received considerable attention in freshwater and marine ecology (review of rocky intertidal communities: Dayton 1994; reviews of freshwater communities: McAuliffe 1984; Peckarsky 1984; Sih *et al.* 1985; Power *et al.* 1988; Power 1990). Understanding how species interactions are affected by contaminants has been largely ignored probably because most of the concerns about the effects of contaminants in soft-substrate environments have proved much less tractable to experimental investigation than those on hard substrata described in the references cited above. Studies of macrofaunal communities in marine sediments, for example, have so far produced ambivalent results, and there are no general models of how these communities are structured (*e.g.*, Posey 1990; Hall *et al.* 1994).

The effects of contaminants on species interactions will likely depend on the magnitude of the stress and the relative sensitivities of the interacting species. Menge and Sutherland (1987) developed a model of community regulation showing that the relative importance of species interactions decreases with increased environmental stress. If environmental contaminants are analogous to environmental stress, then similar responses may occur along contaminant stress gradients. Changes in the number of species present in successively higher trophic levels as a consequence of the alteration of interspecific interactions in lower levels by contaminants have been modeled by Landis *et al.* (1989). Warner *et al.* (1993) found that the outcome of interspecific competition between anurans was changed by acidification. Wipfli and Merritt (1994) observed that reduced black fly density in a larvicide-treated stream altered species interactions and indirectly affected community structure.

The specific effects of contaminants differ depending on whether consumers or their prey are more susceptible to the stress (Menge and Olson 1990). For example, if predators are more sensitive to stress than their prey, the consumer-stress model predicts that predation rates will be lower in high-stress environments. This pattern was observed in rocky intertidal habitats subjected to extreme wave action (Menge and Olson 1990), in high gradient streams (Peckarsky *et al.* 1990), and in acidified lakes (Locke and Sprules 1994). In contrast, if prey are more sensitive to stress, the prey-stress model predicts that predation rates will be greater in stressful environments. This pattern was observed in the predator-prey experiments between stoneflies and caddisflies (Clements *et al.* 1989).

These authors reported that susceptibility of net-spinning caddisflies to stonefly predation increased in streams dosed with copper, compared to control streams. Understanding the importance of indirect effects of contaminants may help reconcile differences between laboratory and field responses. For example, failure to show direct effects in laboratory toxicity tests, despite altered benthic community structure observed in the field, may be a result of indirect effects or the benthos is responding to noncontaminant factors.

The lack of studies on indirect effects of contaminants (but see Hurlbert 1975) is probably related to the difficulty of conducting the necessary field experiments to separate direct effects on species from indirect effects on species interactions. First, it must be shown that species interactions are important, and then it must be demonstrated that these interactions are affected by contaminants. For example, caging experiments in reference and polluted locations could be used to evaluate the effects of predation in both situations. Greater stonefly predation on benthic invertebrate communities occurred in metal-polluted locations than reference locations as measured in experimental streams (Clements, unpublished data). This result supports the hypothesis that contaminants increase the susceptibility of prey organisms to predation. The same experiment could be done in artificial stream mesocosms.

For benthic communities structured by bottom-up processes, perturbations caused by either natural or anthropogenic stress on primary production will be observed as indirect effects on population sizes of consumers and higher trophic levels. Microbial decomposers can be similarly affected (Cairns *et al.* 1992). Effects of contaminants in sediments in bottom-up systems are more likely to be significant in assemblages that are dominated by detritus and deposit feeders than when filter feeders are dominant because the food supply of filter feeders is less likely to be affected. Posey's (1990) review of the success of functional models in predicting the structure of communities in marine sediments suggests that assemblages in which both trophic groups are abundant are common (*cf.* the "trophic-group amensalism" hypothesis proposed by Rhoads and Young [1970]) to explain the structure of communities in marine sediments). Estuaries, however, provide an exception in that assemblages are often dominated by deposit/detritus feeders and their predators, presumably because suspension feeders are intolerant of the large loads of suspended material occurring in these environments. Estuaries are particularly prone to the accumulation of sediment-associated contaminants because they have high rates of sedimentation and are recipients of river-borne contamination. Thus, estuaries may represent areas where contaminants that adversely affect the lowest trophic levels are of greatest ecological significance, assuming that bottom-up factors structure the community. The susceptibility of soft-sediment communities in estuaries to natural disturbances may vary geographically and may be large.

Predators may be able to influence the temporal and spatial distribution, species-density and abundance of their prey (Ambrose 1991). Thus, in top-down systems, control of population abundance may result from "trophic cascades" of predator–prey interactions. If contaminants affect the predator more than the prey, the prey may be released from trophic control and bloom as an indirect effect of contamination. For example, in Norwegian fjords, two predatory polychaetes, *Glycera rouxii* and *Lumbrineris* sp., were intolerant of copper contamination (Rygg 1985). Thus, copper contamination could release their prey from predation by these species, assuming that the prey were relatively less affected. However, other polychaete groups in the same system, *Glycera alba*, *Goniada maculata*, and *Nephtys* spp., were more tolerant of copper, so that if one species of predator was affected, another could simply take its place. Such "functional redundancy" has been cited as an argument against the use of the structure of assemblages as a reliable surrogate for the protection of ecological integrity (Cairns *et al.* 1992). Alternatively, it could be argued that such redundancy makes structural changes more sensitive indicators of change (Rapport *et al.* 1985). The ecological effects are, in any case, likely to depend on the nature of the stressor. When one predator species is substituted for another, the trajectory of the system will be affected and such changes in trajectory may well be both sensitive and important indicators of ecological change.

The trajectory of a community through time will likely be diverted as a consequence of these various effects of contaminants at different scales and levels of organization. For example, the recolonization of disturbed areas in Manukau Harbour by bivalves (see above) can depend not only on spatial and temporal variation in immigration but also on degree of contamination in the environment. Juveniles of the bivalve *M. liliana* avoided laboratory sediments contaminated with copper at 5 g.g^{-1} or greater (Roper and Hickey 1994), so contamination of the sediments by copper could retard or eliminate the potential for recolonization despite the availability of propagules. This would disturb the normal trajectory of the system. However, copper concentrations of sediments in the harbor are in the range of 20 to 30 g.g^{-1} and populations of *M. liliana* persist, suggesting that other environmental factors or bioavailability issues mitigate the inhibitory effects of copper *in situ*.

The persistence of *M. liliana* populations despite high copper concentrations in the sediments raises the problem of spatial and temporal scale in studies of the effects of contaminants. For example, if the structure of faunal assemblages is reset by large-scale physical disturbances such as storms, then the impact of contaminants on smaller-scale processes such as competition and predation may be relatively unimportant. In reality, the impact of contaminants on the potentially more sensitive larval stages may dictate the success of recolonization.

12.3.4 Future work

Ecosystems must be recognized as spatially and temporally variable, and this natural variability and how it relates to the structure of benthic communities needs to be established. Information is also needed about the relative variabilities and sensitivities of benthic communities to contaminants under different environmental conditions and about their usefulness as measurement endpoints for different assessment endpoints. Methods should be developed that allow determination of the trajectory of natural assemblages (communities) in light of recognized temporal and spatial structures or that allow appropriate comparisons of trajectories among reference and contaminated sites (Underwood 1993). Although their development is still in an early stage, manipulative field experiments (Cooper and Barmuta 1993; Hare *et al.* 1994; Morrisey *et al.* 1995 and in review) offer a means of studying the effects of contaminants on structural and functional variables under environmentally realistic conditions. Manipulative field experiments also offer a high potential for identification of cause and effect.

12.4 Cause and effect

12.4.1 Why is causation important to sediment ecological risk assessment?

The demonstration of causal relationships between ecological responses and anthropogenic stressors is important in SERAs in order to correctly assign an underlying reason for change in an ecological measurement. Expensive remediation and litigation decisions are based on the assumption of causality, so it is essential that SERAs attempt to establish a cause-and-effect relationship between stressors and responses. In addition, it is important to know the identity of the stressor, so its effects can be mitigated.

Direct demonstration of causality is possible only when all relevant factors which may mitigate change are accounted for and controlled. Therefore, in risk assessments that are based on traditional toxicity tests using spiked sediment, and in some cases of modeling, causality is not in doubt. However, predictions about the nature and level of effects under natural conditions may be doubtful due to many uncontrolled but mitigating factors (see Section 12.3.2 and Chapter 10). When conditions are not under the control of the investigator, such as in biological surveys, associations with toxicity can be demonstrated but causality can only be inferred. In such cases, causality must be established by demonstrating consistency among the uncontrolled variability in control versus impacted areas. In addition, some measurement endpoints for uncontrolled field studies can be useful for inferring causality because they have a well-defined relationship to standard test endpoints. Similarly, some test endpoints are useful for demonstrating causality because they have well-defined or relatively consistent exposure-response relationships for the contaminants of concern or because they have well-defined relationships to endpoint effects in the field. Demonstration of the association between measurement endpoints from controlled and uncontrolled studies is a component of a weight-of-evidence analysis, which also considers the quality of the various lines of evidence. The analysis of consistency is typically based on rules such as Koch's (1966) postulates and Hill's (1965) criteria but may be statistical or use problem-specific inference procedures (Suter 1993).

Every measurement endpoint must have an element of causality to link the response to the magnitude of the stressor. Depending on the goals of the risk assessment, the specific identity of the stressor may nor may not be important to know. It may just be necessary to know that a characteristic of the stressor causes the effect. Demonstrating direct causation will not be possible in many situations, particularly those in which benthic communities are surveyed; however, integrating field observations with laboratory experiments and developing weight-of-evidence approaches will help support the hypothesis that changes in benthic populations or communities are a direct result of sediment contaminants.

The ability to demonstrate a cause-and-effect relationship between specific stressors and responses is influenced by 1) understanding a stressor's mode of action; 2) complexity of the suite of stressors (*i.e.*, multiple, possibly interacting, modes of action); 3) ability to measure exposure at scale and form, relevant to an endpoint; 4) ability to discriminate between natural variability (in space and time) and contaminant-induced response; and 5) other confounding factors.

12.4.2 Mode of action

The strength of the cause-effect relationship will be affected by response- and stressor-related factors. Some assessment endpoints are directly associated with a specific type of stressor, so these endpoints may be useful for showing cause-and-effect relationships. For example, increased production of metallothionein is generally indicative of metal exposure, whereas increased primary productivity is often associated with nutrient additions or other nonspecific factors. However, theses endpoints may not respond to other stressors that underlie ecological changes in the sediment of concern. Other endpoints commonly used in sediment studies may respond to a wide variety of stressors (*e.g.*, growth can change in response to contaminants, food limitation, temperature, or other factors; community metrics may be responsive to contaminants, habitat characteristics, physical disturbance, interactions among species). These general responses can provide a broad measure of stress, but inferring causation may be difficult. Thus, a trade-off exists between endpoints wit high sensitivity to specific stressors and those that are sensitive to a wide variety of stressors.

12.4.3 Complexity of stressors

Certain types of stressors may be directly tied to specific effects, especially if only a single stressor is present or the stressors are spatially distributed along a linear gradient. This situation is usually the exception rather than the rule. Most sediment assessments have the difficult task of differentiating among multiple stressors (*i.e.*, mixtures of chemicals), between chemical contamination and habitat modification, or among stressors distributed patchily in space and time. Suter (1993) introduces several terms to classify the cumulative effects of multiple stressors:

*   *Nibbling:* "The cumulative effects of a number of actions which have similar small incremental effects..." (p. 372)
*   *Time-crowded:* "...actions [that] are so close in time that the system has not recovered from effects of one before the next one occurs..." (p. 373)
*   *Space-crowded:* "...actions [that] are so close in space that the areas within which they can induce effects overlap." (p. 373)
*   *Indirect*: "The cumulative effects that occur when the direct effects of actions are not space- or time-crowded, but their indirect effects are." (p. 373)

Separating the component impacts of multiple stressors is a difficult task, and methods to address this issue are discussed in Chapter 16, Bedford and Preston (1988), and Gosslink *et al.* (1990).

12.4.4 Measurement of exposure

Measurement of exposure is the most problematic issue in defining cause and effect of sediment-associated contaminants. The problem lies in defining the dose to the organism in the dose-response relationship. Sediment-associated contaminants, in contrast to aqueous contaminants, are subject to a greater magnitude and range of factors that influence their bioavailability to organisms. The toxicity paradigm for ecological testing has generally used environmental concentration as the dose to which the organism is exposed. For aquatic toxicity tests, this stems from the proportional relationships between the dose at the receptor, the dose in the organism, and the external concentration. These relationships are more difficult to establish for sediment exposures because of many factors that can alter contaminant bioavailability. In addition, behavioral responses of organisms to contaminants may alter exposure (Kukkonen and Landrum 1994; Landrum *et al.* 1994). However, in these cases, the dose in the organism (*i.e.*, the tissue concentration) often is proportional to the observed response and may be a better indicator of exposure (Landrum *et al.* 1994).

Factors that influence the bioavailability of sediment-associated contaminants include character of the sediment, characteristics of the contaminants and behavior of the organism (Hamelink *et al.* 1994). Thermodynamic equilibrium models have proven useful in predicting the bioavailability of hydrophobic organic contaminants (Di Toro *et al.* 1991) and some metals in sediment (Di Toro *et al.* 1990; Ankley *et al.* 1994; DeWitt *et al.*, in press), particularly for homogeneous and temporally undisturbed sediments. Under more variable conditions, simple equilibrium relationships may break down (Harkey, Landrum, and Klaine 1994; Landrum *et al.* 1994, 1995). For example, if concentrations of contaminants or normalizing factors (*i.e.*, OC or AVS) are vertically or horizontally stratified within the sediment, organisms burrowing through such strata will likely experience differences in exposure over relatively small spatial scales that may not be reflected in the usual bulk measurements of sediment properties or chemistry (Lee 1991). Additionally, the behavior of benthic organisms may affect their exposure as a consequence of 1) feeding mode, which can affect the routes of exposure (*i.e.*, deposit versus filter-feeding; Harkey, Lydy *et al.* 1994); 2) bioturbation, which can change local sediment geochemistry and thus alter local contaminant bioavailability; or 3) dispersal ability, which can alter the duration of exposure to contaminated sediments. These factors can confound the estimation of exposure for organisms in the field when exposure is based simply on either a whole sediment or some normalized concentration. Consequently, assessments of exposure within sediments would be improved by including tissue-level concentrations as well as measurements (or predictions) of the bioavailable fraction of contaminants in sediments.

12.4.5 Separating natural background variation from anthropogenic variation

Natural systems are inherently variable in space and time. In general, if we look closely enough or take enough samples, we will almost always be able to detect differences between two locations, regardless of the degree of anthropogenic stress. Therefore, one of the most important issues in establishing causal relationships between sediment contaminants an ecological effects is to separate natural spatial-temporal variation from anthropogenic variation. For example, benthic macroinvertebrate communities are strongly influenced by substrate composition (grain size, percent OC; Sanders 1960). On a larger scale, it is well established that structure and function of streams vary naturally from upstream to downstream (Vannote *et al.* 1980). Although this longitudinal variation is predictable, it also must be accounted for when designing SERAs.

Understanding natural variation is the key to separating it from anthropogenic variation. The best understanding is gained form long-term studies over relatively broad spatial scales, such as those conducted at the Experimental Lakes Area (ELA) in northern Ontario (Heckey *et al.* 1994) or at the National Science Foundation's Long-Term Ecological Research (LTER) sites (Callahan 1984).

Experimental manipulation, either in the laboratory or in the field, is the most direct method of separating natural from anthropogenic variation. For example, laboratory studies can be conducted to measure the direct effects of a particular confounding variable (*e.g.*, grain size, OC) on survival, growth, or reproduction (DeWitt *et al.* 1988). Similarly, microcosm or mesocosm experiments can be used to measure the influence of confounding variables on community structure or ecosystem processes.

Well-designed field studies, with appropriate reference sites and true replicates, can also be used to understand the relative importance of natural variation. Multiple regression models and multivariate procedures can be developed to estimate the importance of potential confounding variables on benthic community endpoints (*e.g.*, Clements 1994). One potential limitation of these field procedures is that large amounts of data are generally necessary to produce robust models.

12.4.6 Confounding factors

Noncontaminant factors can confound the response of a measurement endpoint and lead to incorrectly ascribing high environmental risk to a stressor of concern. For example, survival of some marine amphipods can be significantly affected by the grain size of the sediment (DeWitt *et al.* 1988) or ammonia concentration in the pore water (Ankley *et al.* 1990; Kohn *et al.* 1994). Failure to recognize these factors can lead to erroneous attribution of toxicity to persistent contaminants present in the sediments at sublethal concentrations. Other confounding factors within sediment habitats include both abiotic (*i.e.*, salinity, temperature, pH, dissolved oxygen, hydrogen sulfide, alkalinity, hardness, light) and biotic parameters (*i.e.*, food limitation, predators, competitors, disease, biogenic disturbance). Typically, confounding factors depress the measurement endpoint response (*i.e.*, reduce survival, decrease growth); however, the opposite can also occur, such as when a test organism fails to burrow into sediment because of anoxic conditions or is less exposed to sediment-associated contaminants and therefore experiences greater than expected survival or growth. In laboratory studies, use of appropriate controls and monitoring of water quality conditions can greatly reduce the influence of confounding factors. In field studies, however, confounding factors may be much more difficult to account for because of spatial and temporal variability. Knowledge of the physiological tolerances of the species of interest and their interactions with other organisms at the study site can provide inferential guidance to the potential influence of confounding factors in the field, but manipulative field experiments may be required to unequivocally measure the impact of these confounding factors.

12.4.7 Assessment of causation

Several laboratory and field approaches may be used to establish cause-and-effect relationships between stressors and endpoints (Table 12-1). Of particular relevance is the use of TIE procedures which can be defined as "a stepwise process that combines toxicity testing and analysis of the physical and chemical characteristics of samples to identify potentially causative toxicants" (USEPA 1985; and see Chapter 16). Laboratory toxicity tests provide the greatest control over confounding variables and are highly replicable. However, laboratory toxicity tests have greatly limited spatial and temporal scales and lack ecological realism. Field biomonitoring approaches may be conducted over broader spatial and temporal scales but are poorly controlled with regard to confounding variables. Controlled mesocosm studies provide the highest degree of environmental realism and

Table 12-1 Usefulness of laboratory and field approaches for assessing cause-and-effect relationships between stressors and endpoints

Approach	Control of confounding variables	Replicability	Spatial-temporal scale	Ecological realism
Laboratory toxicity tests	high	high	low	med
Site-specific biomonitoring	low	med	med	high
Large-scale biomonitoring studies	med	med	high	high
Paleoecology	low	low	high	med
Natural experiments	med	low	med	high
Ecosystem models	med	high	high	med
Microcosms/mesocosms	med	med	med	med
Ecosystem manipulations	low	low	high	high

may be conducted over relatively broad spatial and temporal scales, but these approaches also lack control over confounding variables. As strength of inference is related to a weight-of-evidence approach, consistency of relationships among replicates in laboratory studies; correspondence between field evidence, laboratory experiments, and published literature; and biomonitoring information should all be utilized to make sound decisions in SERAs.

12.5 Predictive and retrospective inference

12.5.1 Introduction

Assessment endpoints in SERAs usually have some relevance in terms of legal mandates, regulatory policies, stakeholder interests, and/or broad societal values because risk managers are involved in the process of selecting such endpoints. However, during the risk assessment process, the question becomes one of ecological relevance: In what ways are the measurement endpoints that are available from the literature or potentially available from *ad hoc* testing and measurement activities relevant to the chosen assessment endpoints?

Types of measurement endpoints from which inferences are made and the types of inference to be made can be either predictive or retrospective. Inferences can be made from single measurement endpoints or combinations of measurement endpoints. Inference from a single measurement relies on models relating the measurement endpoint to the assessment endpoint, whereas statistical, mathematical, or conceptual inferences from multiple lines of evidence use logical or rule-based weight of evidence.

12.5.2 Inference from biological survey endpoints

Biological surveys are activities that sample biological populations or assemblages to estimate their properties. For example, the mean number of species per m², the total number of species, and the mass of macroinvertebrates per m² are potential endpoints from a survey of benthic macroinvertebrates conducted by collecting multiple Ponar samples at sites on a transect. It is often assumed that the organisms collected from sediment have been chronically exposed to the contaminants in those sediments because the invertebrate biota are sedentary.

All inferences from biological surveys involve some comparison to a reference state. This may be done in a number of ways, and the choice influences the types of inferences that can be made.

- Transect methods are quire useful when contaminants have a distinct source (*e.g.*, an outfall or tributary) or location (*e.g.*, a spoil mound). In such cases, the reference is the portion of the transect in which the measurement endpoint achieves the asymptotic condition. This approach requires reasonable uniformity of the physical-chemical properties of the sediment in order to avoid confusion. In theory, one could identify relevant physical-chemical properties and eliminate their influence if their distribution was independent of the distribution of contaminants, but such an approach is not likely to succeed in practice.

- Where transects are not practical, community properties from the contaminated site must be compared to those of less-contaminated sites. If biological surveys are to provide the sole or primary line of inference, then it is important to have multiple reference sites in order to answer the question "Is the contaminated site similar to or different from uncontaminated sites?" rather than "Are the two groups of sites similar or different?" The latter question is of little interest since two sites may differ for any number of reasons other than contamination. If the assessment is to rely primarily on sediment toxicity testing or single-chemical toxicology, then less rigorous designs may be used to support the results of those inferences. For example, if the sediment is toxic to the test organisms and is contaminated to levels that are estimated to result in toxic exposures, then the fact that the species richness of the contaminated site and a reference site differs by 30% serves to support those inferences even though it has little utility alone.

- A survey of reference communities is sometimes not necessary. If a sediment supports only a few taxa that are known to be resistant to a contaminant, then little need exists to expend resources on sampling and characterizing reference communities. It is sufficient to assure, based on knowledge of habitat, that the sediment is suitable for a different sort of community in the absence of contamination.

- Paleoecological references may be used if sediments are bedded so that current community composition can be compared to predisturbance composition at the same location. Inference is limited by confounding historical changes such as changes in sediment texture or frequency of anoxia.

Biological survey results are seldom used in predictive assessments, and it is unlikely that they would be used as the sole line of evidence for inferring effects. However, inference from similarly contaminated sites to a site that is proposed for contamination is prudent, given the many uncertainties associated with predicting the responses of benthic communities to contaminated sediments. For example, the best evidence concerning the risks to the benthic community of a proposed disposal site could be surveys of the effects of disposal of sediments from prior dredging of that harbor. The relevance in this case is limited by changes in the contaminant composition of the sediment and the ability of available survey techniques to measure properties that are closely related to the assessment endpoint.

Measurement endpoints that are direct estimators of the assessment endpoint can be selected when planning biological surveys. For example, if the assessment endpoint is a reduction in species richness of benthic invertebrates in a river reach, the measurement endpoint may be mean species richness in samples from that reach. The inference is largely statistical (*e.g.*, How representative are the samples, and what summary statistics best describe the sampled community?) but also includes assumptions about methods (*e.g.*, the consistency of taxonomic identification). Inferences from these measurement endpoints to assessment endpoints are relatively simple and direct in most cases because assessment endpoints correspond to conventional measurement endpoints. However, recall that inferences about causation cannot be made from these measurement endpoints alone.

It is also possible to measure an intermediate stage in an indirectly induced effect and to model the effects on the assessment endpoint (*e.g.*, measure invertebrate abundance and model the effects of the loss of food on benthic fish).

12.5.3 Inference from biomarkers and gross pathologies
Biomarkers include various biochemical and histological measures of the internal state of organisms (Melancon 1995). They are seldom used with benthic invertebrates but are used with fish, including epibenthic species. Gross pathologies include deformities, lesions, and tumors. Tumors and fin erosion have often been associated with contaminated sediments.

When predicting effects of future contamination, it is more useful to do a toxicity test that measures responses such as mortality, growth, and reproduction (endpoints relevant to population and community properties) than to test for biomarker responses because the relationship of biomarkers to organismal, population, or community effects is not clear. Gross pathologies are not usually included in toxicity tests, although developmental abnormalities are relevant to population effects if it is assumed that abnormal organisms will not reach maturity.

Biomarkers and pathologies have no known ecological relevance, so they cannot be used alone to make inferences about ecological risks. However, they may be used as supporting evidence. For example, the presence of gross or histological pathologies characteristic of a particular contaminant or class of contaminants will support the inference that toxicity is a cause of a depauperate fish assemblage. In addition, acetylcholinesterase inhibition may be used to diagnose death resulting from organophosphate pesticides.

12.5.4 Inference from sediment toxicity test endpoints

Toxicity tests that expose organisms to whole sediments that have been contaminated *in situ*, or to pore waters or elutriates derived from whole sediments, are considerably less well developed than those for water. Inference from the existing standardized tests is limited by the small number of sublethal test endpoints, particularly reproductive endpoints, and the lack of systematic validation relative to field effects. Currently, we know that sediments that kill organisms in the laboratory are likely to show depauperate communities in the field, but we do not know that benthic communities that are significantly affected in the field inhabit sediments that would be toxic in the standard tests. Like biological surveys, inference is limited by sampling issues (*i.e.*, the ability of the tested samples to represent the sediments to which the endpoint community is exposed) and by the need to use reference sediments in whole sediment toxicity tests.

These tests use contaminated sediments so they are not used in predictive assessments. However, they can be useful in specific circumstances. For example, such sediments could be used to determine a site-specific exposure-response relationship if a contamination gradient occurs in space. That relationship could then be used to predict changes in time if concentrations are changing as a result of degradation or other processes.

Sediment toxicity tests may be used alone to infer that a sediment poses unacceptable risks. That is, if amphipods die in a test, then it is reasonable to assume that organisms in the field will be killed by the sediment. Inferences could be improved if well-designed validation studies established statistical models relating test endpoint to field measurement endpoint that correspond to common assessment endpoints. Alternatively, demographic models or individual-based models could be used to infer population-level effects from measurement endpoints that are relevant parameters of the models (*e.g.*, survivorship and fecundity for the demographic models; DeWitt 1994). If biological survey data are available, then the sediment toxicity data need not be used to make inferences about the effects on the assessment endpoint. Instead, sediment toxicity data can be used to confirm that the effects detected by a survey are associated with toxicity, so causation can be inferred.

12.5.5 Inference from single chemical toxicity test endpoints

Traditional toxicity tests involve addition of individual chemicals to a clean sediment or water to generate test endpoints such as LC50s, NOECs, or exposure-response functions. The inference problems of contaminated sediment toxicity tests are also applicable here. In addition, inference from these tests involves assumptions concerning combined toxic effects of mixtures (unless one contaminant dominates) and exposure (*i.e.*, bioavailability and speciation) that are avoided by contaminated sediment tests (see Chapter 18).

Most efforts to develop models to extrapolate to assessment endpoints have been devoted to endpoints from single chemical tests. Various types of models have been developed for this purpose, but they are only beginning to be implemented for sediment community endpoints (Suter 1993). Species-sensitivity distributions, regression models, toxicodynamic models, and other types of extrapolation models that have been developed for estimating effects on aquatic populations and communities from conventional test endpoints can be applied to benthic species as well. Therefore, inference from this class of measurement endpoints alone currently relies on conceptual models (*e.g.*, in the field, if the LC50 is exceeded, organisms will die, some benthic species are likely to go extinct, and species richness will be reduced).

Use of single chemical test endpoints alone in retrospective assessments is subject to the same limitations as in predictive assessments. If the contaminated sediments have been tested or the community has been surveyed, then single chemical test endpoints can be used to identify the chemicals that are responsible for the toxicity and to infer whether community degradation is likely as a result of a particular chemical or source.

12.5.6 Inference from contaminants in biota
Benthic organisms accumulate contaminants from sediments and transfer them along trophic webs to fish and wildlife. Hence, benthos are a source of exposure to other endpoint populations and communities, as well as being an endpoint community in themselves (discussed above). Some models are available for estimating contaminant uptake from sediments, but the complexity and diversity of sediments makes the predictions highly uncertain and not applicable to all classes of chemicals. Therefore, concentrations of contaminants of concern should be measured in both the biota and the sediments to which they are exposed along with the various laboratory and field measurements that serve to generate measurement endpoints for the benthic biota. Concentrations of contaminants in the biota can be used to estimate risks to avian and mammalian wildlife using conventional dietary exposure models and oral toxicity endpoints (Suter 1993, USEPA 1994). Risks to fish that feed on benthic organisms are difficult to estimate from concentrations in food because that mode of exposure has received very little consideration (Woodward *et al.* 1994). Thus, concentrations in benthic organisms can only be used as supporting information to establish that a route of exposure exists.

12.6 Recommendations and critical issues

12.6.1 Reference conditions
Sediment ecological risk assessments involve comparison of potential stressors with one or more reference conditions. Establishment of appropriate reference conditions has been problematic for many studies, largely because of a lack of knowledge concerning natural ecological processes. The following recommendations concern the establishment of reference conditions:

- Contributions to the systematics (classification) of marine and freshwater benthic organisms. Systematics is at the base of our understanding of the natural world. Unfortunately, systematics has become an unfashionable activity in today's highly technical world, and our knowledge of biodiversity is languishing (Wilson 1992). This area of basic biology needs more — not less — attention in SERAs. In addition, the taxonomic levels necessary to detect in ecological assessments need to be determined.

- Develop regional reference bases for benthic community structure in fine-grained depositional sediments. The establishment of adequate reference or control sites has long been a difficulty in field-based investigations. Three different approaches have recently been developed to circumvent this problem. The regional reference site approach (*e.g.*, Plafkin *et al.* 1989) uses a biocriteria approach, whereas the RIVPACS model from the UK (Wright *et al.* 1984, 1988) and the BEAST model from Canada (Reynoldson *et al.* 1995) use multivariate approaches. Similar results can be achieved by conceptual clustering (Matthews and Hearne 1991; Landis *et al.* 1994), a nonmetric technique. Whatever method is used, regional sites should be established and sampled before more anthropogenic disturbance occurs and their usefulness as "non-impacted" sites is eliminated.

12.6.2 Exposure

Exposure to sediment-associated contaminants is complicated, and our current level of understanding limits the probability of an accurate risk assessment for sediment-associated contaminants. The following recommendations would improve our overall understanding of exposure to contaminants:

- Guidance should be developed to assist regulators in selecting ecologically relevant endpoints to reflect the appropriate routes and levels of exposure.

- Better understanding of exposure, appropriate for each level of trophic organization needs to be developed, to account for differences in the interactions of benthic species (both animals and plants) with sediments.

- Methods are needed to characterize exposure to individual stressors from the many different natural and anthropogenic stressors to which benthic organisms are exposed. Methods are also needed to improve our ability to predict exposure to contaminant mixtures.

- Better ability is needed to predict the effects of body residues on benthic biota. Improving relationships between accumulation of contaminants and expected effects will improve both interpretation of bioaccumulation data and our confidence to explain the cause of observed ecological disturbances.

- Better methods are required to assess the influence of benthic community structure and function (and variation in both) on contaminant concentrations and bioavailability. Benthic organisms (especially deposit feeders and bioturbators) can substantially alter the biogeochemistry of sediments they inhabit, yet little is know about the activity of theses organisms on the fate of sediment-bound contaminants.

12.6.3 Effects

Ecologically relevant responses to sediment-associated contaminants is complicated, and our current level of understanding limits the probability of an accurate ERA for sediment-associated contaminants. The following recommendations would improve our overall understanding of responses to contaminants:

- A need exists to determine whether SQGs (*e.g.*, SQC) are protective or ecologically relevant within the context of SERAs.

- The current suite of sediment toxicity tests and their ability to predict effects on populations and communities in diverse biogeographic locations should be systematically validated. The results of theses studies should be used to direct the development of new sediment toxicity tests and to assess and develop methods to extrapolate from sediment toxicity tests to endpoint properties in the field.

- Existing QSARs should be systematically created and evaluated for phylogenetically diverse sets of benthic animals and plants.

- Mechanisms underlying contaminant effects at different levels of ecological organization and linkages among them need to be understood. Furthermore, a need exists to understand how contaminants affect ecological change through time (trajectory).

- Better methods need to be developed to measure spatial and temporal variation in structural and functional properties of benthic communities and to measure and predict the trajectory of change (temporal) of ecological variables, especially at community and ecosystem levels. Studies are needed to better understand how spatial and temporal patchiness in ecological responses affect prediction and detection of impacts.

- Better understanding is needed of direct effects of contaminants on inter- and intraspecific interactions, succession, and other community-level processes, and on ecosystem-level processes (nutrient or energy cycling, metabolism, productivity).

- Methods need to be developed to evaluate the effects of mixtures of toxicants in laboratory experiments and in field studies, and the relative importance of an individual stressor from suites of multiple stressors in field studies.

12.7 References

Adams WJ. 1995. Aquatic toxicology testing methods. In: Hoffman DJ, Rattner BA, Burton Jr GA, Cairns Jr J, editors. Handbook of ecotoxicology. Boca Raton FL: Lewis. p 25–46.

Ambrose Jr WG. 1991. Are infaunal predators important in structuring marine soft-bottom communities? *Amer Zool* 31:849–860.

Ankley GT, Katko A, Arthur JW. 1990. Identification of ammonia as an important sediment-associated toxicant in the lower Fox River and Green Bay, Wisconsin. *Environ Toxicol Chem* 9:313–322.

Ankley GT, Leonard EN, Mattson VR. 1994. Prediction of bioaccumulation of metals from contaminated sediments by the oligochaete, *Lumbriculus variegatus*. *Water Res* 28:1071–1076.

[ASTM] American Society for Testing and Materials. 1995 Standard test methods for measuring the toxicity of sediment-associated contaminants with freshwater invertebrates. In: Volume 11.05, Annual book of ASTM standards. Philadelphia: ASTM. p 1204–1285.

Barnthouse LW. 1992. The role of models in ecological risk assessment: a 1990's perspective. *Environ Toxicol Chem* 11:1751–1760.

Bedford BL, Preston EM, editors. 1988. Cumulative impacts to wetlands. *Environ Manage* 12(5).

Burton Jr GA, MacPherson C. 1995. Sediment toxicity testing issues and methods. In: Hoffman DJ, Rattner BA, Burton Jr GA, Cairns Jr J, editors. Boca Raton FL: Lewis. p 70–103.

Cairns Jr J. 1983. Are single species toxicity tests alone adequate for estimating environmental hazard? *Hydrobiologia* 100:47–57.

Cairns Jr J, Neiderlehner BR, Smith EP. 1992. The emergence of functional attributes as endpoints in ecotoxicology. In: Burton GA, editor. Sediment toxicity assessment. Boca Raton FL: Lewis. p 111–128.

Callahan JR. 1984. Long-term ecological research. *BioScience* 34:363–367.

Canfield TJ, Kemble NE, Brumbaugh WG, Dwyer FJ, Ingersoll CG, Fairchild JF. 1994. Use of benthic invertebrate community structure and the sediment quality triad to evaluate metal-contaminated sediment in the Upper Clark Fork River, Montana. *Environ Toxicol Chem* 13:1999–2012.

Clements WH. 1994. Benthic community responses to heavy metals in the Upper Arkansas River Basin, Colorado. *J North Amer Benthol Soc* 13:30–44.

Clements WH, Cherry DS, Cairns Jr J. 1989. The influence of copper exposure on predators-prey interactions in aquatic insect communities. *Freshwater Biol* 21:483–488.

Cooper SD, Barmuta LA. 1993. Field experiments in biomonitoring. In: Rosenberg DM, Resh VH, editors. Freshwater biomonitoring and benthic macroinvertebrates. New York: Chapman and Hall. p 399–441.

Corkum LD. 1989./ Patterns of benthic invertebrate assemblages in rivers of northwestern North America. *Freshwater Biol* 21:195–205.

Corkum LD, Currie DC. 1987. Distributional patterns of immature Simuliidae (Dioptera) in northwestern North America. *Freshwater Biol* 17:210–221.

[CWA] Clean Water Act. 33 U.S.C. §1251 *et seq.* (June 30, 1948). Also titled Federal Water Pollution Control Act.

Davis WS. 1995. Biological assessment and criteria: building on the past. In: Davis WS, Simon TP, editors. Biological assessment and criteria. Tools for water resource planning and decision making. Boca Raton FL: Lewis. p 15–30.

Day KE, Dutka BJ, Kwan KK, Batista N, Reynoldson TB, Metcalfe-Smith JL. 1995. Correlations between solid-phase microbial screening assays, whole-sediment toxicity tests with macroinvertebrates and *in situ* benthic community structure. *J Great Lakes Res* 21:192–206.

Dayton PK. 1994. Community landscape: scale and stability in hard bottom marine communities. In: Giller PS, Hildrew AG, Raffaelli DG, editors. Aquatic ecology - scale, pattern and process. Oxford: Blackwell Scientific. p 289–332.

DeWitt TH, Ditsworth GR, Swartz RC. 1988. Effects of natural sediment features on the phoxocephalid amphipod, *Rheposynius abronius*: implications for sediment toxicity bioassays. *Mar Environ Res* 25:99–124.

DeWitt TH, Ozretich RJ, Swartz RC, Lamberson JO, Schults DW, Ditsworth GR, Jones JKP, Hoselton L, Smith LM. 1992. The influence of organic matter quality on the toxicity and partitioning of sediment-associated fluoranthene. *Environ Toxicol Chem* 11:197–208.

DeWitt TH, Swartz RC, Hansen DJ, Berry WJ, McGovern D. AVS regulation of cadmium bioavailability in a life-cycle sediment toxicity test using *Leptocheirus plumulosus*. *Environ Toxicol Chem*: in press.

Di Toro DM, Mahony JD, Hansen DJ, Scott KJ, Hicks MB, Mayr SM, Redmond MS. 1990. Toxicity of cadmium in sediments: the role of acid volatile sulfide. *Environ Toxicol Chem* 9:1487–1502.

Di Toro DM, Zarba CS, Hansen DJ, Berry WJ, Swartz RC, Cowan CE, Pavlou SP, Allen HE, Thomas NA, Paquin PR. 1991. Technical basis for establishing sediment quality criteria for nonionic organic chemicals using equilibrium partitioning. *Environ Toxicol Chem* 10:1541–1583.

Drake JA, Flum TE, Witteman GJ, Voskuil T, Hoffman AM, Creson C, Kenny DA, Huxel GR, Larue CS, Duncan JR. 1993. The construction and assembly of a ecological landscape. *J Animal Ecol* 62:117–130.

Elliott JM. 1971. Some methods for the statistical analysis of samples of benthic invertebrates. Freshwater Biol Assoc Sci Publ No 25.

Enslein K, Tuzzeo TM, Borgstedt HH, Blake B, Hart JB. 1987. Prediction of rat oral LD50 from Daphnia magna LC50 and chemical structure. In: Kaiser KLE, editor. QSAR in environmental toxicology: II. The Netherlands: D. Reidel, Dordrecht. p 91–106.

Enslein K, Tuzzeo TM, Blake BW, Hart JB, Landis WG. 1989. Prediction of *Daphnia magna* EC50 values from rat oral LD50 and structural parameters. In: Suter GW, Lewis MA, editors. Volume 11, Aquatic toxicology and environmental fate. Philadelphia: American Soc for Testing and Materials (ASTM). STP 1007. p 397–409.

Gosslink JG, Lee LC, Muir T. 1990. Ecological process and cumulative impacts: illustrated by bottomland hardwood ecosystems. Boca Raton FL: Lewis.

Gottman JM. 1981. Time series analysis: a comprehensive introduction for social scientists. Cambridge: Cambridge Univ Pr.

Hall SJ, Rafaell D, Thrush SF. 1994. Patchiness and disturbance in shallow water benthic assemblages. In: Giller PS, Hildrew AG, Raffaelli DG, editors. Aquatic ecology: scale, pattern and process. Oxford: Blackwell Scientific. p 333–375.

Hamelink JL, Landrum PF, Bergman HL, Benson WH. 1994. Bioavailability: physical, chemical and biological interactions. Boca Raton FL: Lewis.

Hare L, Carignan R, Huerta-Diaz MA. 1994. A field study of metal toxicity and accumulation by benthic invertebrates: implications for the acid volatile sulfide (AVS) model. *Limnol Oceanogr* 39:1653–1668.

Harkey GA, Landrum PF, Klaine SJ. 1994. Comparison of whole-sediment, elutriate and pore water exposures for use in assessing sediment-associated contaminants in bioassays. *Environ Toxicol Chem* 13:1315–1329.

Harkey GA, Lydy MJ, Kukkonen J, Landrum PF. 1994. Feeding selectivity and assimilation of PAH and PCB in *Diporeia* spp. *Environ Toxicol Chem* 13:1445–1455.

Heckey RE, Campbell P, Rosenberg DM. 1994. Introduction to experimental lakes and natural processes: 25 years of observing natural ecosystems at the Experimental Lakes Area. *Can J Fish Aquat Sci* 51:2721–2722.

Hill AB. 1965. The environment and disease: association or causation. *Proceed Royal Soc Medicine* 58:295–300.

Huggett RJ, Kimerle RA, Mehrle PM, Bergman HL. 1992. Biomarkers: biochemical, physiological and histopathological markers of anthropogenic stress. Chelsea MI: Lewis.

Hunter MD, Price PW. 1992. Playing chutes and ladders: heterogeneity and the relative roles of bottom-up and top-down forces in natural communities. *Ecology* 73:724–732.

Hurlbert SH. 1975. Secondary effects of pesticides on aquatic systems. *Residue Rev* 58:81–148.

Johnson RK, Wiederholm T. 1989. Classification and ordination of profundal macroinvertebrate communities in nutrient poor, oligo-mesohumic lakes in relation to environmental data. *Freshwater Biol* 21:375–386.

Johnson RK, Wiederholm T, Rosenberg DM. 1993. Freshwater biomonitoring using individual organisms, populations and species assemblages of benthic macroinvertebrates. In: Rosenberg DM, Resh VH, editors. Freshwater biomonitoring and benthic macroinvertebrates. New York: Chapman and Hall. p 40–158.

Kemble NE, Brumbaugh WG, Brunson EL, Dwyer FJ, Ingersoll CG, Monda DP, Woodward DF. 1994. Toxicity of metal contaminated sediments from the Upper Clark Fork River, Montana, to aquatic invertebrates and fish in laboratory exposures. *Environ Toxicol Chem* 13:1985–1998.

Kennedy JH, Johnson ZB, Wise PD, Johnson PC. 1995. Model aquatic ecosystems in ecotoxicological research: considerations of design, implementation and analysis. In: Hoffman DJ, Rattner BA, Burton Jr GA, Cairns Jr J, editors. Boca Raton FL: Lewis. p 117–162.

Koch AL. 1966. The logarithm in biology 1: mechanisms generating the log-normal distribution exactly. *J Theoret Biol* 12:276–290.

Kohn NP, Word JQ, Niyogi KD, Ross LT, Dillon T, Moore DW. 1994. Acute toxicity of ammonia to four species of marine amphipod. *Mar Environ Res* 38:1–15.

Kukkonen J, Landrum PF. 1994. Toxicokinetics and toxicity of sediment-associated pyrene to *Lumbriculus variegatus* (Oligochaeta). *Environ Toxicol Chem* 13:1457–1468.

Landis WG, Chester NA, Haley MV, Johnson DW, Muse Jr WT, Tauber RM. 1989. Utility of the standardized aquatic microcosm as a standard method for ecotoxicological evaluation. In:

Suter II GW, Lewis MS, editors. Volume 11, Aquatic toxicology and environmental fate. Philadelphia: American Soc for Testing and Materials (ASTM). STP 1007. p 353–367.

Landis WG, Ho-Yu M. 1994. An introduction to environmental toxicology: impacts of chemicals on ecological systems. Boca Raton FL: Lewis.

Landis WG, Matthews RA, Markiewicz AJ, Matthews GB. 1993. Multivariate analysis of the impacts of the turbine fuel JP-4 in a microcosm toxicity test with implications for the evaluation of ecosystem dynamics and risk assessment. *Ecotoxicol* 2:271–300.

Landis WG, Matthews RA, Markiewicz AJ, Shough NA, Matthews GB. 1993. Multivariate analyses of the impacts of the turbine fuel Jet-A using a microcosm toxicity test. *J Environ Sci* 2:113–130.

Landis WG, Matthews GB, Matthews RA, Sergeant A. 1994. Application of multivariate techniques to endpoint determination, selection and evaluation in ecological risk assessment. *Environ Toxicol Chem* 13:1917–1927.

Landrum PF, Dupuis WS, Kukkonen J. 1994. Toxicokinetics and toxicity of sediment-associated pyrene and phenanthrene in *Diporeia* spp.: examination of equilibrium partitioning theory and residue based affects for assessing hazard. *Environ Toxicol Chem* 13:1769–1780.

Landrum PF, Harkey GA, Kukkonen J. 1995. Evaluation of organic contaminant exposure to aquatic organisms: the significance of bioconcentration and bioaccumulation. In: Newman MC, Jargoe C, editors. Ann Arbor MI: Lewis. p 85–131.

Lee II H. 1991. A clam's eye view of the bioavailability of sediment-associated pollutants. In: Baker RA, editor. Organic substances and sediments in water. III. Biologicals. Ann Arbor MI: Lewis. p 73–93.

Liber K, Kaushik NK, Solomon KR, Carey JH. 1992. Experimental designs for aquatic mesocosm studies: a comparison of the "ANOVA" and "regression" design for assessing the impact of tetrachlorophenol on zooplankton populations in limnocorrals. *Environ Toxicol Chem* 11: 61–78.

Locke A, Sprules WG. 1994. Effects of lake acidification and recovery on the stability of zooplankton food webs. *Ecology* 75:498–506.

MacArthur RH, Wilson WO. 1967. The theory of island biogeography. Princeton: Princeton Univ Pr.

Matthews GB, Hearne J. 1991. Clustering without a metric. *IEEE Transactions on Pattern Analysis and Machine Intelligence* 13:175–184.

Matthews GB, Matthews RA, Hachmoller B. 1991. Mathematical analysis of temporal and spatial trends in the benthic macroinvertebrate communities of a small stream. *Can J Fish Aquat Sci* 48:2184–2190.

Matthews RA, Landis W, Matthews G. 1995. Community conditioning: an ecological approach to environmental toxicology. *Environ Toxicol Chem*: in press.

Matthews RA, Matthews GB, Ehinger WJ. 1991. Classification and ordination of limnological data: a comparison of analytical tools. *Ecol Model* 53: 167–187.

May RM, Oster GF. 1978. Bifurcations and dynamical complexity in simple ecological models. *Amer Nat* 110:573–599.

McAuliffe JR. 1984. Competition for space, disturbance, and the community structure of a benthic stream community. *Ecology* 65:894–908.

McCarty LS. 1991. Toxicant body residues: implications for aquatic bioassays with some organic chemicals. Volume 14, Aquatic toxicology and risk assessment. STP 1124. p 183–192.

McCarty LS, Mackay D. 1993. Enhancing ecotoxicological modeling and assessment. *Environ Sci Technol* 27:1719–1728.

McCarty LS, Ozburn GW, Smith AD, Dixon DG. 1992. Toxicokinetic modeling of mixtures or organic chemicals. *Environ Toxicol Chem* 11: 1037–1047.

McKinney JD, Waller C. 1994. Polychlorinated biphenyls as hormonally active structural analogues. *Environ Health Persp* 102:290–297.

Melancon MJ. 1995. Bioindicators used in aquatic and terrestrial monitoring. In: Hoffman DJ, Rattner BA, Burton Jr GA, Cairns Jr J, editors. Handbook of ecotoxicology. Boca Raton FL: Lewis. p 220–240.

Menge BA, Olson AM. 1990. Role of scale and environmental factors in regulation of community structure. *Trends Ecol Evol* 5:52–57.

Menge BA, Sutherland JP. 1987. Community regulation: variation in disturbance, competition, and predation in relation to environmental tress and recruitment. *Am Nat* 130:730–757.

Morrisey DJ, Underwood AJ, Howitt L. 1995. Development of sediment-quality criteria - a proposal form experimental field-studies of the effects of copper on benthic organisms. *Mar Pollut Bull*: in press.

Morrisey DJ, Underwood AL, Howitt L. Effects of copper on the faunas of marine soft-sediments: a field-experimental study. *Outlet*: in review.

Norris RH, Georges A. 1993. Analysis and interpretation of benthic macroinvertebrate surveys. In: Rosenberg DM, Resh VH, editors. Freshwater biomonitoring and benthic macroinvertebrates. New York: Chapman and Hall. p 234–286.

Oreskes N, Shrader-Frechette K, Belitz K. 1994. Verification, validation and confirmation of numerical models in the earth sciences. *Science* 263:641–646.

Peckarsky B. 1984. Predator-prey interactions among aquatic insects. In: Resh VH, Rosenberg DM, editors. The ecology of aquatic insects. New York: Praeger Scientific. p 196–254.

Peckarsky BL, Horn SC, Statzner B. 1990. Stonefly predation along a hydraulic gradient: a field test of the harsh-benign hypothesis. *Freshwater Biol* 24:181–191.

Plafkin JL, Barbour MT, Porter KD, Gross SK, Hughes RM. 1989. Rapid bioassessment protocols for use in streams and rivers: benthic macroinvertebrates and fish. Washington DC: U.S. Environmental Protection Agency (USEPA). EPA/444/4-89-001.

Posey MH. 1990. Functional approaches to soft-substrate communities: how useful are they? *Rev Aquat Sci* 2:343–356.

Power ME. 1990. Effects of fish in river food webs. Science 250:811–814.

Power ME, Stout RJ, Cushing CE, Harper PP, Hauer FR, Mathews WJ, Moyle PB, Statzner B, Wais de Bagden IR. 1988. Biotic and abiotic controls in river and stream communities. *J North Amer Benthol Soc* 7:456–479.

Pratt JR, Bowers NJ, Balczon JK. 1993. A microcosm using naturally derived microbial communities: comparative ecotoxicology. In: Landis WG, Hughes JS, Lewis MA, editors. Environmental toxicology and risk assessment. Pittsburgh: American Soc for Testing and Materials (ASTM). STP 1179. p 178–191.

Rapport DJ, Regier HA, Hutchinson TC. 1985. Ecosystem behavior under stress. *Amer Nat* 125:617–640.

Resh VH, Jackson JK. 1993. Rapid assessment approaches to biomonitoring using benthic macroinvertebrates. In: Rosenberg DM, Resh VH, editors. Freshwater biomonitoring and benthic macroinvertebrates. New York: Chapman and Hall. p 195–233.

Resh VH, McElravy EP. 1993. Contemporary quantitative approaches using benthic macroinvertebrates. In: Rosenberg DM, Resh VH, editors. Freshwater biomonitoring and benthic macroinvertebrates. New York: Chapman and Hall. p 159–194.

Reynoldson TB, Bailey RC, Day KE, Norris RH. 1995. Biological guidelines for freshwater sediment based on BEnthic Assessment of SedimenT (the BEAST) using a multivariate approach for predicting biological state. *Aust J Ecol* 20:198–219.

Rhoads DC, Young DK. 1970. The influence of deposit-feeding organisms on sediment stability and community structure. *J Mar Res* 28:150–178.

Robbins JA, Kielty TJ, White DS, Edgington DN. 1989. Relationships between tubificid abundances, sediment composition and accumulation rates in Lake Erie. *Can J Fish Aquat Sci* 46:223–231.

Roper DS, Hickey CW. 1994. Behavioural responses of the marine bivalve *Macomona liliana* exposed to copper-and chlordane-dosed sediments. *Mar Biol* 118:673–680.

Rygg B. 1985. Effect of sediment copper on benthic fauna. *Mar Ecol Prog Ser* 25:83–89.

Sanders HL. 1960. Benthic studies in Buzzards Bay. III. The structure of the soft-bottom community. *Limnol Oceanogr* 5:138–153.

Sih A, Crowley P, McPeek M, Petranka J, Strohmeier K. 1985. Predation, competition, and prey communities: a review of field experiments. *Ann Rev Ecol Syst* 16:269–311.

Suter II GW. 1993. Ecological risk assessment. Boca Raton FL: Lewis.

Swartz RC, Cole FA, Lamberson JO, Ferraro SP, Schults DW, DeBen WA, Lee II H, Ozretich RJ. 1994. Sediment toxicity, contamination and amphipod abundance at a DDT- and dieldrin-contaminated site in San Francisco Bay. *Environ Toxicol Chem* 13:949–962.

Taylor LR. 1961. Aggregation, variance and the mean. *Nature* 189:732–735.

Turner SJ, Thrush SF, Pridmore RD, Hewitt JE, Cummings VJ, Maskery M. 1995. Are soft-sediment communities stable? An example from a windy harbour. *Mar Ecol Prog Ser* 120:219–230.

Underwood AJ. 1993. The mechanics of spatially replicated sampling programmes to detect environmental impacts in a variable world. *Aust J Ecol* 18:99–116.

Underwood AJ, Denley EJ. 1984. Paradigms, explanations, and generalizations in models for the structure of intertidal communities on rocky shores. In: Strong Jr DR, Simberloff D, Abele LG, Thistle AB, editors. Ecological communities: conceptual issues and the evidence. Princeton: Princeton Univ Pr. p 151–180.

[USEPA] U.S. Environmental Protection Agency. 1985. Technical support document for water-quality-based toxics control. Washington DC: Office of Water. EPA 440/4-85-032.

[USEPA] U.S. Environmental Protection Agency. 1992. A framework for ecological risk assessment. Washington DC: Risk Assessment Forum. EPA/630/R-92-001.

[USEPA] U.S. Environmental Protection Agency. 1994. Methods for measuring the toxicity and bioaccumulation of sediment-associated contaminants with freshwater invertebrates. Duluth MN: USEPA. EPA 600/R-94/024.

Vannote RL, Minshall GW, Cummins KW, Sedell JR, Cushing CE. 1980. The river continuum concept. *Can J Fish Aquat Sci* 37:130–137.

Warner SC, Travis J, Dunson WA. 1993. Effect of pH variation on interspecific competition between two species of hylid tadpoles. *Ecology* 74:183–194.

Whitlatch RB, Zajac RN. 1985. Biotic interactions among estuarine infaunal opportunistic species. *Mar Ecol Prog Ser* 21:299–311.

Wilson EO. 1992. The diversity of life. Cambridge MA: Belknap Pr of Harvard Univ Pr.

Wipfli MS, Merritt RW. 1994. Disturbance to a stream food web by a bacterial larvicide specific to black flies: feeding responses of predatory macroinvertebrates. *Freshwater Biol* 32:91–103.

Woodward DF, DeLonay AJ, Little EE, Smith CE. 1994. Effects on rainbow trout fry of a metal-contaminated diet of benthic invertebrates from the Clark Fork River, Montana. *Trans Am Fish Soc* 123:51–62.

Wright JF, Armitage PD, Furse MT, Moss D. 1988. A new approach to the biological surveillance of river quality using macroinvertebrates. *Int Ver Theor Angew Limnol Verh* 23:1548–1552.

Wright JF, Moss D, Armitage PD, Furse MT. 1984. A preliminary classification of running-water sites in Great Britain based on macroinvertebrate species and the prediction of community type using environmental data. *Freshwater Biol* 14:221–256.

SESSION 6
CRITICAL ISSUES IN
METHODOLOGICAL UNCERTAINTY

_____Chapter 13

Issues of uncertainty in
ecological risk assessments

William J. Warren-Hicks, Jonathan B. Butcher

13.1 Introduction

Uncertainty analyses seek to describe and interpret lack of knowledge that may be present in the implementation or interpretation of a risk analysis. Depending on the specific goals of the risk analysis and on the availability of information, the uncertainty analysis can take many forms. For example, the analysis can be quantitative or qualitative, predictive or retrospective, model-based or empirical, experimental or survey-oriented. The number of available approaches for documenting and assessing the magnitude of uncertainty is large. Given the array of goals for risk analyses and the need to explicitly address uncertainty in a decision-making context, developing methods to effectively document, calculate, and communicate uncertainty in risk analyses is an area of major research for the future. With the current emphasis on documenting uncertainty, we hope that the use and consideration of uncertainty in risk-based decisions will continue to evolve and improve. In this paper, we address some of the issues in risk uncertainty analysis and attempt to cover a broad spectrum of issues and approaches appropriate for addressing uncertainty in risk methods. The issues discussed in this paper apply to many areas of risk analysis, including risk methods used to assess contaminants associated with sediments.

13.2 The problem of uncertainty

Because risk assessment is a decision-oriented process, the USEPA and other organizations have promoted uncertainty analysis as a tool for addressing the expected variation in environmental response to risk-based decisions. As an example, USEPA endorses uncertainty analysis in the Framework For Ecological Risk Assessment (USEPA 1992). Uncertainty analyses will become ingrained within the scientific community when decision-makers become aware of the importance of explicitly addressing measures of accuracy, precision, and reliability of scientific information available for risk-based decisions (Chapter 17). The information can take the form of either qualitative or quantitative measurements of chemical, biological, or toxicological data, including information on the accuracy and precision of risk estimates.

Uncertainty analyses can be straightforward or complicated; they can be easily implemented or oriented toward solving complex equations requiring significant computing power. There are no typical or accepted methods for conducting an uncertainty analysis, and proponents exist on both sides of the argument for uncertainty analysis standardization. Such confusion can frustrate the investigator, possibly resulting in badly conducted, or simply ignored, uncertainty analyses. These problems affect both the regulatory and regulated communities. For example, in the United States, the need for and interpretation of uncertainty analyses have been the basis of legal actions (*e.g.*, uncertainty in toxicity testing endpoints used in NPDES permits; Warren-Hicks and Parkhurst 1992).

The goal of an uncertainty analysis is to provide the risk manager with the most complete information available on the expected outcomes of management options. Thus, the analysis provides information on not just the most likely outcome, but the whole range of potential outcomes resulting from a management decision. Uncertainty in risk methods, procedures, and data should not negate their use; rather, assessing the degree of uncertainty should provide information and insight useful for selecting alternative management actions using these tools. A correct treatment of uncertainty provides risk managers with a stronger scientific basis for making risk-based decisions. Risk analysis is concerned with evaluating current and future risk caused by environmental perturbations, and the response of the environment to candidate remediation options. Within the risk characterization process (USEPA 1992), an uncertainty analysis may be used to identify particular components and sources of error for the purpose of risk reduction, ultimately leading to the selection of a remediation alternative in the risk management phase of the risk assessment.

Analysis of uncertainty provides valuable information that can aid risk abatement decisions, for example, estimating the probability of success (or risk of failure) of a remediation strategy. The amount of uncertainty in the expected outcome can be balanced against the need to ensure environmental protection with a relatively high degree of confidence. For example, in some cases the uncertainty in a biological impact may be too large to implement management options. As a consequence, the manager may defer action and gather additional information rather than initiate costly pollution controls that may or may not achieve the desired level of environmental protection. If both the uncertainty and costs of controls are high, adopting relatively low-cost remediation strategies for an interim period may be wise while additional monitoring occurs. On the other hand, delaying implementation of the remediation may indeed result in increased environmental damage, and possible public outcry. Large uncertainty in the available data and information does not mean environmental damage is not occurring. Damage may be occurring, but we cannot isolate the degree of damage in the available information. Choosing the level uncertainty that is acceptable for the decision at hand is a task for the risk manager.

In any case, overlooking uncertainty in risk-based decisions can lead to the illusion that the scientific information upon which the decisions are based is adequate, when in fact the data exhibit larger than acceptable variability, or can result in wrongly delaying remediation action when in fact the variability in the risk estimates are unexpectedly small. In either case, risk managers may exhibit surprise, or disappointment, when the environmental outcomes are substantially different from expectations (Reckhow and Chapra 1983).

The amount of uncertainty in risk estimates that is acceptable within a risk analysis framework can reflect a policy decision for a regulator or specific industry (*e.g.*, the data quality objective process; USEPA 1986) or can reflect an individual's level of comfort. For example, a risk manager (or organization) may be a risk taker (*i.e.*, willing to accept a high level of uncertainty in outcome in the selection of an optimal management decision) or may be risk averse (*i.e.*, requiring a significant margin of safety, so that the uncertainty in achieving the desired outcome is small, despite potentially higher costs). This conceptual development of risk as a decision science is extensively applied in business and engineering departments in the United States. With the emphasis on environmental decision-making in the ecological risk framework, the concept of risk aversion/risk acceptance (see Winkler 1972) is growing in popularity (Morgan and Henrion 1990).

In risk analysis, scientific uncertainty derives from many sources, including inadequate scientific knowledge, natural variability, measurement error, sampling error, and incorrect assumptions. Uncertainty also arises from model mis-specification, including errors in statistics, parameters, initial conditions, and failure to appropriately capture expert judgment. The outcome from an analysis of uncertainty can be a probability distribution on the quantity of interest or a ranking scheme based on expert judgment. In some cases, simple estimates of uncertainty, such as the standard error of the mean, may be relevant; more complex measures, such as prediction intervals, odds ratios, marginal density functions, Bayes factors (see Berger 1985), and robust estimates of centrality and variance (Gilbert 1987) may also be appropriate. Unfortunately, this large array of possible approaches and methods to assessing uncertainty can inhibit the understanding and interpretation of the analysis results, thereby effectively reducing the utility of the uncertainty analysis in risk-based decision-making.

In the final analysis, we must overcome the dilemma of uncertainty if the risk process is to succeed and become an integral part of our scientific assessments of the environment. As always, we can look toward better education and increased scientific awareness of the importance of uncertainty in risk-based decision-making. But, uncertainty analysis will become ingrained in our risk paradigms only when decision-makers find value in the process and realize the usefulness of the information gained in the analysis.

13.2.1 Some viewpoints on uncertainty

Given the variety of statistical techniques, models, and approaches available for uncertainty analysis, it is easy to understand why measures of uncertainty are frequently overlooked or minimized in practice. In fact, some assessment of uncertainty enters any decision process, and it may be treated either explicitly or implicitly. In some cases, a risk manager may be presented with a qualitative analysis of uncertainty which enables an informed evaluation of candidate management options. In other cases, a risk manager presented with a point estimate of risk might decide that the risk analysis has low (or high) reliability and might assign it little (or great) weight in the formation of a management decision (implicitly assigning an evaluation of uncertainty associated with the point estimate).

An assumption frequently underlying statistical measures of uncertainty (*e.g.*, mean square error) is that the lack of precision about the measurement endpoint or model predictions can be reduced with more data. This derives from classical statistical theory where the variance of the sample mean is a function of the number of replicate samples. With large sample sizes (greater than 20 to 30), the expected reduction in uncertainty about the mean with larger sample sizes is usually found in practice. However, with small or moderate sample sizes, additional data can easily increase the realized sample variance in the mean. In small samples, each individual value is relatively influential, therefore an extreme data point (a sample far from the calculated mean) can easily result in a large calculated variance and a comparably large shift in the mean value. This problem emphasizes the need for calculating and interpreting standardized measures of variance and centrality (like the mean and variance of a normal distribution). In response to such issues, investigators should realize that a variety of statistical measures exists, each with their own advantages, for describing the characteristics of data (*e.g.*, robust measures of centrality such as the trimmed mean [Gilbert 1987], shrinkage estimators of variance [Berger 1985], and Bayesian methods for combining data [Warren-Hicks and Wolpert 1994]).

In general, uncertainty is the absence of information that may or may not be obtainable. In this regard, the concept of uncertainty is categorized into those components that are reducible and those components that are not reducible. The reducible component of uncertainty is associated with the ability to predict or measure a single value, say benthic species richness at a single site (here, true replicate measurements would provide a measure of sampling error). This uncertainty can be reduced with a better model, improved method, or more information. In practice, we never reduce all sources of uncertainty in an assessment, and some sources of variation may not be reducible. For example, when testing and validating a selected exposure model, natural variability in flow and other hydrological components will result in the presence of model prediction error, regardless of the quality of the model or site-specific measurements used in model testing. The investigator must establish the degree of model prediction error that is appropriate for the decisions to be made with the model predictions. This viewpoint of uncertainty is consistent with USEPA's data quality objectives process (USEPA 1986; Warren-Hicks 1989).

13.2.2 Is uncertainty analysis always needed?

Situations exist where a formal uncertainty analysis may not be useful prior to obtaining additional data. In general, if little knowledge exists about the environmental process under investigation, or if little empirical data are available, then the information generated from a prospective uncertainty analysis is limited in scope and interpretation. A good strategy is to increase the amount of available information (*e.g.*, gather data or consult experts) prior to conducting a formal analysis. In the extreme, if you are completely uncertain, it may not be possible to quantify your lack of understanding. Of course, the extreme case where no information of any kind is available is not likely. As part of a comprehensive strategy, however, a risk analyst should judge whether enough information is available to warrant a formal uncertainty analysis. For example, if a mechanistic computer model is built completely from a conceptual understanding of an environmental process (*e.g.*, uptake of PCBs by flounder in southern estuaries) for which no data exist, employing Monte Carlo analysis to estimate the model prediction uncertainty is not informative as to the variance of actual environmental concentrations. The model cannot be used to estimate actual environmental concentrations without calibration to on-site conditions, which require actual measurements. If remediation decisions require estimates of the mean and range of PCB concentrations in the water, then the uncertainty analysis will provide only misleading information. When little empirical data are available, qualitative or expert judgment assessment of uncertainty can focus the data collection effort. The risk assessor could be better served spending resources to gather empirical data that can be used to calibrate and validate the model. After model calibration and validation with empirical evidence, estimating the relative ability of the model to accurately predict the risk to the environment is appropriate.

13.3 Methods

Approaches to uncertainty are briefly discussed below, classified into three areas:

- Margin of safety (safety factors)
- Statistical paradigms
- Qualitative approaches to uncertainty

This discussion attempts to provide insight into the various approaches to dealing with uncertainty that are generally used in risk assessments.

13.3.1 The margin of safety (or safety factor) approach

The margin of safety approach in environmental assessment is frequently used in regulatory settings (MacDonald *et al.* 1995). The terminology results from analogy to engineering design practice, where, for instance, a bridge is designed to accommodate significantly more than the greatest expected load, thus incorporating a margin of safety into the specification. For example, USEPA uses safety factors when establishing water quality criteria for the protection of aquatic life (Stephan and Erickson 1988); the computed concentration of a toxic chemical estimated to protect 95% of the aquatic community is divided in half to adjust for the use of LC50 data in the analysis. The CWA requires a margin of safety in the estimation of total maximum daily load (TMDL) for the attainment of designated uses, which is "normally incorporated into the conservative assumptions within the calculations or models" (USEPA 1991).

In practice, the margin of safety approach is generally environmentally protective and is useful when little or no data are available for more quantitative estimates of variance, but provides insufficient basis for assessing the tradeoffs between costs and risk. In addition, when there are many sources of uncertainty (*e.g.*, sources of uncertainty in toxicity testing or exposure concentrations), safety factors provide a relatively easy way to implicitly aggregate the multiple sources of uncertainty. In this case, a complicated statistical variance components analysis may be extremely costly and impractical. In a first-tier risk analysis attempting to screen toxic chemicals, safety factors may be appropriate under the assumption that the screening criteria should be conservative (Parkhurst *et al.* 1995). The biggest drawback of blindly applying a margin of safety approach is that it may result in significant over-design and waste of resources.

If the goal of the environmental assessment is to accurately reflect the probability of environmental impact, then safety factors are usually inappropriate. Here, the investigator is focused on accuracy and precision, and not on conservative estimates of risk. In a risk assessment, we are generally interested in describing the potential impact with as much accuracy and precision as possible (expect possibly during a screening-level assessment in which conservative assumptions are frequently employed). Therefore, the data requirements are higher, and the requirements for high prediction accuracy are greater. Conservative safety factors should not be used with these goals in mind. However, conservative safety factors are generally appropriate for assessments attempting to screen nontoxic chemicals from further analysis. The drawbacks to using safety factors are the loss of information and a tendency to reduce the amount of effort in obtaining additional data. Incumbent upon the risk assessor is to identify the goals of the assessment and the amount of uncertainty he/she is willing to accept before using the safety factor approach to uncertainty.

13.3.2 Statistical paradigms
Below, we discuss two statistical paradigms for uncertainty analysis. The classical statistical approach, or frequentist approach, is most familiar to environmental scientists. The second paradigm is based on Bayes theory and has a long history in engineering and business applications. Bayesian methods are associated with decision theory and provide a rigorous framework for combining multiple types of data in a risk analysis (*e.g.*, laboratory toxicity data, benthos data, and measures of value such as cost).

13.3.2.1 Frequentist paradigm
Most environmental scientists are familiar with the frequentist approach to uncertainty estimation. Frequentist theory is based on the long-run expectations of statistical estimators. For example, the standard estimator for the mean of a distribution

$$\frac{\sum_{i=1}^{N} X_i}{N}$$

can be shown with asymptotic theory to converge to the population mean (which is assumed to be fixed, but in reality is unknown). This theory is used to assign the above estimator with the statistical properties of being both unbiased (convergence in expectation to the population mean) and efficient (smallest variance). These desirable properties apply only to the statistical estimators, not to the actual data that is collected. Unfortunately, any single biological measurement or sample is neither unbiased nor efficient. In fact, the mean from a small random sample could be very different from the population mean. Standard measures of uncertainty for the frequentist are usually associated with the normal distribution, for example, the normal variance, normal standard deviation, and standard error of the normal mean. Popular methods for estimating variability include variance estimates based on the normal distribution, Monte Carlo techniques, and first-order error analysis.

Most statistical computer software includes frequentist estimates of variance. Unfortunately, the proliferation of statistical software serves to promote the use of statistics by untrained individuals not familiar with the sometimes subtle, but important, statistical theories upon which the results depend. For example, Monte Carlo methods are frequently misused and their underlying assumptions violated: Figure 13-1 summarizes two Monte Carlo analyses of the same risk assessment endpoint and shows that a change in the specification of distributions can result in a large difference in the estimation of risk. Hypothesis testing procedures also are a popular frequentist method for determining differences between population means. Again, opinions concerning the appropriate use of hypothesis testing methods vary widely (Casella and Berger 1987; Suter 1996).

Uncertainty in model predictions can also be directly evaluated based on uncertainty or variability in model inputs. A popular, and relatively user-friendly, method for propagating uncertainty through equations or models is first-order error analysis.

First-order analysis can be written as

$$S_y^2 \approx \sum_{i=1}^{n} \left(\frac{\partial f}{\partial x_i} \right)^2 S_{x_i}^2 + 2 \sum_{j=1}^{n-1} \sum_{i=j-1}^{n} \left(\frac{\partial f}{\partial x_i} \right) \left(\frac{\partial f}{\partial x_j} \right) S_{x_i} S_{x_j} \, \rho_{x_i x_j}$$

where

x_i = a model parameter estimated from the data (*e.g.*, a regression coefficient),

ρ_{x_i, x_j} = the correlation between the *i*th and *j*th model parameter,

S_{x_i}, S_{x_j} = the standard deviation of the individual model parameters, and

f = a function (*e.g.*, the regression model, or mechanistic model used for prediction).

First-order analysis (like Monte Carlo analysis) requires an explicit estimation of the correlation between the uncertain parameters. In most cases, first-order analysis can be implemented with a calculator and provides a relatively straightforward means of propagating error from multiple sources (*e.g.*, parameters in a multivariate regression model) into the error in model prediction. Applications of first-order analysis can be found in Reckhow and Chapra (1983).

Although there are many arguments for and against the classical frequentist approach to uncertainty estimation, the most compelling reason for using the frequentist measures is that most of the scientific community is familiar with the methods. The methods have been adopted in most environmental programs, are generally understood, and usually are accepted as an appropriate means of data analysis.

Estimated   **Actual**

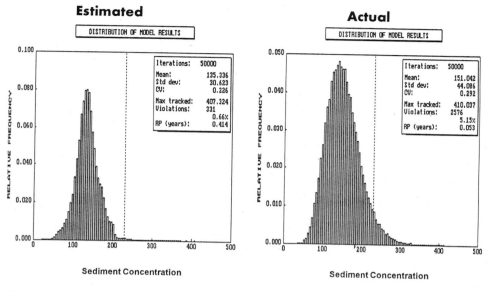

Figure 13-1 Use and abuse of Monte Carlo analysis

13.3.2.2 Bayesian paradigm

Another approach to the evaluation of uncertainty is the Bayesian approach (Press 1989). Bayesian methods are based on Bayes' theorem, which can be written as follows:

$$P(\theta|y) = \frac{L(y|\theta)\,\pi(\theta)}{\int_{\theta} L(y|\theta)\,\pi(\theta)\,d(\theta)}$$

where

$P(\theta|y)$ = the distribution of the parameters conditional on the observed data, referred to as the *posterior distribution*, which contains all current information;

$L(y|\theta)$ = the likelihood function based on the sample data, containing all information available in the current sample; and

$\pi(\theta)$ = the prior distribution of the parameters before observing the data; the distribution can be based on subjective judgment or actual data.

Here, parameters (*e.g.*, measurement endpoints, or parameters in a predictive model) are treated as random variables (unlike classical statistics in which the population mean is fixed but unknown). In the Bayesian concept, the posterior probability distribution expresses a "degree of belief" about the variable of interest, based on prior information obtained before data collection (the prior), and the sample data (the likelihood) collected during the experiment or monitoring study. The prior and likelihood function are combined using Bayes theorem to form a revised estimate of the parameters (the posterior distribution), which contains all of the available information about the parameters (unlike classical statistics, which infers information about the characteristics of the parameter based on the expected value of future data, which is the long-run frequency concept). Bayesian methods can be used to incorporate expert judgment and information from historical sources into the estimation process via the prior distribution. Bayesian methods form the basis of the decision sciences, which have historically been used and developed in engineering and business applications. In addition, Bayesian methods can be used to assess the risk associated with alternative actions such as site remediation strategies (Reckhow 1994), to explicitly incorporate costs of various alternative strategies into the decision process, and to evaluate the value of collecting further information (Dakins *et al.* 1994).

Figure 13-2 presents a conceptual output from a Bayesian decision approach. While this figure may be difficult to develop in practice (mostly because of the lack of appropriate information), the information depicted on the graph is implicitly incorporated into most risk management decisions. Decision analysis combines the probability distribution of the risk endpoint of interest with a function representing the net benefits (or net loss, or utility) that occur across the range of the risk endpoint. The benefits are evaluated for several management options. In our example, decision analysis seeks to maximize the net benefits of management options, given the uncertainty in the risk endpoint. The probability distribution needed from the analysis is effectively the denominator of Bayes theorem (called the *predictive distribution*). The benefits function need for the analysis is typically difficult to define. For example, valuing the benefit of "no species loss" can be difficult. In fact, the benefit for a fisherman is probably different than that for an industry using the water for cooling. Generally, risk managers do not explicitly generate benefit functions when making decisions. However, implicitly, managers weigh the amount of information, the degree of uncertainty, and the relative benefits of management options when making decisions. Decision theory provides a formalized way to incorporate all of this information in a rigorous fashion.

Given the data, decision analysis provides a tool for rigorously combining information into a probabilistic assessment that incorporates uncertainty and is useful for aiding management decisions. Figure 13-2 shows the probability of risk (X-axis; *e.g.*, the probability of species loss in an area of contaminated sediment) and a measure of net benefits or value (Y-axis; the sum of positive benefits, *e.g.*, revenue due to recreational use including good public relations, and negative benefits, such as costs of treatment or fines imposed by regulatory agencies). In addition, from the figure, a risk manager can find the remediation strategy that minimizes the expected loss, given the uncertainty in the risk estimate. Note that the uncertainty in the site-specific species loss leads to several treatment strategies depending upon the risk the manager is willing to take. In the figure, no treatment appears to be optimal at expected species loss; however, for a risk-averse decision (right-hand tail of the distribution of species loss), the treatment option gives greater net benefits. Practical applications of this approach are available in Reckhow (1985).

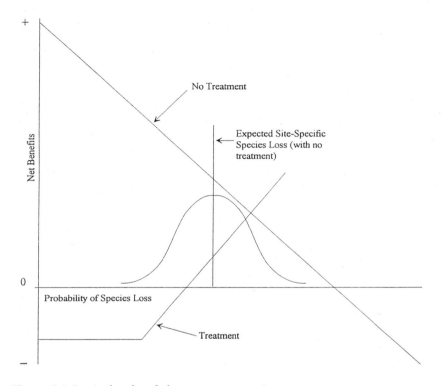

Figure 13-2 Expected net benefit from management actions

13.3.3 Qualitative estimates of uncertainty

An important source of information in most risk assessments is expert judgment. In those cases where little monitoring data are available, expert judgment plays an ever-increasing role in uncertainty estimation. Expert judgment can be incorporated into the risk process in many places (Meyer and Booker 1991), including exposure assessment, effects assessment, and risk characterization. We encourage risk assessors to incorporate expert judgment in a rigorous fashion, either through Bayes theory (Winkler 1967, 1977, 1978, 1980; Kadane *et al.* 1980) or through rigorously designed experiments. While expert judgment is implicitly a part of all risk-based decisions, a formal approach provides a level of credibility that is frequently lacking in many applied applications of risk assessment. For example, Winkler (1977, 1978, 1980) developed interview methods for eliciting expert opinion. Because of the form of the questions, he is able to develop quantitative estimates of variance surrounding the endpoint of interest.

13.4 Conclusions

In most ecological applications, uncertainty analysis can provide useful information to risk decision managers with respect to the quality of information available for making decisions. Many of the issues in uncertainty are briefly discussed, or referenced, above. We hope that uncertainty analysis becomes an integral part of the risk framework and that the concept of uncertainty and variability become institutionalized in the risk assessment process as practiced by both industry and the regulated community.

13.5 References

Berger JO. 1985. Statistical decision theory and Bayesian analysis. 2nd ed. New York: Springer-Verlag.

Casella G, Berger RL. 1987. Reconciling Bayesian and Frequentist evidence in the one-sided testing problem. *J Am Statistic Assoc* 82(397).

[CWA] Clean Water Act. 33 U.S.C. §1251 *et seq.* (June 30, 1948). Also titled Federal Water Pollution Control Act.

Dakins ME, Toll JE, Small MJ. 1994. Risk-based environmental remediation: decision framework and role of uncertainty. *Environ Toxicol Chem* 13:1907–1915.

Gilbert RO. 1987. Statistical methods for environmental pollution monitoring. New York: Van Nostrand Reinhold.

Kadane JB, Dickey JM, Winkler RL, Smith WS, Peters SC. 1980. Interactive elicitation of opinion for a normal model. *J Am Statistic Assoc* 75:845-854.

MacDonald DD, Carr RS, Calder FD, Long ER, Ingersoll CG. 1996. Development and evaluation of sediment quality guidelines for Florida coastal waters. *Ecotoxicology* 5:253–278.

Meyer MA, Booker JM. 1991. Eliciting and analyzing expert judgment: a practical guide. San Diego: Academic Pr.

Morgan MG, Henrion M. 1990. Uncertainty: a guide to dealing with uncertainty in quantitative risk and policy analysis. New York: Cambridge Univ Pr.

Parkhurst BR, Warren-Hick WS, Cardwell RD, Volosin J, Etchison T, Butcher JB, Covington SM. 1995. Methodology for aquatic ecological risk assessment. Washington DC: Cadmus Group Inc. Water Environment Research Foundation Contract No. RP91-AER-1.

Press SJ. 1989. Bayesian statistics: principles, models, and applications. New York: J Wiley.

Reckhow KH, Chapra SC. 1983. Engineering approaches for lake management. Volume 1, Data analysis and empirical modeling. Woburn MA: Butterworth.

Reckhow KH. 1985. Decision theory applied to lake management. Lake and Reservoir Management: Practical Applications; Proceedings of the Fourth Annual Conference and International Symposium; 1984 Oct 16–19; McAfee NJ.

Reckhow KH. 1994. A decision analytic framework for environmental analysis and simulation modeling. *Environ Toxicol Chem* 13:1901–1906.

Stephan CE, Erickson RJ. 1988. Calculation of the final acute value for water quality criteria for aquatic organisms. Duluth MN: U.S. Environmental Protection Agency (USEPA) Environmental Research Laboratory. EPA/600/3-88/018.

Suter II GW. 1996. Abuse of hypothesis testing statistics in ecological risk assessment. *Hum Ecol Risk Assess*: In press.

[USEPA] U.S. Environmental Protection Agency. 1986. Development of data quality objectives: description of stages I and II. Washington DC: USEPA.

[USEPA] U.S. Environmental Protection Agency. 1991. Guidance for water quality-based decisions: the TMDL process. Washington DC: Office of Water. EPA 440/4-91/001.

[USEPA] U.S. Environmental Protection Agency. 1992. Framework for ecological risk assessment. Washington DC: Office of Research and Development. EPA/630/R-92/001.

Warren-Hicks WJ, Wolpert RL. 1994. Predictive models of fish response to acidification: using Bayesian inference to combine laboratory and field measurements. In: Cothern CR, Ross NP, editors. Environmental statistics, assessment, and forecasting. Boca Raton FL: CRC Pr.

Warren-Hicks WJ. 1989. Quality assurance and data quality objectives. In: Ecological assessment of hazardous waste sites: a field and laboratory reference. Corvallis OR: U.S. Environmental Protection Agency (USEPA), Environmental Research Laboratory. EPA/600/3-89/013.

Warren-Hicks WJ, Parkhurst BR. 1992. Performance characteristics of effluent toxicity tests: Variability and its implications for regulatory policy. *Environ Toxicol Chem* 11:793–804.

Winkler RL. 1967. The assessment of prior distributions in Bayesian analysis. *J Am Statistic Assoc* 62:776-800.

Winkler RL. 1972. An introduction to Bayesian inference and decision. New York: Holt Rinehart Winston. 563 p.

Winkler RL. 1977. New developments in the applications of Bayesian methods. Amsterdam: North-Holland.

Winkler RL. 1978. Adaptive forecasting models based on predictive distributions. *Management Sci* 24:997-986.

Winkler RL. 1980. Prior information, predictive distributions and Bayesian model-building. In: Zellner A, editor. Bayesian analysis in econometrics and statistics. Amsterdam: North-Holland. 474 p.

SESSION 6
CRITICAL ISSUES IN
METHODOLOGICAL UNCERTAINTY

_____*Chapter 14*
Uncertainties in assessing
contaminant exposure from sediments

Samuel N. Luoma, Nicholas Fisher

14.1 Introduction

Assessing biological exposures to contaminants is an essential step in understanding the toxicity risks from contaminated sediments. Total contaminant concentrations in sediments can be directly determined and are the simplest measure of exposure. However, the bioavailability of the contaminant controls the dose that organisms actually receive. Predictions of risk from contaminated sediments will be most relevant when exposure and dose are both understood. This requires accurate methods for predicting bioavailability in a site-specific manner and for linking bioavailability to toxicity, as the latter is manifested in nature.

In this chapter, we evaluate uncertainties in the methods and models available for describing contaminant bioavailability and their applicability to site-specific risk assessment. We use correspondence, response, and mechanisms in the evaluation: Do bioavailability predictions generally correspond with observations in nature (*i.e.*, in broad datasets)? Do organisms respond in specific instances as predicted (*e.g.*, short-term responsiveness to bioavailability changes in a field experiment or after an event)? Can the methods and models be justified on the basis of mechanistic understanding? We conclude that geochemically robust methods have been developed to date, but important uncertainties in the models and methods stem from incomplete consideration of biological processes. Including biological processes in the hydrophobicity model has advanced risk predictions for organic chemicals. For metals, kinetic models that consider both external geochemical controls and internal biological processes will be necessary. Better understanding of tissue-residue–based toxicity may be a necessary supplement for linking bioavailability (dose) and toxicity.

14.2 Predicting exposure from total concentrations

Concentration provides a first order control on the ecological risk from contaminants in sediments. Total concentration is recommended in some methods as the basis for estimating exposure (Long *et al.* 1995), but mechanistic and correlative studies have long shown substantial uncertainties in this approach (Neely *et al.* 1974; Veith *et al.* 1979; Luoma 1983, 1989). For example, no relationship is found between sediment chemical concentrations and bioavailability or biological effects when experiments are conducted with sediments that differ widely in critical geochemical characteristics (Luoma and Jenne 1977; Di Toro *et al.* 1990, 1991).

Quantitative uncertainties in the relationship between concentration and bioavailability can be variable in nature. Significant, even strong, correlations between bioavailability and total concentration are found within simple gradients or among geochemically similar environments (Bryan 1985). It is when comparisons are made across environments (or across complex geochemical gradients) that less correspondence is found. The large datasets from estuaries of the United Kingdom assembled by Bryan and co-workers illustrate the effects of complex geochemistry (Luoma and Bryan 1981; Bryan 1985; Bryan *et al.* 1985; Bryan and Langston 1992). For these studies, concentrations of metals were determined in fine-grained surface sediments (judged to be oxidized by appearance) as well as in bivalve and polychaete tissues (three of the most commonly employed species are shown in Table 14-1). The estuaries included a wide range of physical, biogeochemical, and pollution conditions. Data ranges for all elements were several orders of magnitude, and confounding cross-correlations among independent variables were minimized (Luoma and Bryan 1981). Obvious sources of bias such as particle size, large reduction/oxidation (redox) differences, or dilution of tissue concentrations by reproductive tissue were carefully controlled.

Predictions of bioaccumulation from sediment concentrations in the English estuaries varied among elements and among species (Table 14-1). No significant correlation was observed between tissue residues and sediments in a few instances (*e.g.*, Cd in the polychaete *Nereis diversicolor*; Cu in the bivalve *Scrobicularia plana*). Bioavailability in these cases was completely unpredictable from total metal concentrations in sediments. In other cases, the uncertainty in predicting bioaccumulation from only sediment concentrations was not as great as some mechanistic studies would suggest. Silver and lead concentrations in sediments explained about half the variance in bioaccumulation in all three species. Results for Cd and Cu were variable among species, suggesting a significant biological contribution to the uncertainties. Tissue residues commonly differed by tenfold at median sediment concentrations, even where half of the variance was explained by the correlation (Bryan 1985; Bryan and Langston 1992). Thus, for all metals, predictions of bioavailability from concentration alone were usually well outside the two-fold criteria for accuracy suggested by Landrum *et al.* (1992).

14.3 Incorporating bioavailability into exposure assessment

It is difficult to quantify the uncertainty in many estimates of bioavailability because critical processes are not fully understood. Knowledge is growing along a similar path for metals and organic chemicals, although the latter is more fully developed. Below we describe, briefly, the state of the existing methods applied to exposure assessments. Equilibrium partitioning (Di Toro *et al.* 1991) is the present method of choice. We will consider some of the most important areas of uncertainty in EqP and use it as a basis for comparison to other methods. Organic chemical bioavailability and the associated models have been widely reviewed, most recently by Landrum *et al.* (1992), Hamelink *et al.* (1994),

Table 14-1   Percent variance in bioaccumulation of Ag, Cd, Cu, and Pb explained by variation in sedimentary concentrations of these metals (1N HCl extractions)

Organism	Percent variance in bioaccumulation			
	Ag	Cd	Cu	Pb
Nereis diversicolor (polychaete)	41–62	0–16	72	54
Scrobicularia plana (bivalve)	49	29	0	50
Macoma balthica (bivalve)	69	21	10	47

Data from a broad range of estuaries in the United Kingdom.
Regressions published in Luoma and Bryan 1981, Bryan 1985, Bryan and Langston 1992.

and Meador *et al.* (1995a); we will try to minimize redundancies with those insightful discussions.

14.3.1 Organic chemicals

Mackay (1982) suggested that, although biologically based kinetic models were the traditional approach to describing organic contaminant bioconcentration, an alternative was to "view the biotic phase (*e.g.*, fish) as an inanimate volume of material that is approaching thermodynamic equilibrium with its medium (*e.g.*, water) as defined by the chemical potential or fugacity of the bioconcentrating-persistent solute." This remains the central principle guiding risk assessments for organic chemicals in aquatic habitats. Experimentally derived bioconcentration factors (where bioconcentration is defined as uptake from dissolved phase) for different organic chemicals were related to the hydrophobicity of the chemical (Mackay 1982). Bioaccumulation (uptake from all sources) from sediments was not necessarily predictable from chemical hydrophobicity alone (Pereira *et al.* 1988; Swackhamer and Hites 1988), but predictions were improved by normalization to the lipid content of the organisms and the TOC content of the sediments (*e.g.*, Lake *et al.* 1990), consistent with the fugacity theory. Equilibrium partitioning predicted porewater concentrations, based upon hydrophobicity and concentrations of the chemical normalized to OC in sediment (Di Toro *et al.* 1991). Toxicity was predicted by comparison of porewater concentrations with dissolved concentrations that cause toxicity in bioassays.

Bioassays that manipulated OC in sediments provided the mechanistic basis for predictions of bioavailability to benthos by EqP. For example, Swartz *et al.* (1990) found that EqP predictions explained porewater concentrations in three sandy marine sediments spiked with fluoranthene. Toxicity to two amphipods correlated with porewater concentrations, although the species differed in sensitivity. Swartz *et al.* (1990) cited the range of toxicities of fluoranthene to be 8 to 62 µg/l; comparable to 12 to 41 µg/l observed in their study — a maximum uncertainty of 3X to 4X.

Dissolved organic carbon (DOC) confounds EqP predictions of porewater concentrations and toxicity of organic chemicals. Equilibrium partitioning underestimated porewater concentrations of fluoranthene and overestimated toxicity in sediment bioassays where DOC accumulated in pore waters (DeWitt *et al.* 1992; Suedel *et al.* 1993). The uncertainty in these treatments was 10X or more. Lower porewater concentrations of PAHs than predicted by EqP are found in field sediments in contrast to spiked sediments; the source of the PAHs may be one cause of these uncertainties (Meador *et al.* 1995b).

Correlative studies of organic chemical bioavailability to benthos across broad geochemical gradients have received little attention. In one of the few field studies of EqP predictions, Lake *et al.* (1990) compared PCB concentrations in three species of benthos with EqP predicted accumulation factors (AF was lipid-normalized tissue PCB/TOC-normalized sediment PCB). Animals were collected from New Bedford Harbor, Long Island Sound, and Narragansett Bay. The maximum variability in the aggregated concentration data was ~3X, and AFs varied by 3X to 5X between the molluscan and polychaete species, apparently due to species-specific influences on bioaccumulation.

14.3.2 Metals
Some consensus exists about principles that guide prediction of trace element bioavailability, but knowledge uncertainties are probably greater than for organic chemicals. It is clear that element speciation or form, not total metal concentrations, influence bioavailability of both dissolved and sediment-bound trace elements. However, geochemical controls are more diverse than with organic contaminants. The primary influence on bioconcentration (from solution) appears to be the activity of the free metal ion, for metals that behave as cations (Cd, Cu, Pb, Zn) (Sunda and Guillard 1976; Anderson and Morel 1978; Sunda *et al.* 1978; Campbell and Tessier 1989). Ag does not follow this principle in marine waters because its chloro-complex is bioavailable (Engel and Fowler 1989). Methylation of some elements (*e.g.*, Hg and Sn) greatly changes bioavailability. Oxidation state controls the bioavailability of elements that behave as anions (Se, As, Cr, V).

Tissue accumulation does not appear to be driven by any single principle analogous to hydrophobicity. Chemical potential (*fugacity*) in food and within biotic tissues is controlled by diverse biochemical reactions and a variety of metal forms (*e.g.*, complexes like metallothioneins occur in cell solution; membrane-bound forms occur within and outside cells, as do granules and other insoluble bodies). Fisher (1986) showed that differences in bioconcentration by phytoplankton among cationic-behaving metals was explained by differences in hydroxyl ligand association constants. This is not a good predictor of bioaccumulation in animals, however, probably because of the diversity of the above biochemical reactions within organisms and in the food sources from which they assimilate metals.

Geochemical processes governing metal bioavailability from sediments are more complex than for organic chemicals; redox state is important, as are the diversity of surfaces that can sorb metals (Jenne 1977). In the laboratory, bioavailability from oxidized sediments changes with the sediment component to which the metal is bound (iron oxides, manganese oxides, organic materials, carbonates; Luoma and Jenne 1977). Oxidized estuarine sediments are sufficiently variable in nature that such differences in metal partitioning among such components might be expected (Luoma and Davis 1983). Formation of sulfides also affects metal bioavailability. Bioavailability is reduced when moles of AVS exceed moles of weak acid-extractable metals in the sediments to which an organism is exposed (Di Toro *et al.* 1990).

Normalizations or correction factors were derived with metals, as with organic contaminants, to account for effects of sediment geochemistry on bioavailability. Simple, single-factor normalizations of sedimentary Pb (by extractable Fe), Hg (by TOC), As (by extractable Fe), and Cu (by extractable Fe) allowed surprisingly accurate predictions in large datasets from geochemically diverse field conditions (Luoma and Bryan 1978; Langston 1980, 1982; Tessier *et al.* 1984). In many circumstances, however, factors affecting bioaccumulation were more complex than accounted for by simple normalizations (*e.g.*, Luoma and Bryan 1982; Amyot *et al.* 1995).

Di Toro *et al.* (1990, 1991) suggested that normalization by AVS might explain varying metal bioavailability from sediments. Short-term sediment bioassays showed that the molar ratio of AVS/SEM provided a first-order control on activities of at least some metals (Cd, Ni) in pore water (*e.g.*, Di Toro *et al.* 1990; Ankley *et al.* 1991). Activities were reduced to very low levels at ratios <1, because of the high stability of the metal sulfide. Toxicity and bioaccumulation were correlated with porewater metal activities in the bioassay. Thus, very different doses resulted if two sediments had the same total metal concentrations but different AVS concentrations; the differences were corrected by AVS normalization.

A recent field experiment in a lake in Quebec illustrated that porewater concentrations of Cd were responsive to AVS changes in nature (Hare *et al.* 1994). Pretreated sediments were transplanted in trays to the lake, and bioaccumulation was studied in organisms that colonized the trays of sediment. Porewater Cd concentrations increased as excess Cd increased relative to AVS, as would be predicted from equilibrium partitioning. The effect on Cd bioavailability was different for species with different life cycles, however. Organisms that diurnally migrated from sediments to the water column of lakes were the least influenced by enriched porewater metal concentrations; chironomids that live buried in sediments were the most influenced.

Uncertainties in the use of AVS as the universal sediment normalizer are difficult to quantify at the present state of knowledge. Studies to date have not defined how to determine AVS concentrations that are biologically relevant in the field. Nor have they confronted the challenge of ensuring that the AVS concentrations used in experiments are representative of the system (or risk assessment site) of interest. Reduction/oxidation reactions, and thus sulfide concentrations, are patchy on biologically relevant micro-scales within reduced sediments. AVS varies widely with depth in a different manner in every sediment and with time in the same sediment (see Luoma and Ho 1993 and Luoma 1995 for reviews of the geochemical literature relevant to bioassays). Sediments collected for bioassays are typically taken from a constant depth, such as the surface 3 to 6 cm, and natural redox gradients are homogenized to reduce experimental variability. Typically, limited numbers of samples are collected, sometimes from a single type of depositional area in otherwise complex systems (*e.g.*, Brumbaugh *et al.* 1994). Accurate representation of nature will require systematic characterization of AVS variability in a variety of circumstance (Table 14-2). Ultimately, standardized methods should include sampling from a depth defined by site-specific redox conditions and by the redox conditions favored in the sedimentary microhabitat of the species of interest.

Mechanistic knowledge of sediment geochemistry suggests factors in addition to AVS should influence porewater metal concentrations and metal bioavailability from sediments. Most macrofauna have an obligate requirement for oxygen. Macrofaunal and meiofaunal life concentrates in the oxidized zones of sediments, where thermodynamics do not favor occurrence of sulfides. On the other hand, the relatively slow oxidation kinetics of sulfides suggest some sulfides could be present at times in oxidized sediments. Multi-ligand equilibrium models have successfully predicted porewater concentrations and bioavailability of metals in Canadian lakes (Campbell and Tessier 1989; Belzile *et al.* 1990; Tessier *et al.* 1993; Amyot *et al.* 1994). These models incorporate binding constants for iron oxides and organic materials but do not consider AVS. Tessier *et al.* (1993) showed a strong correlation between Cd activity in pore waters (as predicted from the model) and Cd concentrations in a bivalve. Amyot *et al.* (1994) used multivariate analysis to predict metal bioavailability to an amphipod from two lakes using the oxic equilibrium model to define Cd activities in solution. They could explain 61% to 81% of the variability in concentrations of Cd, Cu, Pb, Ni, and Zn in the amphipods from Cd activity, pH, and Ca^{+2}. The maximum differences between predicted and observed concentrations were 3X to 4X. Seasonal variability was poorly explained, possibly because of the influences of biotic variables and changes in diet (Amyot *et al.* 1994). It seems likely that AVS and oxic surfaces should both be included if equilibrium bioavailability models are to be employed across a range of conditions.

14.4 Are equilibrium models sufficient to predict bioavailability?

Where multiple factors affect a process like bioavailability (Table 14-3), models are an effective explanatory and predictive tool. The simplicity of EqP models make them an attractive choice for risk assessment. However, they are also characterized by incomplete or incorrect descriptions of some fundamental processes and a limited capability for dealing with important environmental complexities (Landrum *et al.* 1992; Iannuzzi 1995). These mechanistic limitations suggest that uncertainties in the use of EqP in risk assessments could be substantially greater than the 3X to 10X differences found in the limited assessments of variability conducted to date. Farrington (1989) concluded that "numerous field data and experimental studies demonstrate the need to extend the equilibrium theory model...." Landrum *et al.* (1992) state "...kinetic models are needed to

Table 14-2 Some uncertainties and research needs in defining the usefulness of AVS normalization for estimating metal bioavailability in sediments

Uncertainty	Problem	Research question
Biological relevance of AVS	Behavior/ecology influences exposure	Does AVS effect vary among species in equilibrated sediment systems?
		Does AVS vary among micro-habitats in a sediment?
Spatially representative sampling	Redox and sulfide concentrations are patchy: • Sediment characteristics • Depth in sediment • Season • Among depositional zones within a system	Are there inaccuracies due to sampling constant depth?
		What is variability within a system (rivers, estuaries, coastal zones, lakes)?
		How does AVS vary with time and depth within a year?
		What sediment characteristics most influence AVS in oxidized surface sediments? In surface 3 to 6 cm?

predict non-steady state, nonequilibrium accumulation from temporally and spatially varying exposures when the simplifying assumptions of equilibrium partitioning models are inappropriate, for example, when multiple sources contribute significantly to accumulation."

In a kinetic bioaccumulation model, a set of rate expressions is used to relate chemical fluxes among biological "compartments" and the environment (McKim and Nichols 1994). Kinetic models are more flexible and specific in dealing with biological complexity than are equilibrium models. The focus is on the biological processes that influence contaminant exposures, and responses of such processes are employed to interpret geochemical complexities. The greatest drawback of kinetic models (and other alternatives to EqP) is that they could require extensive data for predictive application to specific circumstances (Landrum *et al.* 1992; McKim and Nichols 1994). A critical question is whether it is reasonable to develop such data. Landrum *et al.* (1992) evaluated and discussed the mathematical details of both equilibrium and kinetic models available to aid exposure assessments. Physiological, pathway, and food web models with both kinetic and equilibrium components have also been described (Clark *et al.* 1988; Thomann 1989; Clark *et al.* 1990; Landrum *et al.* 1992; Bremle *et al.* 1995; Kidd *et al.* 1995; LeBlanc 1995).

The choice of model for exposure assessment should account for at least the following general sources of uncertainty that plague the simpler approaches: 1) nonequilibrium conditions (the assumption of equilibrium is critical to EqP but not necessary for all modeling); 2) chemicals for which the simple EqP approach is unsuitable (*e.g.*, sulfides are not a factor affecting bioavailability of some trace elements; bioavailability of some organic chemicals is not predictable from hydrophobicity); 3) different pathways of uptake (at least, models should assess whether multiple pathways of uptake add uncertainty

Table 14-3   Some biological attributes and geochemical factors affecting metal bioavailability from sediments

Biological attributes	Geochemical factors
Feeding rate	Water column chemistry
Filtration rate across gills	Type of surfaces in sediment
Food selection	Oxidation/reduction conditions
Life-cycle characteristics	Particle/solution partitioning
Microhabitat	Porewater DOC (or other factors affecting speciation)
Behavior with regard to sediment	Diffusion and other reactions at microhabitat scale
Digestive processing	Biological productivity

to exposure assessment); 4) food web transfer of contaminants (sediment risk assessment should extend to higher trophic levels in the sedimentary food web); and 5) biological attributes that affect bioavailability to a target or sensitive species. A diversity of experimental designs, beyond the short-term bioassay paradigm, will also be necessary to convincingly assess the above.

14.4.1 Nonequilibrium conditions: How common?

The assumption of equilibrium inherently limits the circumstances in which EqP models can be applied (Landrum *et al.* 1992). There is no consensus on how frequently contaminant distributions among phases are governed by nonequilibrium conditions nor about the quantitative error that arises in these circumstances. Two of the circumstances in nature that can cause distributions of contaminants to differ from those predicted at thermodynamic equilibrium are high temporal variability and biological control of distributions.

Key processes in many ecosystems are variable on several time scales. In estuaries, for example, metal inputs are from myriad sources (rivers, atmosphere, local anthropogenic) and physical-chemical factors that affect metal distributions and bioavailability (salinities, dissolved organic matter, resuspension, hydrodynamics, and biological cycles) change on several scales in time and space. Intensive field studies in estuaries have clearly demonstrated the highly dynamic nature of metal exposures (Luoma 1976; Fisher and Frood 1980; Cain and Luoma 1990; Luoma *et al.* 1990). Rivers also undergo changes in physical conditions, contaminant inputs, and internal physicochemical characteristics on diurnal, weekly, and seasonal scales. Again, highly unstable metal dynamics can be typical in such ecosystems. In contrast, it seems possible that some lakes and coastal environments, while not stable, may be subject to a regularity in physical-chemical change more amenable to maintenance of near-equilibrium conditions.

Biological processes can result in steady-state or varying contaminant distributions that deviate substantially from those predicted by thermodynamic equilibrium. Wangersky (1986) suggested that "control of concentrations of most trace metals in surface waters appears to be by physical and chemical adsorption on biological materials, and is related to local primary productivity." Distributions will eventually be predictable from thermodynamic principles if biological controls exert their effect by changing the abundance and nature of surfaces (although biological change is often rapid). For example, the cycling and flux of particle-reactive, non-essential metals is predictably linked to production in oceanic systems (Coale and Bruland 1985; Fisher *et al.* 1987, 1988; Moore and Dymond 1988). However, thermodynamic equilibrium will not predict distributions and reactions where a) biological processes add new intracellular biochemical sinks, b) biota create and release products not favored thermodynamically, or c) biota incorporate contaminants into structures, compounds, or complexes that are either stable or break down slowly.

For example, concentrations of Se species in the oceans are explained by biological processing, not by thermodynamics. Phytoplankton strip selenite from solution, transform it to organo-selenium, and release the latter both to solution and to their consumers (Wrench and Measures 1982; Cutter and Bruland 1984; Reinfelder and Fisher 1991; Luoma *et al.* 1992). It is well known that dissolved metals (Cd, Zn, and perhaps Ni and Cu) are "depleted" by biological productivity (*i.e.*, concentrations are lower than expected on the basis of thermodynamics) in the oceans and, at least at times, in lakes (Bruland 1980; Sigg 1985; Reynolds and Hamilton-Tayler 1992). Sunda and Huntsman (1995) showed that Cu concentrations (particulate and dissolved) in oceanic nutriclines are regulated by phytoplankton uptake and regeneration, not by thermodynamics alone. Lee and Fisher (1992) demonstrated that Cd and Zn are accumulated in the cytosol (*i.e.*, within the cells) of phytoplankton, while Cu and Ag are predominantly sorbed to particle surfaces. As a result of such reactions, Cd and Zn concentrations were rapidly depleted from solution and increased in suspended materials during a phytoplankton bloom in South San Francisco Bay (van Geen *et al.*, USGS, in prep.). Passage of metals from phytoplankton to their consumers (Reinfelder and Fisher 1991) or recycling of the metals from decaying phytoplankton (Lee and Fisher 1992) are also controlled by the proportion of metal partitioned to the cytosol of the plants, a purely biological phenomenon. Consumer organisms also transform contaminants. For example, detoxification of metals in the tissues of animals occurs by conjugation into stable forms like granules that may be of low bioavailability to predators (Nott and Nicolaidou 1990; Wallace and Lopez, in press).

Equilibrium-based models inherently assume a stability in all of nature. They may misstate exposures where biota control contaminant partitioning between dissolved and particulate phases, where trophic transfer is influenced by biochemically transformed contaminants or where physical-chemical conditions are highly dynamic. Comparisons of predictions between models that assume equilibrium and those that do not could be fruitful in testing the quantitative importance of this source of uncertainty.

14.4.2 Biological attributes that affect bioavailability: Is metal availability biologically generic?

Differences in organismic and ecological attributes of the ecosystem at risk are often excluded from consideration in the most geochemically robust equilibrium models (*e.g.*, Figure 14-1). Multiple pathways of uptake, bioenergetics, species-specific differences in mode of exposure to sediments and associated waters, physiological control of uptake, differences in elimination rates, and food web structure are potentially important but often not considered (Clark *et al.* 1988, 1990; Thomann 1989; Landrum *et al.* 1992). While the conceptual model guiding exposure assessments must include first-order geochemical principles, substantial uncertainties will remain in specific predictions until a robust appreciation of biology is also incorporated.

Biological attributes could affect whether AVS is a relevant control on bioavailability to a species, for example. Benthic species obtain oxygen and their nutrition differently from sediments and are exposed to different microenvironments. Some oligochaetes feed "head-down" in reduced sediments and "breathe" by periodically returning to the oxidized surface of the sediments (G. Lopez, SUNY Stony Brook, personal communication). These organisms would be predominantly exposed to AVS-rich, reduced sediments in many environments, probably sediments with low metal bioavailability. In contrast, most meiofauna are restricted to oxidized layers of sediments, where metals are likely to be in their most bioavailable forms. Many macrofauna use tubes or burrows to feed and obtain oxygen or particulate food from oxygenated waters above or near the interface between oxidized and reduced sediments, even though their bodies are buried in anoxic sediments. As cited earlier, Hare *et al.* (1994) found that AVS-controlled porewater concentrations of Cd in spiked sediments transplanted to a lake, but differences in life history, behavior, and/or ecology determined differences in the Cd exposure of different species.

Different species ingest different foods, and feeding can change within species in response to their environment or life stage. Differences in food sources affect trace element assimilation (Decho and Luoma 1991, 1994; Luoma *et al.* 1992; Wang *et al.* 1995; Wang and Fisher 1996); less is known about effect of feeding on assimilation of organic chemicals. For example, Cr was generally not assimilated by bivalves from sediment particles but was assimilated with ~90% efficiency from ingested bacteria (Decho and Luoma 1991). Assimilation efficiencies of ingested metals varied among 7 different algal diets in the mussel (*Mytilus edulis*) (Wang and Fisher 1996) and changed approximately two fold with the abundance of available food (greater assimilation at lower feeding rates) (Wang *et al.* 1995). Decho and Luoma (1996) found that Cd assimilation varied from 20 to 80% between different phytoplankton or bacteria; only 5% to 17% of Cd was assimilated from sediments (probably with a minimal living component) or from nonviable particles (iron oxides, humic substances, fulvic substances). Deposit feeding clams assimilated ~90% of ingested Se from diatoms but only 20% of the Se when they fed on sediments containing microbially deposited elemental Se (Luoma *et al.* 1992).

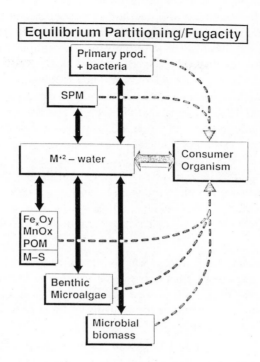

Figure 14-1 Forms of metal to which benthic organism might be exposed and their contributions to dose experienced. All potential pathways of uptake do not need to be directly considered (dotted lines) in EqP model because all sources are in equilibrium. Fugacity theory implies that common process controls chemical potential in all equilibrated phases that are sources of exposure. This is more appropriate for organic chemicals than for trace elements. Alternative models (e.g., kinetic) assume all pathways are additive; therefore, contribution of each must be quantified to correctly estimate dose.

Contaminant (and especially trace-element) bioavailability is not biologically generic. It is not practical to understand all biological factors for all species (*e.g.*, contaminant assimilation from all combinations of food sources available to all benthos). But understanding, for example, assimilation from end members in a continuum of food sources, and generalizations about how biology and ecology affect exposures for key species, may be necessary for reliable exposure assessments. The models employed should be sufficiently flexible to include such considerations.

14.4.3 Pathways of uptake: Additive or equilibrated?
The EqP model assumes that only one pathway of exposure needs to be defined. "It is assumed that the organism receives an equivalent exposure from a water-only exposure or from any equilibrated phase; either from pore water via respiration; from sediment

carbon via ingestion; or from a mixture of the routes" (Di Toro *et al.* 1991). Alternative models assume that pathways of uptake are additive (Thomann 1989; Clark *et al.* 1990; Landrum *et al.* 1992). This fundamental difference in conceptual models is important. For example, the applicability of water-only bioassays implicitly depends upon the assumption of equilibrium. If pathways are additive, then it is not appropriate to extrapolate to nature from single-pathway bioassays. Such extrapolations will underestimate the full exposure of organisms ingesting bioavailable contaminants in their food (Luoma 1995; Meador *et al.* 1995a). Luoma *et al.* (1992) used a pathway model and environmental data to show that Se contamination problems in San Francisco Bay occurred at much lower concentrations than predicted by toxicity or uptake from the dissolved phase alone. The dose of Se that organisms received from the dissolved vector was very low at concentrations of available forms (selenite) typical of the Bay, but the dose received via the pathway selenite-phytoplankton-clam-birds was sufficient to explain the environmental contamination (Brown and Luoma 1995). Whether or not the system was in equilibrium, bioassay-based criteria could not account for the additive threat achieved via the food web.

The choice of models also can affect what is monitored for exposure assessment. If EqP is used for risk assessment, then only sediments need be analyzed for the parameters necessary for modeling. If pathways are additive then the different sources (water, suspended particulates, sediments, or prey organisms) must be monitored, unless their insignificance is demonstrated.

Contrary to the observations of Luoma *et al.* (1992), Kemp and Swartz (1988) found that Cd bioavailability was predictable from the EqP concepts. Their paper is one of the important references supporting the proposal that pore water is the significant pathway of Cd bioaccumulation by amphipods. While the experiment was rigorous, a close inspection of the approach suggests the results were affected by the sediment-water distribution chosen. A Cd bioaccumulation model for *Macoma balthica* (Table 14-4) was applied to the data of Kemp and Swartz (1988), who used *Macoma nasuta* as an experimental animal. Porewater concentrations of Cd were enhanced by 1000X to 10,000X over the concentrations found in contaminated estuarine waters, while concentrations in food were enhanced by 16X to 72X compared to moderately contaminated sediments. The pathway model predicts the same conclusion as the experiment; that is 99% of Cd bioaccumulation by *M. nasuta* uptake should be from pore waters with these experimental conditions. However, it was because of the difference in relative enhancement that food was a trivial source of bioaccumulation. Using conditions typical of a moderately contaminated estuary (Table 14-4), the pathway model predicts that both food and water are sources of Cd bioaccumulation. The experiment does not refute the hypothesis that sediments are not a direct source for contaminant bioaccumulation under the conditions that occur in nature. The point is that the conditions of exposure determine the relative importance of pathways, and experiments that do not mimic conditions typical of nature (*e.g.*, distributions between dissolved and sedimentary concentrations) will not yield results that can be widely extrapolated to nature. Experiments that greatly enrich (pore) water concentra-

Table 14-4 Cadmium bioaccumulation via different pathways by *Macoma balthica* predicted from a pathway model

Experiment	[Cd]sed (µg/g)	[Cd]soln. (µg/l)	AE	Css-soln. (µg/g)	Css-food (µg/g)	Css-total (µg/d)	% from soln.	% from food
Kemp & Swartz	16	1000	0.1	2000	16	2016	99.2	0.8
Kemp & Swartz	72	5000	0.1	10,000	72	10,072	99.3	0.7
Estuary: some contamination	1	0.2	0.4	0.4	4	4.4	9	91

Steady-state bioaccumulation (Css) is compared between the experimental conditions reported in Kemp and Swartz (1988) and conditions typical of an estuary with moderate Cd contamination.
Data for the model derived from experiments reported by Luoma *et al.* (in prep.).
Elimination rate was 0.01 d^{-1}; feeding rate was 0.1 g sediment d^{-1} g^{-1} tissue; assimilation efficiencies were 0.1 for sediment and 0.4 for sediment with living diatoms; uptake rate from solution (Css-soln.) was determined experimentally at 20‰ salinity; Css-soln. and Css-food were determined from influx rate/elimination rate.

tions of contaminants relative to other sources will show dominance by that source term, whatever the conceptual model of bioaccumulation.

The assumption of pathway equilibrium, rather than pathway additivity, has affected the approach of the studies that best support EqP. Most such studies (including the one discussed above) exclude contaminated food sources, especially the living component of natural sediment. Most sediment collection, handling, or storage procedures change or eliminate the meiofauna, diatoms, and bacterial flora of natural sediments. Assimilation of contaminants from these important food sources could change the outcome of experiments.

Correlative studies have not differentiated which conceptual model of bioaccumulation best reflects nature. In nature, bioaccumulation could correlate with dissolved activity of a contaminant if all pathways are in equilibrium, whether pathways are additive or not. The significant correlation between predicted free Cd ion activities and Cd residues in mussel or amphipod tissue in the lakes of Quebec suggested equilibrium distributions controlled bioavailability (Tessier *et al.* 1993; Amyot *et al.* 1994), but neither study claimed that only one pathway of exposure was significant. Empirical correlations could provide a basis for predicting contaminant exposures (avoiding the pathway question), but first more study is necessary of the questions of ecosystem equilibrium (Can similar correlations be found in a range of environments?) and species-specific bioaccumulation (How do such correlations differ among species?).

There is some direct experimental evidence that bioaccumulation pathways are additive. Experiments that compare water exposures to food + water exposures show that when contaminants are available from the type of food employed, total bioaccumulation or toxicity increases over that from water alone; *i.e.*, uptake is additive (Chandler *et al.* 1993; see metals literature cited in Luoma 1989). Recent experiments that provided more quantitative determinations of the efficiencies of assimilation (*e.g.*, Boese *et al.* 1990; Decho and Luoma 1991; Reinfelder and Fisher 1991; Wang and Fisher 1996; Wang *et al.* 1996) may also help resolve the contributions of pathway additivity (Thomann 1989; Luoma *et al.* 1992; Thomann *et al.* 1995; Wang *et al.* 1996).

14.4.4 Food web transfer: Trophic transfer beyond the benthos?

The food web beyond sediment-dwelling benthos should be included if sediment risk assessments are to be relevant to ecosystems. Several factors may affect the distribution of organic chemicals between consumer organisms, their solute environment, and their food sources. Residues in higher trophic-level species are often elevated beyond those predicted from hydrophobicity and simple normalizations (Oliver and Niimi 1988; Pereira *et al.* 1988; Clark *et al.* 1990; LeBlanc 1995). Organic chemical bioconcentration factor (BCF) (for fish, *e.g.*) differs among environments where prediction from laboratory studies forecast the same BCF (Bremle *et al.* 1995). Exciting advances are occurring in studies that combine ecological principles with principles that govern organic chemical bioaccumulation. Differences in organochlorine bioaccumulation by predators, for example, appear to occur among lakes because of differences in food web structure (Kidd *et al.* 1995). These results suggest that the most accurate food web exposure assessments will ultimately come from models incorporating robust geochemical, biological, and ecological concepts relevant to bioaccumulation. Most trace elements do not biomagnify through marine food webs (Fisher and Reinfelder 1995), but their trophic transfer is just starting to be studied quantitatively and may be toxicologically relevant under some conditions.

14.5 Are alternatives to EqP realistic?

Although some studies have shown reliable predictions from simple EqP modeling, the important sources of uncertainty discussed above suggest that such models will eventually prove inadequate for site-specific exposure assessment, especially for trace elements. The best exposure assessments need to be geochemically robust, biologically specific, flexible for a variety of environmental circumstances, and bioaccumulation should be quantitatively predicted, making model predictions verifiable in nature. The ideal model would take advantage of the equilibrium-based geochemical theory to define exposure and would employ biologically robust modeling approaches to predict dose (bioaccumulation).

Models that incorporate geochemical and biological principles are available. Multiple compartment pharmacokinetic models, some of which consider bioenergetics and multiple pathways, have been developed to predict tissue distributions of (especially organic) contaminants in higher order organisms (Norstrom *et al.* 1976; Thomann 1989; Clark *et al.* 1988, 1990; Landrum *et al.* 1992; McKim and Nichols 1994). For assessing exposure from sediment to benthos, model approaches can probably be simplified to assess whole organism exposures (one or two compartment models) (Pentreath 1973; Thomann 1989; Boese *et al.* 1990; Clark *et al.* 1990). Such simple kinetic models were extensively used in radioecology to study exposures, although multiple pathways were not always considered (Ruzic 1972; Pentreath 1973).

One way to reduce extensive data requirements of some of the above models is to incorporate generic biological and geochemical constants (Thomann 1989; Clark *et al.* 1990). Generic choices of concentration factors, distribution coefficients, growth rates, assimilation efficiencies, and efflux rate constants have been used in model projections. Unfortunately, all of these can vary by an order of magnitude or more in specific circumstances. Thus, generic models are instructive with regard to broad principles governing contaminant exposure but may have large uncertainties when employed in specific exposure assessments.

For site-specific exposure assessment, an alternative is what we will call an *empirical* kinetic, multiple-pathway model. Such a model is similar in principle to the generic kinetic models but is made more applicable to specific circumstances by empirically developing physiological parameters representative of one or more key species native to that system and employing environmental data representative of a range of conditions in the system (Figure 14-2). At first glance, obtaining the data for such models appears onerous (Landrum *et al.* 1992). However, model behavior can be reasonably constrained to a restricted subset of parameter inputs (McKim and Nichols 1994) and manageable methods are available for obtaining species-specific biological data, especially if whole organism data is the goal (Luoma *et al.* 1992; Wang *et al.* 1996).

The mathematics of the simplest kinetic model with several pathways illustrates the necessary experiments:

$$dC_m/dt = (I_f + I_w) - Ck_e \qquad (14\text{-}1).$$

For I_f

$$C_{m,t} = I_f/k_e + (1 - e^{-kt}) \qquad (14\text{-}2)$$

and

$$C_{m,ss} = I_f/k_e \qquad (14\text{-}3)$$

where

C_m = the concentration in animal,
t = time,
I_f = the gross influx rate from food,
I_w = the gross influx rate from water,
k_e = the rate constant of loss (slowest compartment), and
$C_{m,ss}$ = the concentration at steady state.

For the influx rate from food,

$$I_f = \text{feed rate} \times \text{concentration} \times \text{assimilation} \qquad (14\text{-}4).$$

Key parameters to determine experimentally include influx rates from solution, influx rates from ingestion, and efflux rates (Equations 14-1, 14-2; Figure 14-2). If correctly determined, rate constants for these parameters describe mechanisms that are directly comparable among species. Thus, in addition to their use in models, these data could

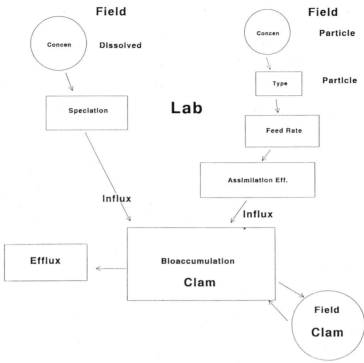

Figure 14-2 Description of experiments and field data necessary to derive empirical pathway model for benthic detritus feeder. Approach also allows quantitative determination of contributions of food and dissolved pathways of bioaccumulation. Circles indicate data that should be obtained from site of interest (data ranges are preferable to single values). Squares indicate experiments or species-specific experimentally derived parameters. Bioaccumulation is calculated from experiments and field data. Prediction can be validated by direct comparison to concentrations observed in resident (or transplanted) individuals from that species at site of interest.

lead to better appreciation of reasons for interspecies differences in bioaccumulation and, perhaps, why some species are more vulnerable to contaminants than others.

It is very important how the physiological parameters are determined. Mechanistically, bioaccumulation results from a combination of gross influx rate as balanced by the gross efflux rate. Gross influx is a species-specific function of the concentration of bioavailable metal. Gross efflux is an instantaneous function of the concentration in tissues and the rate constants of loss (Equations 14-1, 14-2). Typically in past studies, bioaccumulation was determined from exposures that lasted weeks or months. Because the results reflected a mixture of physiological mechanisms, they were not suitable for model application. They could not be quantitatively extrapolated to nature and they were difficult to compare quantitatively to other circumstances. Some studies have purported to achieve a steady-state bioaccumulation, but caution is warranted in concluding that fluxes are in balance (Luoma 1977; Meador *et al.* 1995a). For example, life-long exposure to dissolved ^{65}Zn was necessary to obtain uniform labeling, and thus true steady state, in mosquito fish (Willis and Jones 1977).

Influx from solution can be determined with radionuclides in short exposures (*e.g.*, 1-d) because the goal is to estimate the unidirectional flux. Influx rates can be studied as a function of concentration, by spiking with stable carrier element, or as a function of speciation by varying the geochemical characteristics across conditions characteristic of the ecosystem of interest (or, alternatively, using water from the site of interest). Testing a variety of conditions is quite feasible because the experiments are short and manageable.

Influx rates from ingestion vary with the food source and so are best determined from the product of assimilation efficiencies (from specific types of food), feeding rate, and concentration (Equation 14-4). Assimilation efficiency (AE) is the factor for which the least is known and is the key to understanding contributions of different pathways of uptake. In fact, the final concentrations of a metal contaminant in mussels, for example, can be shown to be a direct function of that metal's AE (Wang *et al.* 1996). Repeatable AEs for a variety of food sources and feeding conditions can readily be determined with invertebrates using radionuclides in pulse-chase experiments (Decho and Luoma 1991, 1994; Fisher *et al.* 1991; Reinfelder and Fisher 1991; Luoma *et al.* 1992; Wang *et al.* 1995; Wang and Fisher 1996). These studies showed that assimilation efficiencies are not generic for all food types; a substantial variation was observed in AE among food sources and feeding conditions for nearly every element as well as among elements and species.

Efflux rates can be described by the rate constants of first-order isotope-substitution kinetics (Ruzic 1972; Cutshall 1974). The loss experiments are best conducted with radionuclides in order to assure that no net influx of metal occurs during elimination. In most invertebrates, efflux can be described by one to three components of first-order exponential decay. Separation of the compartments is essential to obtaining repeatable results (Riggs 1963). The rate constants for each compartment could vary with route of exposure, but such variability appears to be small, at least for mussels. A second parameter of importance is the proportion of metal in each compartment. This is affected by exposure time (Cutshall 1974) and thus can be experiment-specific. Exposure time will not affect the rate constants of loss if they are determined by mathematically stripping compartments, but if only single compartment analysis is (inappropriately) employed, rate constants will artificially vary with exposure. A useful alternative for predictive pathway models could be to assume that exposures in nature are chronic, and thus rate constants from the most slowly exchanging compartments can be employed as the efflux term (Luoma *et al.* 1992).

Pathway models can take advantage of site-specific environmental data for assessing exposures. Reliable methodologies are now available for direct determination of contaminant concentrations in water, suspended materials, sediments and even biological components of sediments, and in fact, reliable data already exist for many ecosystems. Site-specific environmental data can be combined with biological rate constants for some key local species, to assess dose (bioaccumulation) under a variety of conditions typical of the site. Variable contaminant concentrations can be included, as well as variable DOC or salinity, changes in food source, changes in sediment character, *etc.* Exposure assessment might be more realistic if bioaccumulation is projected for the range of possibilities typical of a system, rather than assuming that a single numeric response characterizes an ecosystem.

The model described above is the simplest form of a bioaccumulation model. It lacks, for example, bioenergetic terms, or considerations for seasonal gain and loss of lipid that affect both trace element and organic contaminant bioaccumulation (Cappuzzo *et al.* 1989; Cain and Luoma 1990). Even this simple model approach appears to provide reasonable compatibility with field observations (Luoma 1976; Luoma *et al.* 1992; Wang *et al.* 1996), although further studies undoubtedly will find ways to improve the model predictions. Data needs for expanding even the simple, empirical pathway models are, at present, large. However, as rate constants are defined for common species and as experiments with different geochemical conditions are related to these mechanistic biological responses, adequate data should become available for site-specific exposure assessments.

14.6 Extrapolation from bioavailable dose to toxicity

The empirical pathway model predicts tissue residues directly. An advantage of this approach is that predictions can be unambiguously compared to residues of organisms in nature (*e.g.*, to verify the model approach under one set of conditions). However, the residue output of the pathway model means that tissue-residue–based toxicity extrapolations will be needed for risk assessment (Figure 14-3). Some data suggest that tissue-residue–based toxicity predictions may be reasonable for organic contaminants (Widdows and Donkin 1989), but their mechanistic feasibility has not been demonstrated for metals. This is an important area of research for the future.

An advantage of equilibrium models is that contaminant activities in pore water can be used to extrapolate toxicity from the large, existing database on dissolved toxicity. These data have proven valuable in regulatory issues to date, but they will be less useful if equilibrium models are inadequate estimators of exposure for risk assessments. Perhaps more important is the question of whether the toxicity test approach will be sufficiently powerful as the only estimator of toxicity, if improved accuracy is demanded for site-specific risk assessment. Large, undetermined, and controversial uncertainties lie in extrapolating from environmental concentrations of a chemical to toxicity, based upon such toxicity tests (Cairns 1983, 1986; Moriarity 1988; Depledge 1989; Cairns and Mount 1990; Iannuzzi *et al.* 1995; Luoma 1995). Mechanistic reasons exist to expect that bioassays can either overestimate (Iannuzzi *et al.* 1995) or underestimate (Luoma 1995) toxic concentrations of contaminants. The existing bioassay data are limited with regard to ecologically key species, contaminant activities rather than concentrations, multi-generation exposures, compensatory mechanisms at all levels of organization, interactions with natural stresses, *etc.* Linkages between short-term exposures of organisms via a single pathway in the laboratory, and complex toxicity responses in nature have been difficult

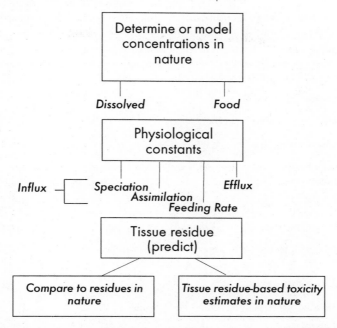

Figure 14-3 Empirical pathway model generates tissue-residue data; therefore, requires understanding tissue-residue–based toxicity

to establish; these difficulties partly explain why field validation of toxicity test results has not been easy to accomplish (Chapman 1986, 1991, 1995). Field validation could be easier if residue-based toxicity could at least be used to supplement existing approaches. Of course, before any approach to toxicity estimation is adequate, we must improve mechanistic understanding of how toxicity is manifested in nature (Luoma and Carter 1991).

14.7 Conclusions

Simple EqP models sometimes proposed for use in risk assessment may not be sufficiently flexible to deal with the range of conditions that can occur in ecosystems (Table 14-5). The greatest uncertainties in assessing contaminant exposure from sediments stem from insufficient understanding of the interplay between geochemistry, biology, and ecology. It may be feasible to reduce some of the crucial uncertainties, however. The key processes in accurate predictive models for organic contaminants will include hydrophobicity, geochemical corrections for factors that strongly influence bioavailability (*e.g.*, DOC), consideration of the influence of multiple pathways, kinetic influences such as variable contaminant elimination rates, and food web complexities. For metals, a better database needs to be developed for key benthic biota describing the kinetic components in bioaccumulation via multiple pathways. These include element-specific uptake rates from solution, elimination rates, and assimilation efficiencies from end members among

Table 14-5 Some areas of knowledge deficit that cause uncertainties in predictions of contaminant bioavailability and exposure assessment

Item	Knowledge deficit	Applicability
1	Dissolved organic materials in pore waters	M & O
2	Partitioning in spiked vs. field-contaminated samples	O (M=?)
3	Obtaining representative sediment samples	M
4	Relevance of the sediment sample to the biological microenvironment	M
5	Role of the living organisms within sediments on exposure and toxicity to their consumers	M & O
6	Frequency of nonequilibrium conditions	M & O
7	Additive pathways or all routes equilibrated	M & O
8	Effect of specific biological, behavioral, or ecological attributes of different species	M & O
9	Food web transfer	M & O
10	Speciation and partitioning models relevant to nature	M

M = applicable to trace elements
O = applicable to organic contaminants

food sources. Feeding rates and modes of exposure to sedimentary contaminants need to be determined for key species. Existing equilibrium models need to be expanded to include both AVS and oxic surfaces. End member effects of dissolved speciation might be linked into bioaccumulation predictions via uptake rate studies. Finally, studies of trace element exposure routes for the upper trophic levels in the benthic food web need to be better understood. This is a large agenda, but for the first time, many of the methodologies are available to accomplish much of it, and experimental work is underway to address at least some of the needs.

Perhaps the most difficult challenge confronting improved risk assessment lies in linking bioavailability predictions to toxicity, where the uncertainties of predictions from the most established approach, toxicity tests, are large. It is possible that the simple toxicity test approach has reached inherent limits with substantial uncertainties remaining unresolvable. The tissue-residue–based toxicity approach has not received much attention to date but offers at least the potential to supplement the toxicity test approach because of its less ambiguous linkage to nature via bioaccumulation. It first needs to be better demonstrated, however, that tissue-residue–based toxicity estimates are feasible at all. Thus while reasonable resolutions to important bioavailability questions may be fast approaching, the great debates of the future about the toxicity–bioavailability linkage could be more difficult to resolve. Until we better understand the limits to exposure assessment imposed by complex factors affecting bioavailability and the limits to risk assessment imposed by the toxicity–bioavailability linkage, science should remain humble about the accuracy of sediment risk assessment.

14.8 Acknowledgments

This work was partially supported by a San Francisco Bay Research Enhancement Program grant administered by the Interagency Ecological Program for the San Francisco Bay/Delta.

14.9 References

Amyot M, Pinel-Alloul B, Campbell PGC. 1994. Abiotic and seasonal factors influencing trace metal levels (Cd, Cu, Ni, Pb and Zn) in the freshwater amphipod *Gammharus fasciatus* in two fluvial lakes of the St. Lawrence river. *Can J Fish Aq Sci* 51:2003–2016.

Anderson DM, Morel FMM. 1978. Copper sensitivity of *Gonyaulax tamarensis*. 23:283–295.

Ankley GT, Phipps GL, Leonard EN, Benoit DA, Mattson VR, Kosian PA, Cotter AM, Dierkes JR, Hansen DJ, Mahoney JD. 1991. Acid-volatile sulfide as a factor mediating cadmium and nickel bioavailability in contaminated sediments. *Environ Toxicol Chem* 10:1299–1307.

Belzile N, Tessier A. 1990. Interactions between arsenic and iron oxyhydroxides in lacustrine sediments. *Geochim Cosmochim* ACTA 54:103–109.

Boese BL, Lee II H, Specht DT, Cahill RC. 1990. Comparison of aqueous and solid-phase uptake for hexachlorobenzene in the tellinid clam *Macoma nasuta* (conrad): a mass balance approach. *Environ Toxicol Chem* 9:221–231.

Bremle G, Okla LL, Larsson P. 1995. Uptake of PCBs in fish in a contaminated river system: Bioconcentration factors measured in the field. *Environ Sci Technol* 29:2010–2015

Brown CL, Luoma SN. 1995. Energy-related selenium and vanadium contamination in San Francisco Bay: effects on biological resources? In: Tenth McKelvey Forum on Mineral and Energy Resources. Reston VA: U.S. Geological Survey (USGS). Circular 1108. p 91–93.

Bruland KW. 1980. Oceanographic distribution of cadmium, zinc, nickel and copper in the North Pacific. *Earth Planet Sci* 47:176–198.

Brumbaugh WG, Ingersoll CG, Kemble NE, May TW, Zajicek JL. 1994. Chemical characterization of sediments and pore water from the Upper Clark Fork River and Milltown Reservoir, Montana. *Environ Toxicol Chem* 13:1971–1984.

Bryan GW. 1985. Bioavailability and effects of heavy metals in marine deposits. In: Duedorf I, Perk B, Kester D, editors. Wastes in the ocean, near shore waste disposal. London: J Wiley. p 41–61.

Bryan GW, Langston WJ. 1992. Bioavailability, accumulation and effects of heavy metals in sediments with special reference to United Kingdom estuaries: a review. *Environ Pollut* 76:89–131.

Bryan GW, Langston WJ, Hummerstone LG, Burt GR. 1985. A guide to the assessment of heavy-metal contamination in estuaries using biological indicators. Plymouth UK: Marine Biol Assn. UK Spcl Publ #4. 92 p.

Cain DJ, Luoma SN. 1990. Influence of seasonal growth, age and environmental exposure on Cu and Ag in a bivalve indicator, *Macoma balthica*, in San Francisco Bay. *Mar Ecol Prog Ser* 60:45–56.

Cairns Jr J. 1983. Are single species toxicity tests alone adequate for estimating environmental hazard? *Hydrobiologia* 100:47–557.

Cairns Jr J. 1986. The myth of the most sensitive species. *BioScience* 36:670–672.

Cairns Jr J, Mount DI. 1990. Aquatic toxicology. *Environ Sci Technol* 24:154–161.

Campbell PGC, Tessier A. 1989. Geochemistry and bioavailability of trace metals in sediments. In: Aquatic ecotoxicology: fundamental concepts and methodologies, Volume 1. Boca Raton FL: CRC Press. Chapter 7.

Capuzzo JM, Farrington JW, Rantamaki P, Clifford CH, Lancaster BA, Leavett DF, Jia X. 1989. The relationship between lipid composition and seasonal differences in the distribution of PCBs in *Mytilus edulis* L. *Mar Environ Res* 28:259–264.

Chapman PM. 1986. Sediment quality criteria from the sediment quality triad: an example. *Environ Toxicol Chem* 5:957–964.

Chapman PM. 1991. Environmental quality criteria. *Environ Sci Technol* 25:1353–1359.

Chapman PM. 1995. Do sediment toxicity tests require field validation. *Environ Toxicol Chem* 14:1451–1453.

Chandler GT, Coull BC, Davis JC. 1993. Sediment- and aqueous-phase fenvalerate effects on meiobenthos: implications for sediment quality criteria development. *Mar Environ Res* 37:313–329.

Clark T, Clark K, Paterson S, Mackay D, Norstrom RJ. 1988. Wildlife monitoring, modeling and fugacity. *Environ Sci Technol* 22:120–127.

Clark KE, Gobas FAPC, Mackay D. 1990. Model of organic chemical uptake and clearance by fish from food and water. *Environ Sci Technol* 24:1203–1213.

Coale KH, Bruland KW. 1985. 234Th:238U disequilibria within the California current. *Limnol Oceanogr* 30:22–33.

Cutter GA, Bruland KW. 1984. The marine biogeochemistry of selenium: a re-evaluation. *Limnol Oceanogr* 29:1179–1192.

Cutshall N. 1974. Turnover of Zinc-65 in oysters. *Health Physics* 26:327–331.

Decho AW, Luoma SN. 1991. Time-courses in the retention of food material in the bivalves *Potamocorbula amurenis* and *Macoma balthica*: significance to the assimilation of carbon and chromium. *Mar Ecol Prog Ser* 78:303–314.

Decho AW, Luoma SN. 1994. Humic and fulvic acids: sink or source in the availability of metals to the marine bivalves *Potamocorbula amurenis* and *Macoma balthica*. *Mar Ecol Prog Ser* 108:133–145.

Decho AW, Luoma SN. 1996. Flexible digestion strategies and trace metal assimilation in marine bivalves. *Limnol Oceanogr* 41:568–572.

DeWitt TH, Ozretich RJ, Swartz RC, Lamberson JO, Schults DW, Ditsworth GR, Jones JKP, Hoselton L, Smith LM. 1992. The influence of organic matter quality on the toxicity and partitioning of sediment-associated fluoranthene. *Environ Toxicol Chem* 11:197–208.

Depledge M. 1989. The rational basis for detection of the early effects of marine pollutants using physiological indicators. *Ambio* 18: 301–302.

Di Toro DM, Mahony JD, Hansen DJ, Scott KJ, Hicks MB, Mayr SM, Redmond MS. 1990. Toxicity of cadmium in sediments: the role of acid volatile sulfide. *Environ Toxicol Chem* 9:1487–1502.

Di Toro DM, Zarba CS, Hansen DJ, Berry WJ, Swartz RC, Cowan CE, Pavlou SP, Allen HE, Thomas NA, Paquin PR. 1991. Technical basis for establishing sediment quality criteria for nonionic organic chemicals using equilibrium partitioning. *Environ Toxicol Chem* 10:1541–1583.

Engel DW, Fowler BA. 1979. Factors influencing cadmium accumulation and its toxicity to marine organisms *Environ Health Perspectives* 28:81–88.

Farrington JW. 1989. Bioaccumulation of hydrophobic organic pollutant compounds. In: Levin SA, Harwell MA, Kelly JR, Kimball KD editors. Ecotoxicology: problems and approaches. New York: Springer-Verlag. p 279–313.

Fisher NS. 1986. On the reactivity of metals for marine phytoplankton. *Limnol Oceanogr* 31:443–449.

Fisher NS, Cochran JK, Krishnaswami S, Livingston HD. 1988. Predicting the oceanic flux of radionuclides on sinking biogenic debris. *Nature* 335:622–625.

Fisher NS, Frood D. 1980. Heavy metals and marine diatoms: influence of dissolved organic compounds on toxicity and selection for metal tolerance among four species. *Mar Biol* 59:85–93.

Fisher NS, Nolan CV, Fowler SW. 1991. Assimilation of metals in marine copepods and its biogeochemical implications. *Mar Ecol Prog Ser* 71:37–43.

Fisher NS, Reinfelder JR. 1995. The trophic transfer of metals in marine systems. In: Tessier A, Turner DR, editors. Metal speciation and bioavailability in aquatic systems. New York: John Wiley. p 363–406.

Fisher NS, Teyssie J-L, Krishnaswami S, Baskaran M. 1987. Accumulation of Th, Pb, U, and Ra in marine phytoplankton and its geochemical significance. *Limnol Oceanogr* 32:131–142.

Hare L, Carignan R, Herta-Diaz MA. 1994. A field study of metal toxicity and accumulation by benthic invertebrates: implications for the acid-volatile sulfide (AVS) model. *Limnol Oceanogr* 39:1653–1660.

Hamelink JL, Landrum PF, Bergman HL, Benson WH, editors. 1994. Bioavailability: physical, chemical and biological interactions. Boca Raton FL: Lewis.

Iannuzzi TJ, Bonnevie NL, Huntley SL, Wenning RJ, Truchon SP, Tull JD, Sheehan PJ. 1995. Comments on the use of equilibrium partitioning to establish sediment quality criteria for nonionic chemicals. *Environ Toxicol Chem* 14:1257–1259.

Jenne EA. 1977. Trace element sorption by sediments and soils - sites and processes. In: Chappel W, Peterson K, editors. Symposium on molybdenum in the environment. New York: Dekker. p 425–553.

Kemp PF, Swartz RC. 1988. Acute toxicity of interstitial and particle-bound cadmium to a marine infaunal amphipod. *Mar Environ Res* 26:135–153.

Kidd KA, Schindler DW, Muir DCG, Lockhart WL, Hesslein RH. 1995. High concentrations of toxaphene in fishes from a Subarctic lake. *Science* 269:240–242.

Lake JL, Rubinstein NI, Lee II H, Lake CA, Heltshe J, Pavignano S. 1990. Equilibrium partitioning and bioaccumulation of sediment-associated contaminants by infaunal organisms. *Environ Toxicol Chem* 9:1095–1106.

Landrum PF, Lee II H, Lydy MJ. 1992. Toxicokinetics in aquatic systems: model comparisons and use in hazard assessment. *Environ Toxicol Chem* 11:1709–1725.

Langston WJ. 1980. Arsenic in U.K. estuarine sediments and its availability to benthic organisms. *J Mar Biol Assn UK* 60:869–881.

Langston WJ. 1982. The distribution of mercury in British estuarine sediments and its availability to deposit feeding bivalves. *J Mar Biol Assn UK* 62:667–684.

LeBlanc G. 1995. Trophic-level differences in the bioconcentration of chemicals: implications in assessing environmental biomagnification. *Environ Sci Technol* 29:154–160.

Lee B-G, Fisher NS. 1992. Degradation and elemental release rates from phytoplankton debris and their geochemical implications. *Limnol Oceanogr* 37:1345– 1360.

Long ER, MacDonald DD, Smith SL, Calder FD. 1995. Incidence of adverse biological effects within ranges of chemical concentrations in marine and estuarine sediments. *Environ Mgt* 19:81–97.

Luoma SN. 1976. Dynamics of biologically available mercury in a small estuary. *Estuarine Coastal Mar Sci* 5:643–652.

Luoma SN. 1977. Physiological characteristics of mercury uptake by two estuarine species. *Mar Biol* 41:269–273.

Luoma SN. 1983. Bioavailability of trace metals to aquatic organisms—a review. *Sci Total Environ* 28:1–22.

Luoma SN. 1989. Can we determine the biological availability of sediment-bound trace elements? *Hydrobiologia* 176/177:379–401.

Luoma SN. 1995. Prediction of metal toxicity in nature from bioassays: limitations and research needs. In: Tessier A, Turner D, editors. Metal speciation and bioavailability in aquatic systems. Sussex England: J Wiley. p 610–659.

Luoma SN, Bryan GW. 1978. Factors controlling availability of sediment-bound lead to the estuarine bivalve *Scorbicularia plana*. *J Mar Biol Ass UK* 58:793–802.

Luoma SN, Bryan GW. 1981. Statistical assessment of the form of trace metals in oxidized estuarine sediments employing chemical extractants. *Sci Tot Environ* 17:165–196.

Luoma SN, Bryan GW. 1982. A statistical study of environmental factors controlling concentrations of heavy metals in the burrowing bivalve *Scrobicularia plana* and the polychaete, *Nereis diversicolor*. *Estuarine Coastal Shelf Sci* 15:95–108.

Luoma SN, Carter JL. 1991. Effects of trace metals on aquatic benthos. In: Newman M, McIntosh A, editors. Metal ecotoxicology: concepts and applications. Boca Raton FL: CRC Pr.

Luoma SN, Dagovitz R, Axtmann E. 1990. Temporally intensive study of trace metals in sediments and bivalves from a large river-estuarine system: Suisun Bay/Delta in San Francisco Bay. *Sci Total Environ* 97/98:685–712

Luoma SN, Davis JA. 1983. Requirements for modeling trace metal partitioning in oxidized estuarine sediments. *Mar Chem* 12:159–181.

Luoma SN, Ho KT. 1993. The appropriate uses of marine and estuarine sediment bioassays. In: Calow P, editor. Handbook of ecotoxicology. Oxford: Blackwell Scientific. p 193–228.

Luoma SN, Jenne EA. 1977. The availability of sediment-bound cobalt, silver, and zinc to a deposit-feeding clam. In: Wildung RW, Drucker H, editors. Biological implications of metals in the environment. Springfield VA: NTIS. ERDA Conf. 750920. p 213–230.

Luoma SN, Johns C, Fisher NS, Steinberg NA, Oremland RS, Reinfelder J. 1992. Determination of selenium bioavailability to a benthic bivalve from particulate and solute pathways. *Environ Sci Technol* 26:485.

Mackay D. 1982. Correlation of bioconcentration factors. *Environ Sci Technol* 16:274–278.

McKim JM, Nichols JW. 1994. Use of physiologically based toxicokinetic models in a mechanistic approach to aquatic toxicology. In: Malins DC, Ostrander GK, editors. Aquatic toxicology: molecular, biochemical, and cellular perspectives. Boca Raton FL: Lewis. p 469–521.

Meador JP, Stein JE, Reichert WL, Varanasi U. 1995a. Bioaccumulation of polycyclic aromatic hydrocarbons by marine organisms. *Rvw Environ Contam Toxicol* 143:79–165.

Meador JP, Casillas E, Sloan CA, Varanasi U. 1995b. Comparative bioaccumulation of polycyclic aromatic hydrocarbons from sediment by two infaunal organisms. *Mar Ecol Prog Ser* 123:107–124.

Moore WS, Dymond J. 1988. Correlation of 210Pb removal with organic fluxes in the Pacific Ocean. *Nature* 331:339–341.

Moriarity F. 1988. Ecotoxicology: the study of pollutants in ecosystems. London: Academic Pr. 289 p.

Neely WB, Branson DR, Blau GE. 1974. Partition coefficient to measure bioconcentration potential of organic chemical in fish. *Env Sci Technol* 8:1113–1115.

Norstrom RJ, McKinnon AE, DeFreitas ASW. 1976. A bioenergetics-based model for pollutant accumulation by fish. Simulation of PCB and methylmercury residue levels in Ottawa River Yellow Perch (*Perca flavescens*). *Can J Fish Aquat Sci* 33:248–267.

Nott JA, Nicolaidou A. 1990. Transfer of metal detoxification along marine food chains. *J Mar Biol Assn UK* 70:905–912.

Oliver BG, Niimi AJ. 1988. Trophodynamic analysis of PCB congeners and other chlorinated hydrocarbons in the Lake Ontario ecosystem. *Environ Sci Technol* 22:388–397.

Pentreath RJ. 1973. The accumulation and retention of 65Zn and 54Mn by the Plaice, *Pleuronectes platessa* L. *J Exp Mar Biol Ecol* 12:1–18.

Pereira WE, Rostad CE, Chiou CT, Brinton TI, Barber LB, Demcheck DK, Demas CR. 1988. Contamination of estuarine water, biota, and sediment by halogenated organic compounds: a field study. *Environ Sci Technol* 22:772–778.

Reinfelder JR, Fisher NS. 1991. The assimilation of elements ingested by marine copepods. *Science* 251:794–796.

Reynolds GL, Hamilton-Taylor J. 1992. The role of planktonic algae in the cycling of Zn and Cu in a productive soft-water lake. *Limnol Oceanogr* 37:1759–1769.

Riggs DS. 1963. The mathematical approach to physiological problems. Baltimore: Williams and Wilkins. 445 p.

Ruzic I. 1972. Two-compartment model of radionuclide accumulation into marine organisms. I. Accumulation from a medium of constant activity. *Mar Biol* 15:105–112.

Sigg L. 1985. Metal transfer mechanisms in lakes: the role of settling particles. In: Stumm W, editor. Chemical processes in lakes. New York: J Wiley. p 283–309.

Suedel BC, Rodgers Jr JH, Clifford PA. 1993. Bioavailability of fluoranthene in freshwater sediment toxicity tests. *Environ Toxicol Chem* 12:155–165.

Sunda WG, Engel DW, Thuotte RM. 1978. Effect of chemical speciation on toxicity of cadmium to grass shrimp, Palaemonetes pugio: importance of free cadmium ion. *Environ Sci Technol* 12:409–413.

Sunda WG, Guillard RR. 1976. The relationship between cupric ion activity and the toxicity of copper to phytoplankton. *J Mar Res* 34:411–422.

Sunda WG, Huntsman SA. 1995. Regulation of copper concentration in the oceanic nutricline by phytoplankton uptake and regeneration cycles. *Limnol Oceanogr* 40:132–137.

Swackhamer DL, Hites RA. 1988. Occurrence and bioaccumulation of organochlorine compounds in fishes from Siskiwit Lake, Isle Royale, Lake Superior. *Environ Sci Technol* 22:543–547.

Swartz RC, Schults DW, DeWitt TH, Ditsworth GR, Lamberson JO. 1990. Toxicity of fluoranthene in sediment to marine amphipods: a test of the equilibrium partitioning approach to sediment quality criteria. *Environ Toxicol Chem* 9:1071–1080.

Tessier A, Campbell PGC, Auclair JC, Bisson M. 1984. Relationships between the partitioning of trace metals in sediments and their accumulation in the tissues of the freshwater mollusc *Elliptio complanata* in a mining area. *Can J Fish Aquat Sci* 41:1463–1471.

Tessier A, Couillard Y, Campbell PGC, Auclair JC. 1993. Modeling Cd partitioning in oxic lake sediments and Cd concentrations in the freshwater bivalve *Anodonta grandis*. *Limnol Oceanogr* 38:1–17.

Thomann RV. 1989. Bioaccumulation model of organic chemical distribution in aquatic food chains. *Can J Fish Aquat Sci* 23:699–715.

Thomann RV, Mahoney JD, Mueller R. 1995. Steady state model of biota sediment accumulation factor for metals in two marine bivalves. *Environ Toxicol Chem* 14:1989–1988.

Veith GD, DeFoe DL, BV Bergstedt. 1979. Measuring and estimating the bioconcentration factor of chemicals in fish. *Can J Fish Aquatic Sci* 36:1040–1048.

Wallace W, Lopez GR. In press. Relationship between subcellular cadmium distribution of prey and cadmium trophic transfer to a predator. *Estuaries*.

Wang W-X, Fisher NS. 1996. Assimilation of trace elements and carbon by the mussel, *Mytilus edulis*: effects of food composition. *Limnol Oceanogr* 41:70–81.

Wang W-X, Fisher NS, Luoma SN. 1995. Assimilation of trace metals by the mussel, *Mytilus edulis*: effects of food quantity. *Mar Ecol Prog Ser* 129: 165–176.

Wang W-X, Fisher NS, Luoma SN. 1996. Kinetic determinations of trace element bioaccumulation in the mussel *Mytilus edulis*. *Mar Ecol Prog Ser* (in press).

Wangersky PJ. 1986. Biological control of trace metal residence time and speciation: a review and synthesis. *Mar Chem* 18:269–297.

Widdows J, Donkin P. 1989. The application of combined tissue residue chemistry and physiological measurement of mussel (*Mytilus edulis*) for the assessment of environmental pollution. *Hydrobiologia* 188/189:455–461.

Willis JN, Jones NY. 1977. The use of uniform labelling with zinc-65 to measure stable zinc turnover in the mosquito fish, *Gambusia affinis*- I. Retention. *Health Physics* 32:381–390.

Wrench JJ, Measures CI. 1982. Temporal variations in dissolved selenium in a coastal ecosystem. *Nature* 299:431–433.

SESSION 6
CRITICAL ISSUES IN
METHODOLOGICAL UNCERTAINTY

Chapter 15
Modeling the transport and fate of hydrophobic contaminants

Wilbert Lick

15.1 Introduction

Contaminated sediments are a major environmental problem in many marine, estuarine, and freshwater systems. This is often due to hydrophobic contaminants which are sorbed to and are transported by suspended solids. These contaminants are removed from the water column by the settling and deposition of the suspended solids onto the bottom sediments. Conversely, the bottom sediments can serve as a source of contaminants to the overlying water and biota because of the resuspension of contaminated bottom sediments or bioturbation and diffusion together with the subsequent desorption of the sorbed contaminants. Since the total amounts of contaminants in the bottom sediments can be much greater than the amounts of contaminants in the overlying water, the bottom sediments can serve as a major source of contaminants to the overlying water and biota long after discharges of contaminants into the system have ended. These contaminant source/sink processes are highly variable in both space and time and cause contaminant concentrations in the sediment bed which are also highly variable in space and time.

Investigations of sediment toxicity have generally been limited by a lack of understanding of the factors controlling contaminant availability in sediments. This availability depends on the transport and fate of the contaminants in the sediments and in the overlying water and can be significantly affected by remedial actions and by natural, large episodic events. A knowledge of the transport and fate of sediments and the contaminants associated with them is therefore necessary in order to quantitatively understand, predict, and minimize the environmental impact and risk of contaminated sediments. In this chapter, the modeling and prediction of the transport and fate of hydrophobic contaminants is discussed. The flux of contaminants between the sediments and the overlying water is emphasized. Our present understanding of specific processes that are essential to this modeling is presented first. These processes are a) the hydrodynamics; b) the resuspension and erosion of bottom sediments and the dynamics of the sediment bed; c) the flocculation, settling speeds, and deposition of suspended sediments; and d) the sorption of hydrophobic contaminants to particles. Recent models of sediment and contaminant transport and fate and applications of these models are then reviewed. The emphasis is on truly predictive models, _i.e.,_ models which are based on parameters that can be and are determined _a priori_ from laboratory tests and simple field process studies and therefore do not need extensive fine-tuning or calibration. These models can be more easily extended to different environmental conditions and different aquatic systems than models significantly dependent on calibration.

As mentioned above, natural systems are highly variable in both space and time. As an example of temporal variability, the flow rate in the Saginaw River from 1940 to 1990 is shown in Figure 15-1. Since the average flow rate is approximately 50 m³/s, it can be seen that the maximum flows are almost 40 times greater than the average. Other rivers have similar highly variable flow rates, while lakes have highly variable currents and wave action due to highly variable wind velocities. These large natural variabilities, together with the nonlinear dependence of sediment resuspension on the shear stress due to currents and wave action, lead to very large temporal and spatial variabilities in sediment and contaminant transport and fate. Because of this, it can be shown that large events such as storms on lakes and floods on rivers, despite their infrequent occurrence, are responsible for much, if not most, of the sediment transport in rivers and lakes. During these events, the flux of contaminants from the bottom sediments to the overlying water due to resuspension/deposition of sediments can be much larger than the contaminant fluxes due to bioturbation and molecular diffusion. As a result, large flow events are also responsible for most of the contaminant transport in surface waters (Lick 1992; Lick *et al.* 1994).

In general, the transport of sediments and contaminants is a very dynamic process, with the fluxes changing continuously in magnitude and direction. There is no steady state. Because of this and the nonlinearity of the processes involved in this transport, an average state is difficult to define and may not be meaningful. It is the time-dependent event, especially the large runoff and/or storm, that must be considered in the modeling and prediction of the transport and fate of sediments and contaminants. This is emphasized in the following sections.

15.2 Hydrodynamics

Currents and wave action are significant processes when considering sediment transport since a) they are directly responsible for the transport of sediments, b) they cause turbulence which disperses sediments both vertically and horizontally, and c) they cause a shear stress (primarily due to turbulence) at the sediment–water interface which is the primary cause of the resuspension of sediments.

Extensive work has been done and is continuing on the dynamics and especially the modeling of currents in rivers, lakes, estuaries, and oceans. The most general of these models are three-dimensional and time-dependent and include the conservation equations for mass, momentum, and energy (*e.g.*, Sheng and Lick 1979; Blumberg and Mellor 1980; Paul and Lick 1985; Mellor 1990). These models have been extensively applied to rivers, lakes, estuaries, and oceans when thermal stratification is not significant and, to a lesser extent, when thermal stratification is important (*e.g.*, Heinrich *et al.* 1981, 1983). Most three-dimensional models now include some form of a turbulence closure submodel (Mellor and Yamada 1982) in order to provide a more realistic parameterization of the vertical mixing processes in stratified and non-stratified situations. The determination of the appropriate parameters to use in these turbulence models, especially during

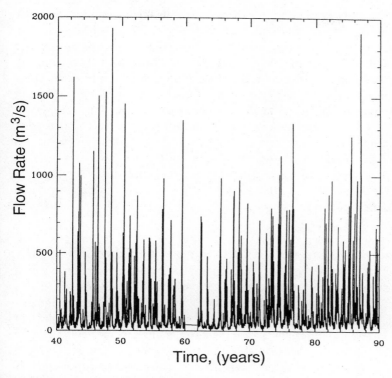

Figure 15-1 Flow rate, Saginaw River 1940–1990

big events, is still a matter of concern. Many of these hydrodynamic models have been coupled to some degree with water quality models, and their use in this context will be referred to below.

Difficulties with three-dimensional, time-dependent models are that they are inherently complex, consume large amounts of computer time, and are difficult to program and use. In many cases, simpler models such as two-dimensional (either vertically integrated or horizontally averaged), time-dependent models are more appropriate. These simpler models are especially useful for the general understanding of phenomena and the determination of the effects of variation in pertinent parameters; in many important situations, they give completely adequate and accurate results for practical applications. In particular, during large storms or runoffs, the waters in shallow bodies of water are very turbulent and well-mixed vertically. In this situation, vertical variations of density and concentrations of sediments and contaminants are not significant, and a vertically integrated model is very accurate. Applications of these two-dimensional, time-dependent models to water quality problems will also be referred to below.

Wave action is primarily responsible for the shear stress at the sediment–water interface and hence the resuspension of bottom sediments, and therefore needs to be known accurately. At present, the most widely used procedures for predicting wave parameters are semi-empirical methods, such as the PNJ method developed by Pierson, Neumann, and James and the SMB method developed by Sverdrup, Munk, and Bretschneider. These procedures give the significant wave height and significant wave period as a function of wind speed, fetch, and mean depth. From these relations and by assuming inviscid flow, one can determine the horizontal periodic flow at the sediment–water interface. Near this interface, a turbulent boundary layer exists (on the order of 10 cm thick for typical surface waves) and produces a shear stress at the interface. Kajiura (1968) has analyzed this problem based on the assumption of a time-independent but spatially varying vertical eddy viscosity. Recent work has extended the analysis to the case when waves and currents are present simultaneously (Grant and Madsen 1979; Christoffersen and Jonsson 1985).

15.3 Resuspension, erosion, and sediment bed dynamics

In order to predict the transport and fate of any hydrophobic contaminant, it is essential to understand the resuspension and transport of bottom sediments. The resuspension of bottom sediments is quite complex with the resuspension rate of bottom sediments strongly and critically dependent on the applied shear stress due to wave action and currents, horizontal location, depth below the sediment–water interface, and consolidation time. The types of sediment can change rapidly from coarse sands (in shallow, near-shore areas or where currents are strong) to fine-grained muds (in deeper areas where currents and wave action are small). Sediment type can also change vertically in the bottom sediments, as for example, when there is a layer of sand in between layers of fine-grained muds. Changes in sediment type can lead to changes in resuspension rates by several orders of magnitude. These rates can also change with time and distance from the sediment water interface as suspended sediments are deposited over previously deposited sediments which are more dense and more compacted. For these reasons, the dynamics of bottom sediments including resuspension, burial by suspended sediments depositing on the surface of bottom sediments, and the consolidation and dewatering of bottom sediments after deposition are all essential to understand.

Considerable work has been done on the transport of coarse-grained sediments by both bed load and suspended load. For summaries of this work, see Dyer (1986) and van Rijn (1984a, 1984b). Many formulas have been derived, but relatively little verification of these formulas has been made.

More recently, the emphasis has been on understanding and predicting the resuspension and deposition of fine-grained sediments. Laboratory experiments on reconstructed sediments (*e.g.*, see Partheniades 1972; Mehta 1973; Fukuda and Lick 1980; MacIntyre *et al.* 1990) have determined the dependence of resuspension on various governing parameters such as the applied shear stress, water content or time after deposition, particle size distribution, mineralogy, and numbers and types of benthic organisms. These experiments are usually done in annular flumes, although other devices are also possible. However, by their very nature, laboratory experiments deal with disturbed sediments. For an adequate understanding of sediment and contaminant transport, the resuspension properties of undisturbed sediments, as they are in the field, need to be known. For this purpose, a portable resuspension device (called a *shaker*) has been devised and applied (Tsai and Lick 1986, 1987; Lick *et al.* 1995).

Many contaminants are buried at depths of up to several meters in the bottom sediments of rivers, lakes, and estuaries. A major question is whether these buried sediments and the contaminants associated with them can be exposed and eroded during large floods and storms. In order to quantitatively understand and predict sediment and contaminant transport and fate during these large events, it is necessary to be able to measure the erosion properties of sediments at high shear stresses, on the order of 100 dynes/cm^2, and with depth, down to a meter or more. However, both the annular flume and shaker are limited to shear stresses below about 10 dynes/cm^2. Because of this, both of these devices are only capable of resuspending the surficial layers of the sediments, usually only the top few millimeters.

In order to determine the erosion properties of relatively undisturbed sediments at high shear stresses and with depth, a unique flume (Sedflume) has been devised, developed, and tested (McNeil *et al.* 1995). This is shown in Figure 15-2 and is such that box cores can be inserted into the bottom of the test section of the flume. By extrusion, the sediments in these cores can then be continuously exposed to a shear flow. Measurements at very high shear stresses, on the order of 100 dynes/cm^2, are possible. Tests have been made on reconstructed sediments in the laboratory and on cores of undisturbed bottom sediments from the Detroit River in Michigan and the Fox River in Wisconsin. Erosion rates of sediments from different locations at low to high shear stresses and as a function of depth down to 1 and 2 m have been measured. Examples of test results for cores from the Detroit River are shown in Figures 15-3a and 15-3b. Figure 15-3a shows the erosion rate as a function of depth for different shear stresses for a core consisting of a dark gray silt near the surface changing gradually to a silt and clay sediment further down in the core. Since the type of sediment changes relatively little with depth, the erosion rates vary relatively smoothly with depth. Figure 15-3b shows the erosion rates as a function of depth for a core which is highly stratified. A surface layer consists of coarse sands and other debris. Below this is a very firm layer of peat, followed by a layer of peat mixed with silt. The core ends in a firm clay layer. The peat layer and the clay layer are extremely difficult to erode, while the coarse surface layer and the layer of peat and silt are relatively easy to erode.

These are just two examples of cores already taken and tested. In general, it can be shown that the shear needed to erode sediments varies by more than an order of magnitude as the depth in the sediment increases and at different locations in these rivers. From these measurements, a reasonably accurate description of the resuspension properties of undisturbed sediments in these two rivers is being obtained.

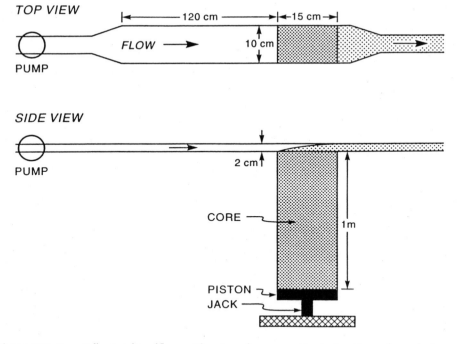

Figure 15-2 Essentially straight sedflume with rectangular cross-section test section and open bottom through which rectangular cross-section coring tube containing sediment can be inserted (By extrusion, sediments in these cores can then be continuously exposed to shear flow.)

15.4 Flocculation, settling speeds, and deposition

Fine-grained sediments and particles always flocculate to some extent. This flocculation affects the transport of the particles due to its effect on settling speeds and also affects the adsorption-desorption process. Extensive work has been done on the aggregation and disaggregation of sediments where the basic particle sizes are from 1 μm to 10 μm. From these basic particles, flocs can form which are up to several centimeters in size and which contain 10^6 to 10^8 basic particles.

In experiments using Couette and disk flocculators, the time-dependent behavior of these flocs has been investigated. In particular, the effects of fluid shear, sediment concentration, salinity (ionic strength), and organic matter on the aggregation and disaggregation rates; the steady-state equilibrium sizes; and the settling speeds of the flocs have been determined (Tsai and Lick 1987; Burban *et al.* 1989, 1990; Lick *et al.* 1993). Figure 15-4 shows some experimental results on the median diameter of flocs as a function of time in the case when differential settling is the dominant mechanism for flocculation. Initially, the flocs are disaggregated into the basic particles, which are about 6 μm in diameter. Due to differential settling, these particles then flocculate with time as shown and even-

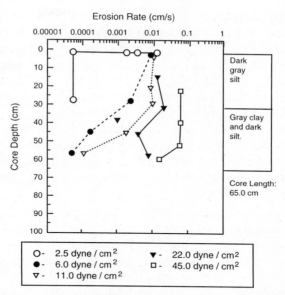

Figure 15-3a Erosion rates as function of depth with shear stress as parameter for sediment core from Detroit River: sediment silts and clays change relatively little with depth

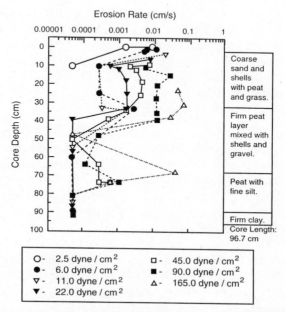

Figure 15-3b Erosion rates as function of depth with shear stress as parameter for sediment core from Detroit River: highly stratified sediments

tually reach a steady state. In general, it can be shown that the diameter of the steady-state flocs and the time to steady state increase as the sediment concentration, fluid shear, and salinity decrease. Theoretical analyses of the experimental results have also been made (Lick and Lick 1988; Lick *et al.* 1993). The analyses are based on a general formula for the time rate of change of the particle size distribution, which includes aggregation due to collisions and disaggregation due to fluid shear and collisions. The theory can accurately reproduce the experimental results of the median diameter of the flocs as a function of time and also the steady-state floc size distribution. The general theory has also been simplified so as to make it much more numerically efficient (Lick *et al.* 1992). From these studies, a fairly complete description of flocculation for a wide and realistic range of parameters (fluid shear, sediment concentration, salinity, concentration of organic matter) is now available.

Although a knowledge of settling speeds is essential in predicting transport in the overlying water, the deposition of particles on the sediment bed is not only determined uniquely by the settling speed but also depends on the shear stress (turbulence) at the sediment–water interface, *i.e.*, only large, dense particles and flocs settle out in a highly turbulent flow while almost all particles, except the finest, settle out in an almost quiescent flow. Little quantitative information is available on this dependence of deposition rate on size and density of the particles and the turbulence of the flow.

15.5 Sorption of hydrophobic organic chemicals to particles

In most previous work on the sorption of hydrophobic organic chemicals to particles, it has been assumed that chemical equilibrium exists and that this equilibrium can be quantified by means of a partition coefficient, K_p. However, in recent work, it has been noted that although the initial adsorption or desorption of a hydrophobic chemical can be quite rapid, with time scales of minutes to hours, the final equilibration may take days to months or even longer. A confusing factor in most experiments is the observation that the rates of both adsorption and desorption seem to depend on the concentration of suspended solids. This is called the solids concentration effect.

Considerable experimental work has been done on the sorption and partitioning of organic chemicals to sediments and saturated soils. In almost all cases, the experiments were short-term, hours to a few days. An extensive review of the work on sediments has been given by Di Toro *et al.* (1991), while work on soils has been recently summarized in two conference proceedings (Sawhney and Brown 1989; Baker 1991). Karickhoff and Morris (1985) present and discuss the results of long-term purge experiments on the sorption dynamics of several different hydrophobic pollutants in several different sediment suspensions. They demonstrated that, for many organic chemicals, sorption is a slow process and equilibrium might take days to months to achieve. In their experiments, a solids concentration effect was present and seemed to be greater for cohesive sediments than for non-cohesive sediments. Coates and Elzermann (1986) studied the desorption of PCBs from sediments. Their data indicated that equilibration times for PCBs in sedi-

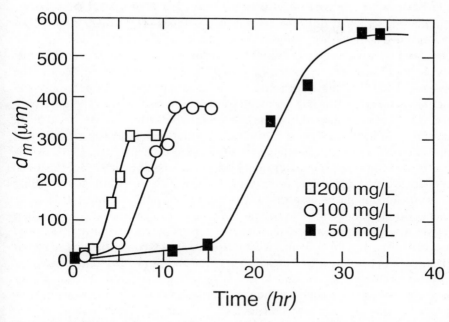

Figure 15-4 Median floc diameter as function of time with sediment concentration as a parameter
(flocculation due to differential settling of particles)

ments could be quite long, probably on the order of six weeks for PCBs with low chlorine
content and months to years for PCBs with a significantly higher chlorine content.
Gschwend and Wu (Gschwend and Wu 1985; Wu and Gschwend 1986) in their sorption
studies have shown that, for many experiments, the solids concentration effect could be
attributed to analytical artifacts caused by incomplete phase separations. Their experiments were short-term (1 to 3 days). Jepsen *et al.* (1995) have investigated the time-
dependent sorption and partitioning of hexachlorobenzene to natural sediments by
means of batch mixing experiments. Experiments were conducted for long times (2 to 5
months) in order to reach sorption equilibrium and at different sediment concentrations
(10, 100, 500, 2000, and 10,000 mg/L). Experiments quantitatively demonstrated the
time-dependent effects on sorption of hexachlorobenzene dissolution, of the flocculation
of colloidal matter in the water, and of the flocculation of the suspended sediments. From
the data, a true equilibrium partition coefficient (independent of solids concentration) as
well as the approximate times to equilibrium were determined. A model for the sorption
of hydrophobic chemicals to suspended sediments has recently been developed (Lick and
Rapaka 1996) which quantitatively describes both adsorption and desorption processes
and gives good agreement with experimental results.

15.6 Models of sediment and contaminant transport and fate

Many models of sediment transport have been developed. These models have been mainly concerned with the description of the suspended solids concentration, and little effort has been made to accurately describe the sediment bed and its properties, especially as the bed changes due to resuspension, deposition, and compaction. None of the models have included accurate descriptions of all the significant processes, especially flocculation. However, through experimental and field work, the processes governing sediment resuspension, transport, and fate have come to be better understood and quantified and are now being incorporated into more accurate and more general sediment transport models. A recent calculation is for the Fox River (Gailani *et al.* 1991). In this calculation, the authors used laboratory and field measurements to determine the resuspension parameters; a quasi-equilibrium model of flocculation for simplicity; settling speeds as measured in the laboratory; a two-dimensional (vertically integrated), time-dependent, hydrodynamic and sediment transport model; an SMB model of wave action when waves were significant; a three-dimensional, time-dependent sediment bed model with properties based on experimental work; suspended solids concentrations measured at the DePere Dam (the upstream boundary) as input; and suspended solids concentrations at the river mouth at Green Bay (the downstream boundary) as verification. Calculations of suspended solids concentrations as a function of time were made for three major runoff events and for steady flows at high, medium, and low flow rates. For the major runoff events, excellent agreement was obtained between the calculated and observed suspended solids concentrations. No data were available to verify changes in bathymetry due to resuspension and deposition.

However, changes in bathymetry over a 6-month period were measured for the Saginaw River by the USACE. For this same period, essentially the same model as described above (but extended to include curvilinear coordinates and bed load) was used to calculate sediment transport in the Saginaw including changes in bathymetry due to resuspension, deposition, and bed load (Cardenas *et al.* 1995). Good agreement between the observations and calculations was obtained. An example of a predictive calculation is given in Figure 15-5. This shows the predicted changes in resuspension and deposition for a 25-year period. Erosion occurs in the channel while deposition occurs in the shallow near-shore areas.

More recently, these calculations have been extended to include an investigation of the transport and fate of PCBs in the Saginaw River. The emphasis was on the effects of large flow events, incoming upstream PCB loads, and burial of contaminated sediments by clean sediments with subsequent erosion of sediments by a large flow event. In previous modeling of contaminant transport and fate, the dynamics of the sediment bed were such that the sediments and contaminants in the bed were generally well mixed vertically. This mixing was primarily due to the numerical procedures and approximations used. Because of this mixing, depositing contaminated sediments were mixed with clean sediments and hence diluted. These sediments, when subsequently resuspended, would then have a lower contaminant concentration and a lesser effect on water quality than if

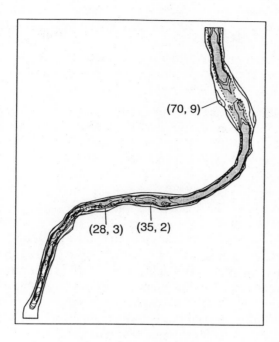

Figure 15-5 Resuspension and deposition: 25-yr simulation. Shaded area is erosional; unshaded areas are depositional. Resuspension contours from 0 g/cm² to 45 g/cm² in 15 g/cm² intervals. Deposition contours from 20 g/cm² to 100 g/cm² in 40 g/cm² intervals. Dashed line is zero erosion/deposition.

they had not been diluted. The extent of this dilution depends on the thickness of the sediment bed, a parameter which was generally somewhat arbitrarily specified. In the absence of benthic organisms, this mixing should not occur. Even when organisms are present and active, this mixing should not occur to a depth of more than a few centimeters, possibly as much as 10 cm. In the latest calculations, discrete layers of sediment (and the contaminants associated with them) can be resuspended and/or deposited without numerical mixing. This capability is essential in accurately predicting the long-term transport and fate of contaminants and nutrients and hence the water quality of rivers and lakes over the long term. Another interesting calculation is that of the resuspension and transport of sediments in Lake Erie (Lick *et al.* 1994). The emphasis in this article was on the effects of major storms. Calculations were made for different constant wind speeds and wind directions and also for the November 1940 storm (see Figure 15-6), one of the largest in the last century. It can be seen that, for this storm, up to a meter or more of sediment can be eroded from near-shore areas, while up to 20 cm of sediment can be deposited in other areas further off shore. In general, the numerical results indicate that major storms, despite their infrequent occurrence, are responsible for most of the resuspension and transport of sediments in Lake Erie. For purposes of verification, the results of numerical calculations were compared with ^{210}Pb and ^{137}Cs data from a sediment core obtained by Robbins *et al.* (1978) at an Eastern Basin location. The data indicates that deposition at this site was very nonuniform with time, with infrequent large depositions caused by major storms which were separated by long periods of time in which very little deposition occurred. The results of the calculated deposition are consistent with this idea.

Sediment-water fluxes occur primarily by a combination of three processes: resuspension/deposition, bioturbation, and diffusion. Each of these processes is distinctly different from the others. In general, these processes occur simultaneously and there are interactions among them. However, in many realistic situations, one of the processes is dominant over the others and so, to a good approximation, can be considered as independent. All of these processes are affected by the adsorption-desorption process for hydrophobic chemicals. Contaminant fluxes due to diffusion are generally small compared to fluxes due to resuspension/deposition and bioturbation. Bioturbation has often been assumed to be the dominant cause for contaminant fluxes from the bottom sediments. The justification for this has been the surficial layers in some areas of lakes which have fairly uniform composition. This uniform composition is deduced from radiometric dating and is attributed to benthic organisms, which presumably mix these layers by furrowing and burrowing. From a consideration of sediment dynamics, an alternative and more plausible explanation for this in many instances is that this layer of uniform properties is due to episodic resuspension/deposition events (Lick 1992).

15.7 Conclusions

Contaminated sediments can cause significant environmental problems in marine, estuarine, and freshwater systems. In order to quantitatively understand, predict, and minimize the environmental impact and risk of these sediments, a knowledge of the transport and fate of the sediments and the contaminants associated with them is necessary. In this chapter, the modeling of this transport and fate and the processes that are essential to this modeling were discussed. These processes are a) the hydrodynamics; b) the resuspension and erosion of bottom sediments and the dynamics of the sediment bed; c) the flocculation, settling speeds, and deposition of suspended sediments; and d) the sorption of hydrophobic contaminants to particles. A brief review of recent models of sediment and contaminant transport and fate was then given. From results of calculations and observations, it can be shown that the sediment–water fluxes of contaminants and the contaminant concentrations in the sediment bed are highly variable in both space and time. This variability must be taken into account in modeling and in assessing the exposure of organisms to contaminants. Because large flow events are responsible for most of the sediment and contaminant transport in surface waters, these large flow events must be considered in detail in the modeling.

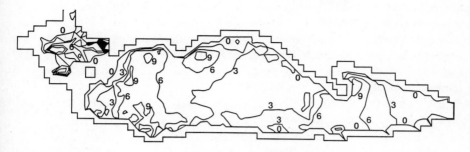

Figure 15-6 Net changes in sediment bed thickness in Lake Erie during major storms, November 1940

15.8 Acknowledgment

This research was supported by the U.S. Environmental Protection Agency.

15.9 References

Baker RA. 1991. Organic substances and sediments in water. Volumes 1, 2, 3. Chelsea MI: Lewis.

Blumberg AF, Mellor GL. 1980. A coastal ocean numerical model. In: Sundermann J, Holz KP, editors. Mathematical Modeling of Estuarine Physics, Proceedings of an International Symposium; 1978 Aug 24–26; Hamburg. Berlin: Springer-Verlag.

Burban P-Y, Lick W, Lick J. 1989. The flocculation of fine-grained sediments in estuarine waters. *J Geophys Res* 94(C6):8323–8330.

Burban PY, Xu YJ, McNeil J, Lick W. 1990. Settling speeds of flocs in fresh water and seawater. *J Geophys Res* 95:182313–18220.

Cardenas M, Gailani J, Ziegler CK, Lick W. 1995. Sediment transport in the lower Saginaw River. *Mar Freshwater Res* 46:337–347.

Christoffersen JB, Jonsson IG. 1985. Bed friction and dissipation in a combined current and wave motion. *Ocean Ingng* 12:387–423.

Coates JT, Elzerman AW. 1986. Desorption kinetics for selected PCB congeners from river sediments. *J Contam Hydrol* 1:191–210.

Di Toro DM, Zarba CS, Hansen DJ, Barry WJ, Schwartz RC, Cowan CE, Pavlov SP, Allen HE, Thomas NA, Paguin PR. 1991. Technical basis for establishing sediment quality criteria for nonionic organic chemicals by using equilibrium partitioning. *Environ Toxicol Chem* 10:1541–1583.

Dyer K. 1986. Coastal and estuarine sediment dynamics. New York: J Wiley. 343 p.

Fukuda M, Lick WJ. 1980. The entrainment of cohesive sediments in fresh water. *J Geophys Res* 85:2813–2824.

Gailani J, Ziegler CK, Lick W. 1991. The transport of sediments in the Fox Rivers. *J Great Lakes Res* 17: 479–494.

Grant WD, Madsen OS. 1979. Combined wave and current interaction with a rough bottom. *J Geophys Res* 84:1797–1808.

Gschwend PM, Wu SC. 1985. On the constancy of sediment-water partition coefficients of hydrophobic organic pollutants. *Environ Sci Technol* 19:90–96.

Heinrich J, Lick WJ, Paul J. 1981. The temperatures and currents in a stratified lake: A two-dimensional analysis. *J Great Lakes Res* 7:264–275.

Heinrich JC, Lick W, Paul J. 1983. Validity of a two-dimensional model for variable-density hydrodynamic circulation. *Math Modeling* 4:323–337.

Jepsen R, Borglin S, Lick W, Swackhamer D. 1995. Parameters affecting the adsorption of hexachlorobenzene to natural sediments. *Environ Toxicol Chem* 14:1487–1497.

Kajiura K. 1968. A model of the bottom boundary layer in water waves. *Bull Earthq Res Int* 46:75–123.

Karickhoff SN, Morris KR. 1985. Sorption dynamics of hydrophobic pollutants in sediment suspension. *Environ Toxicol Chem* 4:469–479.

Lick W. 1992. The importance of large events. Reducing uncertainty in toxic mass balance models, Great Lakes Monograph No. 4. Buffalo: State Univ of New York.

Lick W, Huang H, Jepsen R. 1993. The flocculation of fine-grained sediments due to differential settling. *J Geophys Res* 98(C6):10279–10288.

Lick W, Lick J. 1988. On the aggregation and disaggregation of fine-grained sediments. *J Great Lakes Res* 14:514–523.

Lick W, Lick J, Ziegler CK. 1992. Flocculation and its effect on the vertical transport of fine-grained sediments. *Hydrobiologia* 325:1–16.

Lick W, Lick J, Ziegler CK. 1994. The resuspension and transport of fine-grained sediments in Lake Erie. *J Great Lakes Res* 20:599–612.

Lick W, Xu Y-J, McNeil J. 1995. Resuspension properties of sediments from the Fox, Saginaw, and Buffalo Rivers. *J Great Lakes Res* 21:257–274.

Lick W, Rapaka V. 1996. A quantitative analysis of the dynamics of the sorption of hydrophobic organic chemicals to suspended sediments. *Environ Toxicol Chem* 15:1038–1048.

MacIntyre S, Lick W, Tsai CH. 1990. Variability of entrainment of cohesive sediments in freshwater. *Biogeochemistry* 9:187–209.

McNeil J, Taylor C, Lick W. 1995. Measurements of the erosion of undisturbed bottom sediments with depth. *J Hydraulic Engineering* 122:316–324.

Mehta AJ. 1973. Deposition behavior of cohesive sediments [thesis]. Gainesville: Univ of Florida.

Mellor GL. 1990. User's guide for a three-dimensional, primitive equation, numerical ocean model. Princeton NJ: Princeton Univ.

Mellor GL, Yamada T. 1982. Development of a turbulence closure model for geophysical fluid problems. *Rev Geophys Space Phys* 20:851–875.

Partheniades E. 1972. Results of recent investigations on erosion and deposition of cohesive sediments. In: Shen HW, editor. Sedimentation. Fort Collins CO.

Paul JF, Lick W. 1985. Numerical model for three-dimensional variable density. Rigid-lid hydrodynamic flows, Volume 1. Duluth MN: U.S. Environmental Protection Agency (USEPA) Environmental Research Laboratory.

Robbins JA, Edgington DN, Kemp ALW. 1978. Comparative ^{210}Pb, ^{137}Cs, and pollen geochronologies of sediments from Lakes Ontario and Erie. *Quarternary Res* 10:256–278.

Sawhney BL, Brown K. 1989. Reactions and movement of organic chemicals in soils. Soil Science Society of America (SSSA). SSSA Special Publication Number 22.

Sheng YP, Lick WJ. 1979. The transport and resuspension of sediments in a shallow lake. *J Geophys Res* 84(C4):1809–1826.

Tsai CH, Lick W. 1986. A portable device for measuring sediment resuspension. *J Great Lakes Res* 12:314–321.

Tsai CH, Lick W. 1987. Resuspension of sediments from Long Island Sound. *Nat Sci Tech* 20:155–164.

Van Rijn LC. 1984a. Sediment transport, Part I: Bed load transport. *J Hydraul Engineer, ASCE* 110:1431–1456.

Van Rijn LC. 1984b. Sediment transport, Part III: Bedforms and alluvial roughness. *J Hydraul Engineer, ASCE* 110:1733–1754.

Wu SC, Gschwend PM. 1986. Sorption kinetics of hydrophobic organic compounds to natural sediments and soils. *Environ Sci Technol* 20:717–725.

SESSION 6
CRITICAL ISSUES IN
METHODOLOGICAL UNCERTAINTY

Chapter 16
Sediments as complex mixtures: an overview of methods to assess ecotoxicological significance

Richard C. Swartz, Dominic M. Di Toro

16.1 The reality of mixtures of sediment contaminants

Sediments from many urbanized embayments contain hundreds of individual contaminants. The limit to the number of chemicals that can be detected in a sediment sample is essentially determined by the accuracy and precision of analytical methods. Risebrough (1994) routinely quantified 93 organic chemicals in a sediment survey in San Francisco Bay. Over 100 chlorinated hydrocarbon compounds alone were measured in some sediment samples from Puget Sound, Washington (Malins *et al.* 1980). Thus, the term *complex mixture* is especially applicable to sediment contaminants. The complexity of the mixture may never be completely understood because of limited funds and analytical methods. However, the reality of the mixture and its ecotoxicological significance must always be recognized. The obvious message is that multiple sediment contaminants have a potential for combined effects that might not be predicted by a risk assessment based on individual chemicals.

Two aspects of sediment contaminant mixtures that warrant special attention are co-occurrence and covariance. Certain compounds tend to have universal co-occurrence because of a common source or chemical relation, *e.g.*, DDT and its metabolites, PCB isomers, higher molecular weight PAHs. Co-occurrence of chemicals in a given study area often exists among diverse chemicals because of common, site-specific sources of contamination, *e.g.*, a sewage or industrial outfall. Chemicals with common sources tend to show spatial and temporal covariance in their distribution patterns (*e.g.*, the distribution of DDT and dieldrin in the Lauritzen Channel off Richmond Harbor CA, Figure 16-1; metals on the Palos Verdes Shelf CA, Figure 16-2). In depositional environments, chemicals may covary in sediment depth profiles (*e.g.*, the distribution of Cd, OC, DDE, and toxicity in sediment profiles from the Palos Verdes Shelf CA, Figure 16-3).

Co-occurrence and covariance of sediment contaminants often result in significant correlations of distribution among chemical, biological, and toxicological parameters (Ferraro *et al.* 1991; Canfield *et al.* 1996). These correlations do not signify causal relations. For example, the correlation of DDT and dieldrin with toxicity in the Lauritzen Channel does not provide sufficient evidence, by itself, that either chemical contributed to the observed toxicity (Swartz *et al.* 1994]. The distinction between effects and exposure is inherent in the fundamental risk assessment paradigm. Correlation of chemical con-

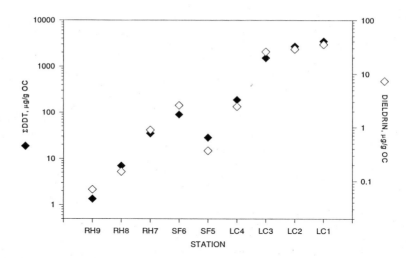

Figure 16-1 Covariance of ΣDDT and dieldrin in the Lauritzen Channel and Richmond Inner Harbor CA (after Swartz *et al.* 1994)

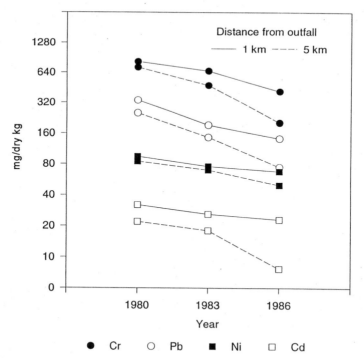

Figure 16-2 Spatial and temporal changes in metals concentrations in sediment on the Palos Verdes Shelf CA (after Ferraro *et al.* 1991)

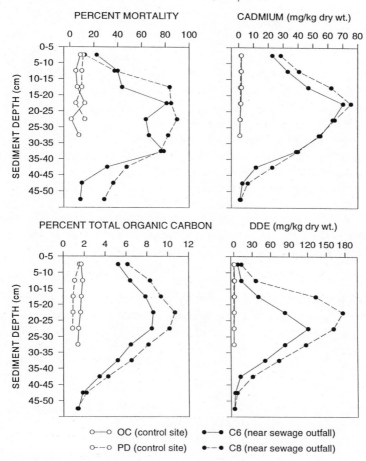

Figure 16-3 Vertical distribution of percent mortality of *Rhepoxynius abronius*, TOC, Cd, and p, p'-DDE in sediment cores from the Palos Verdes Shelf CA (after Swartz *et al.* 1991)

centration and biological response only establishes potential exposure. The effects assessment must be based on independent evaluation of causality.

16.2 Risk assessment based on toxic unit hypothesis

The simplest way to compare the hazard of individual chemicals in sediment contaminant mixtures is to normalize concentrations to an independently derived effects concentration or benchmark. The benchmark might be an SQG, an NOEC, or an LC50. In the latter case, the normalized concentrations are expressed as TUs: number of TU = ambient concentration/LC50 (Sprague 1970). This approach allows comparison of the relative potential contribution of individual chemicals to observed or predicted effects. It does not, by itself, indicate the cumulative effect of the combination of chemicals in the mixture.

The TU concentration of sediment contaminants can be expressed as either porewater TUs (*e.g.*, Kemble *et al.* 1994; Pesch *et al.* 1995) or OC-normalized sediment TUs (*e.g.*, Swartz *et al.* 1994). The latter approach was used to assess the relative contribution of PAHs, PCBs, dieldrin, DDT, and metals in the Lauritzen and Santa Fe Channels, CA. Sediments in both Channels were acutely toxic to marine amphipods (Pinza *et al.* 1992; White *et al.* 1993; Swartz *et al.* 1994). The mean sediment concentration of PAHs, PCBs, dieldrin, and metals in the Lauritzen Channel was $\leq$ 0.28 TU (Table 16-1). DDT and its metabolites had a mean concentration of 5.3 TU. On this basis, DDT was identified as the dominant ecotoxicological factor (Swartz *et al.* 1994). Although dieldrin and DDT were highly correlated with each other (Figure 16-1) and with toxicity, a comparison of their relative hazard in the Lauritzen Channel was made possible by the TU analysis. Similarly, PAH compounds (TU = 1.7) were identified as the major source of toxicity in the Santa Fe Channel because the TUs for ΣDDT, dieldrin, PCB, and metals were $\leq$ 0.24 TU (Table 16-1).

16.3 Metal mixtures: (SEM – AVS)

Acid volatile sulfide is the dominant binding phase for some divalent metals (Cd, Cu, Ni, Pb, Zn) in anaerobic sediment (Di Toro *et al.* 1990). Acid volatile sulfide is composed principally of solid phase, iron monosulfides (FeS(s)) that are soluble in cold acid and thus extracted as AVS. FeS(s) is in equilibrium with aqueous phase sulfide by the reaction

$$FeS(s) \leftrightarrow Fe^{2+} + S^{2-}.$$

Certain metals (Cu, Zn, Pb, Ni, Cd) react with solid phase AVS to form metal sulfide precipitates that are very insoluble in pore water:

$$Me^{2+} + FeS(s) \leftrightarrow MeS(s) + Fe^{2+}.$$

Thus, the iron in FeS(s) is displaced by metal to form soluble iron and solid metal sulfide. This displacement occurs because the metal sulfide solubility parameters of CuS, ZnS, PbS, NiS, and CdS are all less than that of FeS (Di Toro *et al.* 1990). As long as the molar concentration of AVS in sediment exceeds the combined molar concentration of Cu, Zn, Pb, Ni, and Cd, these five metals will form solid sulfides and have very low porewater concentrations.

The appropriate quantification of the combined molar concentration of these five metals is termed *simultaneously extracted metal* (SEM), *i.e.*, the metal which is extracted in the cold acid used in the AVS procedure (Di Toro *et al.* 1990). Thus, when SEM – AVS $\leq$ 0, porewater concentrations of all five metals will be quite low because of the formation and low solubility of their metal sulfides.

Sediment toxicity can be predicted from porewater concentrations of metals (Swartz *et al.* 1986, Kemp and Swartz 1988). Extensive research has shown that when SEM – AVS $\leq$ 0 (or SEM/AVS $\leq$ 1), porewater concentrations of metals are too low to cause toxicity to

Table 16-1   Toxic unit concentrations of organic carbon-normalized sediment contaminants in Lauritzen and Santa Fe Channels

Contaminant	TU value, LC_{50} = 1.0 TU (Reference)	Mean TU in Lauritzen Channel	TU in Santa Fe Channel
ΣDDT[1]	371 μg/g OC (Nebeker *et al.* 1989)	5.3	0.24
Dieldrin	1,955 μg/g OC (Hoke and Ankley 1991)	0.012	0.001
PCB[2]	2,600 μg/g OC (Swartz *et al.* 1988)	0.0026	0.008
ΣPAH[3]	ΣPAH Model (Swartz *et al.* 1995)	0.28	1.7
Metals[4]	SEM – AVS (Di Toro *et al.* 1990)	−52.8 μmol/g	−106.7 μmol/g

Source: Swartz *et al.* 1994

[1] Σ 2,4'-DDE; 4,4'-DDE; 2,4'-DDD; 4,4'-DDD; 2,4'-DDT; 4,4'-DDT
[2] Aroclor 1254
[3] See Figure 16-6 for model description.
[4] Cu, Zn, Pb, Ni and Cd. SEM – AVS is not a TU analysis. Negative values indicate a low probability of effects due to metals. See Section 16.3.

sensitive benthic invertebrates (Figure 16-4; Di Toro *et al.* 1990, 1992; Ankley, *et al.* 1991, 1993; Ankley, Leonard, and Mattson 1994; Ankley, Thomas, *et al.* 1994; Carlson *et al.* 1991; Casas and Crecelius 1994; Hare *et al.* 1994; Pesch *et al.* 1995). SEM – AVS thus provides a geochemical basis for predicting the cumulative effects of mixtures of metals in anaerobic sediment. If SEM – AVS > 0, metal concentrations in pore water will increase greatly and toxicity may occur unless other sediment phases (*e.g.*, OC) bind the metals.

16.4 Sediment toxicity identification evaluation methods

Toxicity identification evaluation is a method to identify contaminant classes and possibly individual compounds that are responsible for the toxicity exerted by effluent, water or sediment samples that contain complex chemical mixtures. Toxicity identification evaluation was originally developed for effluents and later applied to pore water extracted from sediments (Schubauer-Berigan and Ankley 1991). This overview of sediment TIE procedures is based on the recent review by Ankley and Schubauer-Berigan (1995). Toxicity tests by themselves cannot identify the chemical(s) that cause the observed effects. Toxicity identification evaluation uses toxicity-based fractionation procedures to implicate specific contaminants as causative agents; it is conducted in three phases to characterize (Phase I), identify (Phase II), and confirm (Phase III) compounds responsible for observed toxicity.

Phase I characterizes the physical-chemical properties of toxicants in a sample through manipulations designed to alter or render biologically unavailable generic classes of chemicals with similar properties (Ankley and Schubauer-Berigan 1995). The sample extractions and manipulations of Phase I are shown in Figure 16-5. Some of the manipu-

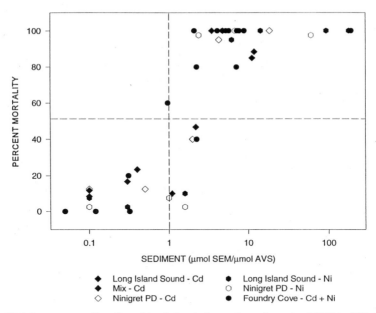

Figure 16-4 Percent mortality of amphipods in relation to the molar ratio of SEM to AVS of the sediment (after Di Toro *et al.* 1992)

lations are the addition of EDTA to implicate cationic metals; solid phase extraction of nonionic compounds in C_{18} columns; pH manipulations to detect toxicity due to ammonia and other ionic chemicals; and aeration to diminish volatile, oxidizable, or sublatable compounds. Changes in toxicity after these treatments indicate the possible contribution of different chemical classes.

Phase II uses a combination of analytical and toxicological methods to identify specific chemicals that may be responsible for observed toxicity (Ankley and Schubauer-Berigan 1995). Phase II chemical analyses are guided by the results of Phase I, *e.g.*, the Phase II analysis would focus on metals if EDTA chelation reduced toxicity. Toxicity tests may be used in single chemical exposures to determine if the measured concentrations in the TIE samples are sufficient to cause toxicity.

Phase III is conducted to confirm that the suspect toxicants identified in Phase II are, in fact, the actual toxicants (Ankley and Schubauer-Berigan 1995). The procedures used in Phase III include correlation between chemistry and toxicity, evaluation of relative species sensitivity to different chemicals, behavioral observations, spiking of samples or controls with suspect toxicants, mass balance of removed and recovered toxicity by filtration or other manipulations, and alteration of pH or other water characteristics that affect the toxicity of specific chemicals. Several of these procedures are typically used to provide the weight of evidence needed for a Phase III confirmation. Use of TIE methods is most effective when a single or limited number of contaminants is responsible for ob-

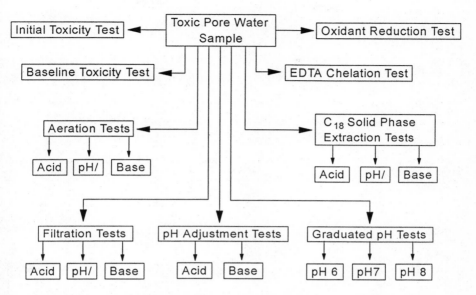

Figure 16-5 Conceptual overview of Phase I of a toxicity identification evaluation (after Ankley and Schubauer-Berigan 1995)

served toxicity. If many chemicals are involved, it can be difficult to discriminate effects of individual contaminants.

The adaptation of effluent TIE methods to sediments was initially based on pore water rather than whole sediments (Ankley and Schubauer-Berigan 1995). Chemical concentrations in pore water are correlated with whole sediment toxicity and the toxicity of extracted pore water can be tested directly. High speed centrifugation without subsequent filtration appears to be the best method for collecting pore water (Schults *et al.* 1992; Ankley and Schubauer-Berigan 1994). Research is currently directed toward the application of TIE methods to whole sediments.

16.5 Models of sediment–contaminant interactions: $\sum$PAH model

The $\sum$PAH model is an initial attempt to predict the sediment toxicity of mixtures of PAH compounds. The model estimates the probability of toxicity using a combination of EqP, QSAR, TU, additivity, and concentration-response models (Figure 16-6). The sediment concentration of OC and 13 PAH compounds are measured. Porewater concentrations (PAH_{iw}) of the 13 compounds are predicted by EqP. The 10-d LC50 of each compound in pore water (10-d $LC50_{iw}$) is predicted by a QSAR regression of 10-d $LC50_{iw}$ (from spiked sediment tests) to K_{ow} (octanol-water partitioning coefficient). Toxic unit concentrations of individual compounds (TU_i) are predicted as $PAH_{iw}/$10-d $LC50_{iw}$. The total number of TUs of the 13 compounds ($\sum TU_i$) is calculated by addition, assuming the additivity of

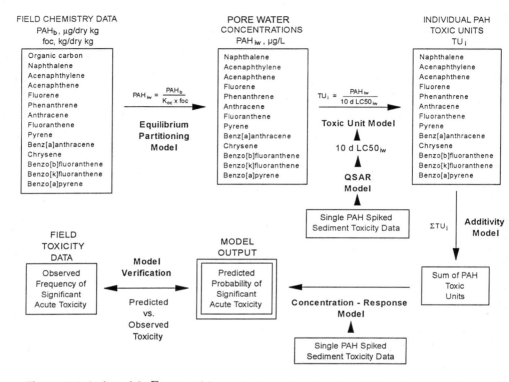

Figure 16-6 Outline of the ΣPAH model to predict the sediment toxicity of PAH mixtures to marine and estuarine amphipods (from Swartz *et al.* 1995)

toxic effects of PAHs. ΣTU_i is used to predict the probability of toxicity of PAH mixtures to marine and estuarine amphipods using a concentration-response model derived empirically from spiked sediment toxicity tests.

Verification of the ΣPAH model was based on comparison of predicted and observed toxicity at sites where PAHs were the major sediment contaminants and at sites where PAHs were relatively minor contaminants (Swartz *et al.* 1995). There was 86.6% correspondence and no significant difference between predicted and observed toxicity at the PAH-contaminated sites. Predicted and observed toxicity were significantly different at sites where PAHs were not the principal contaminants. When toxicity was observed but not predicted, the model provided a useful indication of the relatively minor contribution of PAHs to observed effects.

The ΣPAH model has a number of limitations and sources of error that need further research (Swartz *et al.* 1995). The model is not applicable to chemicals whose interaction is antagonistic or synergistic. The model does not account for changes in mechanisms of toxicity (*e.g.*, photoactivation of certain PAH compounds; Ankley, Collyard, *et al.* 1994). The model concerns acute (10-d) amphipod lethality. With the development of chronic amphipod toxicity tests, the model could be revised to address chronic effects and population dynamics. Additional data on 10-d porewater LC50s are needed to improve the accuracy of the QSAR regression used in the $\sum$PAH model. The principal limitation with respect to mixtures is the present restriction of the model to 13 PAH compounds. In theory, the model could be expanded to include all nonionic, narcotic compounds whose toxicological interaction is additive.

16.6 Models of sediment–contaminant interactions: QSAR/EqP model

Several unpublished models have been used to predict acute and chronic effects of mixtures of narcotic chemicals in sediments. The most parsimonious of these models is based on species-specific QSARs between aqueous LC50s and octanol-water partitioning coefficients (K_{ow}):

$$\log LC50 = a \log K_{ow} + b$$

where

a = slope,
b = y-intercept, and
LC50 is expressed in mol/m^3.

Through multilinear regression analysis, a common slope ($a = -0.83$) was established for the QSARs of 18 aquatic species. At the y-intercept of the QSAR regression, $\log K_{ow} = 0$ ($K_{ow} = 1$) and the equilibrium concentration of a compound in octanol is equal to the equilibrium concentration of a compound in water (*i.e.*, $\log LC50 = b$). Since octanol is a good surrogate for lipid, the y-intercept corresponds to the lipid-normalized LC50 of a narcotic compound. The y-intercept was therefore identified as the toxic endpoint.

Sediment guidelines could be established to protect 95% of aquatic species from acute mortality. The lower 5th percentile in the frequency distribution of toxic endpoints was 15.9 mmol/L. Thus, an acute narcosis baseline toxicity equation can be derived:

$$\text{Log } LC50_{5th} = (-0.83) \log K_{ow} + \log 15.9.$$

Equilibrium partitioning of nonionic organic compounds in sediment is a function of the sediment OC concentration according to the following equation (Adams *et al.* 1985; Di Toro *et al.* 1991):

$$C_s = f_{oc} K_{oc} C_d$$

where
C_s = sediment concentration (kg chem/kg dry weight),
f_{oc} = particle OC weight fraction (kg OC/kg dry weight),
K_{oc} = particle OC partition coefficient, (L/kg organic carbon) [$K_{oc} \sim K_{ow}$], and
C_d = dissolved concentration in pore water (kg chemical/L).

The EqP equation can be used to predict the sediment concentration at which the freely dissolved porewater concentration would equal the acute narcosis baseline concentration; *i.e.*, if $C_d = LC50_{5th}$, the EqP equation becomes

$$C_{s,5th} = f_{oc} K_{oc} LC50_{5th},$$

where

$\quad\quad C_{s,5th} \quad = \quad$ sediment concentration

when

$\quad\quad C_d \quad = \quad LC50_{5th}.$

The SQG for an individual chemical can be expressed as the OC-normalized sediment concentration that is not acutely toxic to 95% of the species:

$$SQG = C_{s,5th}/f_{oc} = K_{oc} LC50_{5th}.$$

Since the toxicological interaction of type I narcotic chemicals is additive, a sediment guideline for a mixture of n such chemicals can be expressed as the sum of the fraction of the OC-normalized sediment concentration divided by the SQG for each chemical:

$$SQG_{mixture} \overset{n}{\underset{i=1}{=}} \Sigma[(C_s/f_{oc})_{ambient}/SQG]_i.$$

When $SQG_{mixture} \geq 1$, the total concentration of type I narcotic chemicals in sediment is expected to exceed the level that will protect 95% of species from acute effects. A chronic SQG can be calculated by assuming an acute-chronic ratio of 10 for each chemical in the mixture.

The initial conceptual development of the QSAR/EqP model of sediment contaminant interactions has been completed. Experimental analysis and field verification are needed to assess the efficacy of this model.

16.7 Sediment guidelines for contaminant mixtures

Sediment toxicity tests are the principal tools currently used in regulatory programs to assess the effects of mixtures of sediment contaminants. The major strength of toxicity tests is that they reflect the cumulative interaction and toxic effect of all contaminant and other stresses imposed by sediment samples on test species. Their major weakness is that they cannot be used alone to identify the factors responsible for the toxic response.

Most numerical SQGs, including the proposed USEPA (1993a, 1993b, 1993c) sediment quality criteria, have been developed for single compounds. However, a few guidelines have been established for mixtures of compounds within chemical classes, especially PAHs (Table 16-2). The sediment mixture guidelines were developed using effects range (ER) (Long *et al.* 1995) and apparent effects threshold (AET) (Washington State Department of Ecology 1990; PTI 1991a, 1991b) methods applied to the sum of the concentrations of compounds in a particular chemical class. For example, the AET guideline for low

Table 16-2 Sediment quality guidelines for chemical mixtures

	Sediment concentration					
	(µg/kg, dry weight)				mg/kg OC	
Mixture	ERL[1]	ERM[2]	AET[3] Microtox	AET Amphipod	WASQS[4]	WACSL[5]
LPAH[6]	552	3,160	5,200	24,000	370	780
HPAH[7]	1,700	9,600	12,000	69,000	960	5,300
TPAH[8]	4,022	44,792				
TDDT[9]	1.58	46.1				
TPCB[10]	22.7	180	130	3,100		

[1] Effects range - Low (Long *et al.* 1995)
[2] Effects range - Median (Long *et al.* 1995)
[3] Apparent effects threshold for amphipod and Microtox toxicity tests (PTI 1991a)
[4] State of Washington Sediment Quality Standards (PTI 1991b, Washington State Department of Ecology 1990)
[5] State of Washington Cleanup Screening Level (PTI 1991b, Washington State Department of Ecology 1990)
[6] Low molecular weight PAHs
[7] High molecular weight PAHs
[8] Total PAHs
[9] Total DDT and its metabolites
[10] Total PCBs

molecular weight PAHs (LPAH) was based on the sum of the concentrations of naphthalene, acenaphthylene, acenaphthene, fluorene, phenanthrene, anthracene, and 2-methylnaphthalene in individual sediment samples.

The AET and ER mixture guidelines average about 75% of the sum of the individual guidelines (Washington State Department of Ecology 1990; PTI 1991a, 1991b; Long *et al.* 1995). In the case of the Cleanup Screening Level of the State of Washington (WACSL, Table 16-2), the mixture guideline for high molecular weight PAHs (HPAH = 5,300 mg/kg OC), actually exceeds the sum of the guidelines for the nine individual HPAH (Σ = 4,189 mg/kg OC; PTI 1991a, 1991b). Similarly, the ERL and ER-median (ERM) guidelines for total PAH exceed the sum of the guidelines for LPAH and HPAH (Long *et al.* 1995). This suggests that sediment contaminants act independently or antagonistically to one another, a conclusion that is inconsistent with evidence supporting the additivity of toxic effects of PAHs, metals, and other sediment-associated chemicals in sediment (Swartz *et al.* 1988, 1995; Di Toro *et al.* 1992; Ankley, Thomas, *et al.* 1994). If cumulative effects are additive, the concentration of individual chemicals in a sediment that exceeds the mixture guideline should be substantially less than the guideline concentration for the individual chemical by itself.

There are two explanations for this apparent paradox. First, the PAH measurements from which the mixture guidelines were derived may include concentrations of compounds for which individual guidelines were not developed. Second, because much (in some cases, all) of the data used to derive PAH guidelines were collected from field samples, the AET and ER guidelines for individual PAHs may reflect the effects of covarying chemicals, including other PAHs. If so, all of the guidelines for individual PAHs do not represent the effects of single chemicals by themselves. Rather, an individual PAH guideline represents the concentration of a single chemical in an unknown mixture of covarying chemicals whose cumulative toxicity is sufficient to exert adverse effects.

16.8 Research issues

We have identified a number of major research needs concerning the development and verification of methods to predict effects of mixtures of contaminated sediments. They including the following:

- Causality/correlation discrimination among covarying factors: Effects attributed to one chemical may actually be caused by another, covarying chemical. This issue must be resolved before reliable guidelines for single chemicals can be developed from field-collected sediment data.
- QSAR models for sediment toxicity: Accurate QSAR regressions are integral components of several models to predict toxicity of sediment contaminant mixtures. The QSARs should be developed directly from porewater LC50s.
- Additivity hypothesis: The assumption of additivity of toxic effects of narcotic and other chemicals in sediment must be verified. The interaction of nonadditive chemicals is poorly understood.
- Concentration-response model: Quantification of concentration-response relations is necessary to predict toxic effects.
- Toxicity identification evaluation methods for whole sediments: TIE methods need to be adapted to whole (solid phase) sediments to improve the direct relevance of TIE to standard sediment toxicity tests.
- Chronic effects of sediment contaminant mixtures: Initial research on the toxicity of mixtures of sediment contaminants has largely been restricted to acute effects. Chronic effects on growth, reproduction, and population dynamics need to be investigated.
- Interactions between chemical contaminants and other stresses: Noncontaminant stress factors like grain size and salinity can interact with chemical contaminants and affect biological responses. Effects of noncontaminant stresses should be examined carefully, especially during chronic exposures.
- Field verification of mixture model: Models of the interactions of contaminant mixtures have to be verified in relation to *in situ* biological and toxicological effects of contaminated sediments.

16.9   Acknowledgments

We thank Beth Power, Jack Word, Chris Ingersoll, Allen Burton, Mike Kravitz, and Peter Landrum for their critical reviews of the manuscript. The information in this document has been funded in part by the U.S. Environmental Protection Agency. It has been subjected to Agency review and approved for publication.

16.10   References

Adams WJ, Kimerle RA, Mosher RG. 1985. Aquatic safety assessment of chemicals sorbed to sediments. In: Cardwell RD, Purdy R, Bahner RC, editors. Aquatic Toxicology and Hazard Assessment: 7th Symposium. Philadelphia: American Soc for Testing and Materials (ASTM). STP 854. p 429–453.

Ankley GT, Collyard SA, Monson PD, Kosian PA. 1994. Influence of ultraviolet light on the toxicity of sediments contaminated with polycyclic aromatic hydrocarbons. *Environ Toxicol Chem* 13:1791–1796.

Ankley GT, Leonard EN, Mattson VR. 1994. Prediction of bioaccumulation of metals from contaminated sediments by the oligochaete, *Lumbriculus variegatus*. *Water Res* 28:1071–1076.

Ankley GT, Mattson VR, Leonard EN, West CW, Bennet JL. 1993. Predicting the acute toxicity of copper in freshwater sediments: evaluation of the role of acid-volatile sulfide. *Environ Toxicol Chem* 12:315–320.

Ankley GT, Phipps GL, Leonard EN, Benoit DA, Mattson VR, Kosian PA, Cotter AM, Dierkes JR, Hansen DJ, Mahony JD. 1991. Acid-volatile sulfide as a factor mediating cadmium and nickel bioavailability in contaminated sediments. *Environ Toxicol Chem* 10:1299–1307.

Ankley GT, Schubauer-Berigan MK. 1994. Comparison of techniques for the isolation of sediment pore water for toxicity testing. *Arch Environ Contam Toxicol* 27:507–512.

Ankley GT, Schubauer-Berigan MK. 1995. Background and overview of current standard toxicity identification evaluation procedures. *J Aquat Ecosys Health* 4:133–149.

Ankley GT, Thomas NA, Di Toro DM, Hansen DJ, Mahony JD, Berry WJ, Swartz RC, Hoke RA, Garrison AW, Allen HE, Zarba CS. 1994. Assessing potential bioavailability of metals in sediments: a proposed approach. *Env Management* 18:331–337.

Canfield PJ, Dwyer FJ, Fairchild JF, Haverland PS, Ingersoll CG, Kemble NE, Mount DR, La Point TW, Burton GA, Swift MC. 1996. Assessing contamination in Great Lakes sediments using benthic invertebrates and the sediment quality triad approach. *J Great Lakes Res* 22:565–583.

Carlson AR, Phipps GL, Mattson VR, Kosian PA, Cotter AM. 1991. The role of acid-volatile sulfide in determining cadmium bioavailability and toxicity in freshwater sediments. *Environ Toxicol Chem* 10:1309–1319.

Casas AM, Crecelius EA. 1994. Relationship between acid volatile sulfide and the toxicity of zinc, lead, and copper in marine sediments. *Environ Toxicol Chem* 13:529–536.

Di Toro DM, Mahony JD, Hansen DJ, Scott KJ, Carlson AR, Ankley GT. 1992. Acid volatile sulfide predicts the acute toxicity of cadmium and nickel in sediments. *Environ Sci Technol* 26:96–101.

Di Toro DM, Mahony JD, Hansen DJ, Scott KJ, Hicks MB, Mayr SM, Redmond MS. 1990. Toxicity of cadmium in sediments: the role of acid volatile sulfide. *Environ Toxicol Chem* 9:1487–1502.

Di Toro DM, Zarba C, Hansen DJ, Berry W, Swartz RC, Cowan CE, Pavlou SP, Allen HE, Thomas NA, Paquin PR. 1991. Technical basis for establishing sediment quality criteria for nonionic organic chemicals using equilibrium partitioning. *Environ Toxicol Chem* 10:1541–1583.

Ferraro SP, Swartz RC, Cole FA, Schults DW. 1991. Temporal changes in the benthos along a pollution gradient: discriminating the effects of natural phenomena from pollution-related variability. *Estuarine Coastal Shelf Sci* 33:383–407.

Hare L, Carignan R, Huerta-Diaz MA. 1994. A field study of metal toxicity and accumulation by benthic invertebrates; implications for the acid volatile sulfide (AVS) model. *Limnol Oceanogr* 39:1653–1668.

Kemble NE, Brumbaugh WG, Brunson EL, Dwyer FJ, Ingersoll CG, Monda DP, Woodward DF. 1994. Toxicity of metal-contaminated sediments from the upper Clark Fork River, Montana, to aquatic invertebrates and fish in laboratory exposures. *Environ Toxicol Chem* 13:1985–1997.

Kemp PF, Swartz RC. 1988. Acute toxicity of interstitial and particle-bound cadmium to a marine infaunal amphipod. *Mar Environ Res* 26:135–153.

Long ER, MacDonald DD, Smith SL, Calder FD. 1995. Incidence of adverse effects within ranges of chemical concentrations in marine and estuarine sediments. *Environ Management* 19:81–97.

Malins DC, McCain BB, Brown DW, Sparks AK, Hodgins HO. 1980. Chemical contaminants and biological abnormalities in central and southern Puget Sound. Boulder CO: National Oceanic and Atmospheric Administration (NOAA). NOAA Technical Memorandum OMPA-2.

Nebeker AV, Schuytema GS, Griffis WL, Barbitta JA, Carey LA. 1989. Effect of sediment organic carbon on survival of *Hyalella azteca* exposed to DDT and endrin. *Environ Toxicol Chem* 8:705–718.

Pesch CE, Hansen DJ, Boothman WS, Berry WJ, Mahony JD. 1995. The role of acid-volatile sulfide and pore water metal concentrations in determining bioavailability of cadmium and nickel from contaminated sediments to the marine polychaete, *Neanthes arenaceodentata*. *Environ Toxicol Chem* 14:129–141.

Pinza MR, Ward JA, Mayhew HL, Word JQ, Niyogi DK, Kohn NP. 1992. Ecological evaluation of proposed dredged material from Richmond Harbor. Sequim WA: Battelle/Marine Sciences Laboratory. PNL-8389.

PTI Environmental Services. 1991a. Pollutants of concern in Puget Sound. Seattle WA: U.S. Environmental Protection Agency (USEPA) Region 10. EPA 910/9-91-003. 107 p.

PTI Environmental Services. 1991b. Sediment management standards - sediment site ranking - SEDRANK guidance document. Olympia: Washington Department of Ecology, Sediment Management Unit.

Risebrough RW 1994. Organic contaminants in sediments and porewaters. San Francisco Bay Regional Water Control Board. Chapter 4, San Francisco estuary pilot regional monitoring program: sediment studies; p 4.1–4.52.

Schubauer-Berigan MK, Ankley GT. 1991. The contribution of ammonia, metals and nonpolar organic compounds to the toxicity of sediment interstitial water from an Illinois River tributary. *Environ Toxicol Chem* 10:925–939.

Schults DW, Ferraro SP, Smith LM, Roberts FA, Poindexter CK. 1992. A comparison of methods for collecting interstitial water for trace organic compounds and metal analyses. *Water Res* 26:989–995.

Sprague JB. 1970. Measurement of pollutant toxicity to fish. 2. Utilizing and applying bioassay results. *Water Res* 4:3–32.

Swartz RC, Cole FA, Lamberson JO, Ferraro SP, Schults DW, DeBen WA, Lee II H, Ozretich RJ. 1994. Sediment toxicity, contamination, and amphipod abundance at a DDT- and dieldrin-contaminated site in San Francisco Bay. *Environ Toxicol Chem* 13:949–962.

Swartz RC, Ditsworth GR, Schults DW, Lamberson JO. 1986. Sediment toxicity to a marine infaunal amphipod: cadmium and its interaction with sewage sludge. *Mar Environ Res* 18:133–153.

Swartz RC, Kemp PF, Schults DW, Lamberson JO. 1988. Effects of mixtures of sediment contaminants on the marine infaunal amphipod, *Rhepoxynius abronius*. *Environ Toxicol Chem* 7:1013–1020.

Swartz RC, Schults DW, Lamberson JO, Ozretich RJ, Stull JK. 1991. Vertical profiles of toxicity, OC, and chemical contaminants in sediment cores from the Palos Verdes Shelf and Santa Monica Bay, California. *Mar Environ Res* 31:215–225.

Swartz RC, Schults DW, Ozretich RJ, Lamberson JO, Cole FA, DeWitt TH, Redmond MS, Ferraro SP. 1995. ΣPAH: A model to predict the toxicity of field-collected marine sediment contaminated by polynuclear aromatic hydrocarbons. *Environ Toxicol Chem* 14:1977–1987.

[USEPA] U.S. Environmental Protection Agency. 1993a. Sediment quality criteria for the protection of benthic organisms: fluoranthene. Washington DC: Office of Science and Technology. EPA 822/R/93/012.

[USEPA] U.S. Environmental Protection Agency. 1993b. Sediment quality criteria for the protection of benthic organisms: acenaphthene. Washington DC: Office of Science and Technology. EPA 822/R/93/013.

[USEPA] U.S. Environmental Protection Agency. 1993c. Sediment quality criteria for the protection of benthic organisms: phenanthrene. Washington DC: Office of Science and Technology. EPA 822/R/93/014.

Washington State Department of Ecology. 1990. Final environmental impact statement for the Washington State Sediment Management Standards. Olympia WA. Chapter 173-204 WAC.

White PJ, Kohn NP, Gardiner WW, Word JQ. 1993. The remedial investigation of marine sediment at the United Heckathorn Superfund site. Sequim WA: Battelle/Marine Sciences Laboratory. Contract Report DE-AC06-76RL01830.

SESSION 6
CRITICAL ISSUES IN
METHODOLOGICAL UNCERTAINTY

_____Chapter 17
Workgroup summary report
on methodological uncertainty

Keith R. Solomon, Gerald T. Ankley, Renato Baudo, G. Allen Burton,
Christopher G. Ingersoll, Wilbert Lick, Samuel N. Luoma, Donald D. MacDonald,
Trefor B. Reynoldson, Richard C. Swartz, William J. Warren-Hicks

17.1 Introduction

In this chapter, a range of issues related to the uncertainty associated with SERAs is described, including an evaluation of 1) uncertainty associated with the overall SERA framework, 2) the effects of false positive and false negative errors associated with sediment toxicity tests, 3) spatial and temporal distributions of sediment contamination, 4) sampling errors, and 5) uncertainties associated with transport and fate models. Chapter 18 describes the uncertainty associated with specific measurement endpoints commonly used in SERAs and discusses approaches for addressing these sources of uncertainty.

The goal of any uncertainty analysis is to describe and interpret knowledge limitations that may be present in the measurement endpoints used to conduct a SERA analysis, for the purpose of incorporating estimates of uncertainty into management decisions. A number of viewpoints were discussed at the Workshop for defining uncertainty, two of which are described below.

In the first viewpoint, uncertainty is considered to be composed of two components: 1) measures of bias (*i.e.*, consistent deviation of measured values from the true value) and 2) measures of precision (*i.e.*, measure of agreement among replicable analyses of a sample). Accuracy is the combination of bias and precision for a procedure which reflects the closeness of a measured value to a true value. Figure 17-1 presents a visual interpretation of these components of uncertainty (Jessen 1978). Note that bias and precision are independent. For example, a method could have low bias and low precision (Figure 17-1b), or high bias and low precision (Figure 17-1c). Either combination leads to a decline in the overall confidence in the measurement.

Strictly defined estimates of accuracy are limited to formal experiments such as interlaboratory testing of a blind (but known) chemical concentration. In contrast, many sediment surveys are conducted without the benefit of knowing the "true" value (*i.e.*, accuracy of sediment toxicity tests with field-collected sediments). In these cases, estimates of field precision are limited and a weight-of-evidence is used as a surrogate to estimating bias. Strictly defined, precision is the observed variance of repeated measurements conducted under the same conditions (*e.g.*, the variance associated with repeated ponar grabs at the

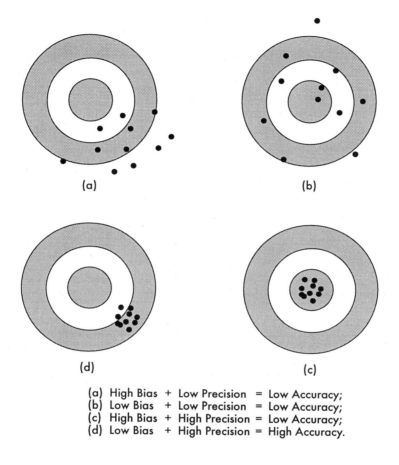

(a) High Bias + Low Precision = Low Accuracy;
(b) Low Bias + Low Precision = Low Accuracy;
(c) High Bias + High Precision = Low Accuracy;
(d) Low Bias + High Precision = High Accuracy.

Figure 17-1 Data precision, accuracy, and bias represented by shot patterns in targets (after Jessen 1978)

same location). In practice however, biological and chemical properties are very dynamic, making rigorous estimates of bias and precision difficult to obtain (see Section 17.2.4.3).

In the second viewpoint, uncertainty can be evaluated in the context of expert judgment and opinion in the analysis of uncertainty. While determination of accuracy and precision of management tools provides direct information for evaluating uncertainty, many methods are not amenable to this type of assessment. For this reason, less quantitative methods are often used to evaluate uncertainty, such as expert judgment (see Chapter 18). Although we may not have definitive numerical measurements on the ecological relevance of a specific measurement endpoint, a well-designed expert opinion survey can be used to generate knowledge relevant to the issue. Similar to numerical analysis, the larger the opinion survey the greater the information we have to assess uncertainty. In practice, expert opinion may be more available than well-conducted numerical analyses of uncertainty and can be a useful source of information. A large statistical literature is available on methods for generating expert opinion in a formal analysis of uncertainty (*i.e.*, Bayes theory; Chapter 13). Bayes theory can be used to combine both subjective and quantitative sources of information in a decision-making process. In the following sections, we use both of the viewpoints described above to discuss the sources of uncertainty and the implication of uncertainty in SERAs. We encourage scientists and policy-makers to consider uncertainty in risk-based decisions. We hope that by addressing the uncertainty issues, decision-makers will have valuable information available for weighing the various options available for risk reduction.

17.2 Uncertainty in the risk assessment framework

Guidelines have been developed for ERAs to promote consistency in analysis (Chapter 1). These guidelines also allow for the establishment of quality standards and consistent terminology for assessments. Consistency in the use of guidelines can help inform all stakeholders as to the relative degree of confidence and scientific knowledge under which the decision was made (Russell and Gruber 1987). Several guidelines are in use with varying degrees of consistency. Many of these methods are based on similar procedures and principles; therefore, the USEPA Framework for Ecological Risk Assessment (USEPA 1992a) was used in this chapter as a guideline (Chapter 1; Figure 1-1). Each of these areas is discussed in more detail below.

17.2.1 Problem formulation

Problem formulation is the planning or experimental design stage of the overall risk assessment process. Uncertainty in the formal statistical sense is of lesser importance at this stage of the process; however, there is qualitative uncertainty in the appropriate choice of assessment endpoints (objectives or purposes of the risk assessment) and measurement endpoints (indicators or tools used to evaluate risk or effects). The best way to reduce these initial uncertainties in the problem formulation is to involve all interested parties through stakeholder input. This involves asking all interested parties (including the risk managers, the scientific community, and the public) to define the problem in the form of a concise narrative. Once the problem has been identified, appropriate assessment and measurement endpoints can be selected. This chapter focuses primarily on uncertainty in relation to evaluating effects of chemical stressors. However, nonchemical stressors could be a dominant process influencing the system (*e.g.*, habitat disturbance) or other chemical or nonchemical stressors potentially will interact to produce perturbations (*e.g.*, ammonia and dissolved oxygen in the lower Fox River, Ankley *et al.* 1992; temperature and metals in the Clark Fork River, Kemble *et al.* 1994).

In the case of retrospective risk assessments (in some instances, termed *impact assessments*), identification of the stressors is a potential source of uncertainty. Identification of stressors should be part of the problem formulation stage and typically consists of 1) a survey of the natural and anthropogenic stressors which may be associated with the test area, 2) an assessment of the available data relative to quality control and quality assurance, and 3) hypothesizing potential stressors. In some cases, it may be necessary to make use of physical and chemical separation techniques (*e.g.*, toxicity-based fractionation methods) to identify specific classes of contaminant stressors. For example, extraction of pore water from sediment may allow partial identification of potential chemical stressors through TIE methods which use physical-chemical manipulations to affect the toxicity of specific contaminants of concern (Chapter 18). Uncertainties also exist in the identification of both assessment and measurement endpoints, an area where good professional judgment is valuable (Chapter 7; section 7.4).

17.2.2 Exposure characterization

The analysis phase of the risk assessment framework consists of two activities: characterization of exposure and characterization of effects (Chapter 1; Figure 1-1). The purpose of characterization of exposure is to predict or measure the spatial and temporal distribution of a stressor and its co-occurrence or contact with the ecological components of concern (USEPA 1992a). These uncertainties may influence the planning, execution, or interpretation stage of the exposure characterization. Primary sources of uncertainty in characterizing exposure include 1) laboratory imprecision, 2) matrix interference errors, 3) sample location biases, 4) sample collection and handling errors, 5) phase distribution of stressor, 6) contamination of the sample with other stressors, 7) spatial and temporal heterogeneity of the stressor, 8) references for comparison of stressor levels, 9) substrate type and interactions with stressor, 10) life history of the organism, 11) nonequilibrium of chemical stressor between sediment and water, 12) response model prediction error, 13) data transformation and normalization errors, 14) exposure pathway analysis errors, 15) sediment transport modeling assumptions, and 16) fate analyses errors.

17.2.3 Effects characterization

The purpose of characterization of effects is to identify and quantify the adverse effects resulting from exposure to a stressor (USEPA 1989). The 16 areas of uncertainties listed above for exposure characterization may also influence the planning, execution, or interpretation stage of the effects characterization. Additional sources of uncertainty in effects characterization may include 1) effects of noncontaminant stressors in toxicity tests, 2) reference comparisons to toxicity or receptor distributions, 3) laboratory-to-field extrapolations, 4) interpretation and definition of natural variability, 5) differences in receptor species sensitivity, 6) differences in physical alterations of the sediment, and 7) differences in stressor-response relationships.

17.2.4 Risk characterization

Risk characterization may either be prospective (*e.g.*, product hazard assessment as described in Chapters 3 and 4) or retrospective (*e.g.*, impact hazard assessment as described in Chapters 6 and 7). Risk characterization at the organismal level has traditionally been done by comparison of the concentration of the stressors found in the environment to the responses reported for those stressors in the laboratory, field, or by use of the literature. This risk characterization can be performed as described in the following sections.

17.2.4.1 Use of quotients for risk assessment

Risk quotients are simple ratios of exposure and effects. For example,

$$\text{Risk } \alpha \; = \; \frac{\text{Exposure concentration}}{\text{Effect concentration}}$$

Traditionally, the quotient method has been used to compare the effect concentrations for the most sensitive species of concern to the average, median, mean, or highest exposure concentration. In addition, these exposure concentrations may be compared to an effect concentration derived from toxicity tests. This assessment can be made more conservative by the use of safety (application) factors, such as division of the effect level by a number such as 20 (Canadian Water Quality Guidelines [CWQG] 1987). Use of safety factors allows for unquantified uncertainty in the effect and the exposure estimations or measurements. Because this uncertainty is unknown and unquantifiable, substantial errors are possible, both in underestimating or overestimating the risk.

In the absence of sufficient information from toxicity tests, these risk assessments may be underprotective. Conversely, where a wide range of toxicity data is available, the variation in receptor response may be well defined and further use of safety factors may be overprotective. Use of the quotient approach is acceptable for early tiers of the risk assessment, but the approach fails to consider the range of variation which may exist in terms of exposures and susceptibility (*i.e.*, Chapter 5 dealing with dredging assessments). Recently a method was proposed for using quotients in Tier I risk assessments, which included the incorporation of uncertainty in both the numerator and denominator of the quotient equation (Parkhurst *et al.* 1995).

17.2.4.2 Probabilistic risk assessment

A second approach for evaluating risk is to express the results of a refined risk characterization analysis as a distribution of toxicity values rather than a single point estimate (*i.e.*, Chapters 3 and 4 dealing with product assessment). For example, this approach has been proposed or is now being used by the Dutch government (Health Council of the Netherlands 1993), Cardwell *et al.* (1993), Baker *et al.* (1994), Solomon *et al.* (1996), and Klaine *et al.* (1996). A major advantage of the probability approach is that it uses all relevant single-species toxicity data and, when combined with exposure distributions, allows for quantitative estimation of the risks to receptors. However, the approach is only valid if endpoints used in the assessment are similar. For example, survival data would not be expected to be protective of reproductive effects. The degree of overlap of the exposure curve (drawn as a log-Pearson Type III distribution; McBean and Rovers 1992) with the effects curve can be used to estimate the probability that a certain percentage of receptors may be adversely affected for a percentage of occasions (*i.e.*, Figure 17-2). A similar approach has been used in the derivation of USEPA Water Quality Criteria (Stephan *et al.* 1985). With the use of overlapping distributions, there is an implicit assumption that protecting a certain percentage of species for a certain proportion of occasions will also preserve ecosystem structure and function.

Although this approach to risk characterization takes into account much of the variability with regard to the range of susceptibility in receptor species, it still embodies several uncertainties and limitations. For example, the choice of protection level (*e.g.*, 90% of species) may not be socially acceptable. Some may view 90% as being overprotective, whereas others may find this level of risk unacceptable, especially if the 10% of potentially affected species includes endangered species or other organisms of ecological, commercial, or recreational importance (see Chapter 11). Additionally, risks of persistent, bioaccumulative chemicals to species at the top of the food chain may not be sufficiently addressed by this approach (Baker *et al.* 1994; Section 18.5). In the situation where there is a desire to protect more sensitive receptors, these species could be identified and appropriate mitigation measures taken.

A further issue requiring consideration in probabilistic risk assessments is the number of data points required to define the distribution of receptor species for either acute or chronic effects. Additional test species and endpoints beyond those now applied for SERAs may be needed (Burton and Ingersoll 1994). In addition, there is a need for methods such as those proposed by Parkhurst *et al.* (1995) for calculating the degree of risk associated with exposures to multiple chemicals.

17.2.4.3 *Retrospective risk assessment*

Risk assessments based on measurement of current conditions are considered retrospective and typically do not forecast the expected change in risk due to remedial or mitigatory options or changes in the ecosystem in the future. Retrospective risk assessments rely on a number of techniques discussed in more detail in Chapter 5 (dredging assessment) and Chapters 6 and 7 (site cleanup assessment). These assessments may include measurement endpoints such as sediment toxicity tests, assessments of structural or functional changes in the benthic communities, cellular and molecular effects in the receptor species, or the presence of tissue residues of contaminants of concern (Chapter 18).

The use of multiple lines of evidence (weight of evidence) is particularly important in retrospective risk assessment (USEPA 1992b) and may also be useful in prospective analysis. For example, the probability that an effect on benthic community structure is the result of exposure to a chemical stressor is made more certain if the concentration of the stressor in the area is high enough to have caused the observed effect and also results in overt toxicity in laboratory toxicity testing.

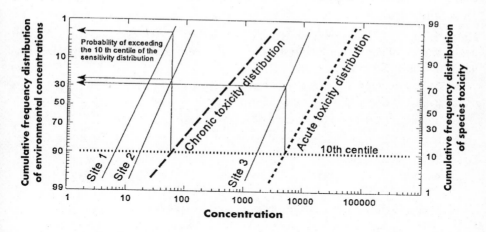

Figure 17-2   Graphical representation of use of probabilistic risk assessment with sediments. Cumulative frequency distributions of concentrations of stressors in sediments are compared with distributions of sensitive benthic organisms. Arrows show probabilities of not exceeding 10th percentile sensitivity concentrations for acute and chronic endpoints at 3 sites (adapted from Solomon *et al.* 1996).

17.2.5 Risk management

The outcome of all risk management actions should be either the acceptance or the reduction of the risk. Risk reduction involves many potential actions which range from the technical through the socioeconomic to the political. In undertaking risk management, it is necessary to take these actions:

- Decide which risks must be managed and in what priority. This requires that some method for measuring and comparing risks must be available (*e.g.,* Chapter 11 on ecological relevance).

- Maximize the reduction of risk for the available resources. This implies that a system must be in place for assessing the degree of risk reduction and for measuring its cost.

17.2.5.1 Uncertainty in prioritizing risks

In general, the first step in ranking risks for management involves evaluation of the harmful effects of the action associated with the production or release of the stressor. In the case of human health, this response may be expressed as a numerical risk. Even though the risk assessment process may have limitations, estimates of relative risk may be comparable if similar processes are used to derive the risks. An additional difficulty is presented by unquantifiable risks. This applies particularly to environmental risk which may have measurement endpoints of an aesthetic nature, such as reduced days of recreational fishing or reduced view. Endpoints of this type cannot be quantified in the same terms as, for example, fish mortality.

Harwell *et al.* (1992) proposed a method for evaluating and prioritizing risk to human health and the environment. The system is based on recognition of the issues raised earlier in this document including the following:

- Acknowledging that ecosystems are diverse
- Knowing that ecosystems respond to stress differently and that this response is governed by the type of ecosystem and the type of stressor
- Recognizing that a wide range of temporal, organizational, and spatial scales are involved
- Knowing that the measurement endpoints are relevant to the selected assessment endpoints
- Knowing the normal baseline behavior of the ecosystem
- Having good extrapolation techniques from laboratory and field measurement endpoints to the selected assessment endpoints
- Considering uncertainty in all of these issues

The risks to be prioritized are then separated into a series of components, which are ranked as follows:

- The potential magnitude of the risk. Magnitude is ranked on an ordinal scale of 5, ranging from low to high as follows: Low < Medium < High < Very High < Extremely High.
- The geographic extent of the risk. Extent is ranked on an ordinal scale of 3, ranging from low to high as follows: Local < Regional < Biosphere.
- The recovery time. Recovery time is ranked on an ordinal scale of 3, ranging from low to high as follows: Short (years) < Medium (decades) < Long (centuries).

These scores can then be combined and used for ranking purposes. However, because these ranks are based on expert assessment, they are subject to uncertainty and bias. As suggested above for problem formulation, uncertainty of qualitative assessments may be reduced by involving expert opinion polls and the stakeholders in the process.

17.2.5.2 *Uncertainty in assessing risk reduction strategies*

Many options for risk reduction may be available to the risk manager; however, there are generally two types of tools: technological and regulatory.

Technological tools for risk mitigation include a wide range of procedures, many of which are specific to the situation. In the case of sediments which are contaminated by effluent discharges, further treatment of the effluent before release is commonly applied in industrial settings. In the case of *in situ* contamination of sediments, many cleanup and disposal options are available (Francinques *et al.* 1985; International Joint Commission [IJC] 1988) once sediment has been identified as containing chemicals at concentrations posing a problem. The sediment can either be removed, stabilized, capped, treated *in situ*, or "no action" may be taken (Lynam *et al.* 1987; Grigalunas and Opaluch 1989). The remediation procedure or combination of procedures chosen is specific to the study area and depends on ecological, chemical, physical, engineering, economic, human health, and political considerations. Furthermore, source control and continued monitoring must be included with any remediation effort to avoid creation of new problem areas.

The regulatory tools which may be used for risk mitigation, in increasing order of effectiveness, are as follows:

- Provide better information and communication to prevent misuse of stressors that may contaminate sediments.
- Better control discharges and releases of stressors to levels which are judged to present a tolerable risk to the benthic community.
- Restrict the use and application of the stressor.
- Impose a total manufacturing ban on the stressor.

Uncertainties exist which affect the selection of technological or regulatory tools that should be used to mitigate risks. Uncertainty of knowledge (which technological options are available) is best addressed through expert opinion surveys and stakeholder consultations. Uncertainties in the degree of risk reduction are best assessed by reiterating the risk assessment procedure for all the appropriate exposure reduction strategies and then ranking these in terms of both the reductions in risk and the uncertainty in achieving these reductions. This matrix will allow informed choices to be made and the "trading off" of costs with risk reductions and the uncertainties of achieving these reductions.

17.2.5.3 *Uncertainty in assessing societal values*

If ecosystems are viewed as providing services to society, these services can be assessed to have an economic value. All components of the ecosystem can be assigned an economic value; however, this view has been criticized, particularly in the assigning of value to concepts such as species richness and diversity. In assigning economic value to ecological services, the implication is that these services, including physical capital (equipment and technology) or human capital (knowledge and skills), are interchangeable and can be traded in the same way as these commodities, for example, writing off the loss of a species for an increase in copper production. In addition, assignment of economic value is often restricted to only a few components of the system at risk and may ignore temporal and spatial interconnectedness of organisms, populations, and ecosystems. Uncertainty with respect to economic issues which should be considered is as follows (Harwell *et al.* 1992):

- Sustainability: Irreversible resource damage will undermine the sustainability of ecosystems (and by extension, human society). Thus, irreversible damage to an ecosystem should not be economically discounted over a period of years (as in the amortization of equipment and capital resources), as this devalues the importance of long-term environmental problems. Thus, regulatory agencies or politicians may relegate a problem to a lower level of importance because the effect will only be felt at some time in the future (*e.g.*, global warming).

- Willingness to pay: This assumes that market prices can be used to assess the tastes and preferences of society. The problem with this approach is that individuals and society may enjoy the services provided by ecosystems (*i.e.*, clean air, water, weather control, food chain maintenance, genetic diversity) without understanding them or even having knowledge of their existence. Thus, their willingness to pay for them or assign values to such services may be incorrect and inappropriate relative to their real ecological value.

- Multipliers: Economic analysis of benefits always includes multipliers (*e.g.*, developing a mine results in jobs for construction workers and a demand for materials). Multipliers should also be used in the risk side of the risk:benefit equation. For example, the loss of a benthic community may result in losses to fisheries, transportation, fishing equipment manufacturers, the accommodation industry, and marina operations.

17.3　Special issues of uncertainty in sediment ecological risk assessment

17.3.1　Decision-making with sediment toxicity measurement endpoints: exploring effects of false positive and false negative errors

Sediment risk assessments can be used in a variety of applications, including the assessment of relative risk between an impacted site and a reference site or the reduction in risk associated with a remediation action. A variety of chemical and biological measurement endpoints can be used in the assessment, including laboratory sediment toxicity tests and SQGs (see Sections 18.2, 18.3, and 18.7). For example, sediment toxicity tests are now used to evaluate the relative difference in organism survival or growth between sediments from reference areas and dredged material (Chapter 5 on dredging assessment; USEPA and USACE 1991, 1994). Test endpoints such as mortality or growth of organisms exposed in the laboratory to field-collected sediments are assumed to reflect the response of organisms in the field exposed to dredged material.

A key issue in the use of sediment endpoints within a regulatory or programmatic environment is the level of confidence in the results of this assessment. Using the above example, in dredging there is uncertainty in determining 1) the probability of stating that the reference area and the dredged material are different with respect to toxicity when in fact they are the same (false positive error) and 2) the probability of stating that the reference area and the dredged material are the same, when in fact they are different (false negative error). The power of the test is the probability of finding an impact when it occurs. In most regulatory applications we are interested only in a single-direction test — whether the toxicity of the dredged material is greater than the reference (*e.g.*, we ignore any information showing the reference toxicity is greater than the dredged area). This type of statistical approach to environmental decision-making achieves the goal of environmental protection. However, some problems do arise. For example, if enough samples are taken, the reference and dredged areas can always be shown to be statistically separable. The degree of separation in toxicity can be very small, but statistically significant. We could frame an alternative null hypothesis to detect a difference of an ecologically significant magnitude. While this has scientific appeal, the degree of difference representing an ecologically significant result could be long debated.

In classical statistical terms, the chance of false positive decisions is termed a Type I error (α), the chance of false negative decisions is considered a Type II error (ß), and power is $1 - ß$. Investigations typically focus on establishing rigorous Type I errors. For example, risk managers are often willing to risk a 5% chance that a Type I error occurs. However, Type II errors are often ignored, or no definitive Type II decision criteria are established. In classical hypothesis testing, balancing the potential occurrence of false positive and false negative results is a function of the number of samples collected and the variance of the sample mean. Type I and Type II errors are mathematically linked, for a fixed sample size and variance, so establishing α in the decision criteria determines ß (see Steel and Torrie 1980).

The interrelationship of Type I and Type II errors requires consideration of the relative importance of the risks associated with making false positive and false negative decisions. Risk assessment usually focuses on reducing the risk of false positive results associated with statistical analysis (by establishing a low α level). However, from an environmental protection perspective, emphasis should be placed on reducing the risk of making false negative decisions (*i.e.*, falsely concluding that an area is not contaminated when it actually is). For example, an environmentally conservative approach would emphasize identifying small differences between the reference and test areas. Therefore, it would be desirable to have a high chance of classifying a site as clean when it is clean (high power), and a small chance of falsely classifying the reference and test site are the same when they are different (small ß). In this conservative approach, one would rather make an error in judging a clean site as contaminated than in misclassifying a contaminated site as clean.

In contrast, an alternative approach would be to classify the test site different from the reference site only when the data provide a large degree of confidence in the decision. Therefore, one would want to reduce the error in stating that the reference site and test site are different when they are not. This is accomplished by establishing a small α, with a higher probability of false negative results.

17.3.1.1 *Case study: interlaboratory variability*
The chance of false positive and false negative results is a function of the number of samples and the variance of the test endpoint. As an example, we will examine variability in sediment toxicity tests. While many sources of error are associated with these tests (see Sections 18.2 and 18.3), the following example focuses only on uncertainty associated with interlaboratory variability. Interlaboratory variance (*i.e.*, round-robin or ring testing) has been extensively studied in whole effluent toxicity testing (Parkhurst *et al.* 1992; Warren-Hicks and Parkhurst 1992), and we will draw on this earlier work in this

analysis. Data for the analysis are obtained from a round-robin study of whole sediment toxicity tests (USEPA 1994a; ASTM 1995a–1995e; Burton *et al.* 1996).

A key issue in the use of any method is the number of tests required for a specified decision criteria. In this example, Burton *et al.* (1996) reported mean survival of *Chironomus tentans* in 10-d whole sediment toxicity tests. Data were generated from eight laboratories, each of which tested split samples of field-collected sediment using the toxicity test method described in USEPA (1994a). For one of the sediments evaluated in the study, the mean survival among the eight laboratories was 76% with a standard deviation of 27% (resulting in a coefficient of variation of 37%, which is considered acceptable interlaboratory precision; Burton *et al.* 1996). The data consisted of survival measurements reported by each of eight laboratories. From this information, the number of laboratories needed to achieve a specified decision criteria can be estimated, based on pre-specified probabilities of either false positive or false negative results. For any one laboratory, the reported survival response was the mean of 8 replicate tests, each replicate test consisting of 10 organisms. If the replicate data were available, we would have calculated the number of replicates required by each of the eight participating laboratories for prespecified decision criteria (the data were not available at the time of this analysis). For discussion purposes only, we use the laboratory mean data and present an analysis of interlaboratory variability. The method for estimating intralaboratory replicates is consistent with the following discussion.

Suppose that an investigator is faced with determining the sediment toxicity of a potentially impacted site. Also, assume that 90% survival has been established as the acceptable control response. Given that the investigator has no prior knowledge of the site toxicity, an assumption was made that the sediment is about as toxic as that in the above referenced Burton *et al.* (1996) study. Alternatively, the investigator could conduct a pilot study of the site instead of making this assumption. Given these data, the investigator wishes to determine the number of laboratories necessary to achieve a specified decision criterion based on the chance of false positive and negative results. (Note: A somewhat related concept is minimum detectable difference [MDD] [USEPA and USACE 1991, 1994]. The MDD is generally used to establish the minimum difference detectable between a control and response solution for a single toxicity test, given a fixed sample size and variance. This concept may be adaptable to our example by evaluating the MDD between a reference site and a dredge site, for a fixed number of laboratories with known interlaboratory variance.)

The following equation provides a means of determining the desired number of laboratories, while balancing the chance of false negative and false positive test results:

$$N = \sigma^2 \left[\frac{Z_{1-\beta} + Z_{1-\alpha}}{C_s - \mu_1} \right]^2$$

where

N = the number of laboratories required to meet specified levels of Type I and Type II errors,

$Z_{1-ß}$ = the critical value of $1 - ß$ for the normal distribution (*e.g.*, 1.64 for a 1-sided test with a $ß = 0.05$ error probability),

$Z_{1-\alpha}$ = the critical value of $1 - \alpha$ for the normal distribution (*e.g.*, 1.64 for a 1-sided test with an $\alpha = 0.05$ error probability),

C_s = the specified standard (*i.e.*, 90% survival), and

μ_1 = the average percent survival across laboratories.

Figure 17-3 presents a plot of the power of a 1-sided test of the null hypothesis

$$H_0: Cs = \mu_1$$

against the alternative hypothesis

$$H_1: Cs > \mu_1.$$

(Note: because we are only concerned if the toxicity of the site is less than the standard, a 1-sided test of the hypothesis is appropriate.)

Notice that fixing either the power of the test $(1 - ß)$ or the Type I error rate (α) establishes the other. For example, with a false positive and false negative error probability of 5%, 33 laboratories are required for testing. With a false positive and false negative error rate of 10%, 20 laboratories are required for testing.

17.3.1.2 Summary

The above example demonstrates a method for estimating the number of laboratories, given Type I and Type II errors. The choice of how much error is acceptable is up to the investigator. The investigator should carefully consider the relative merits and interpretations of Type I and Type II errors when evaluating the results from any sediment measurements used to establish the possibility of contamination.

17.3.2 Uncertainty in estimating spatial and temporal distributions of contaminants in sediment

Sediments may be highly variable on both a spatial and a temporal basis. Therefore, replicate samples need to be collected at each site to determine variance in sediment characteristics. Sediment should be collected with as little disruption as possible; however, subsampling, compositing, or homogenization of sediment samples may be required for some experimental designs (*e.g.*, USEPA 1994a, 1994b; ASTM 1995a–1995e; Environment Canada 1996a, 1996b). Sampling locations might be distributed along a known pollution gradient, in relation to the boundary of a disposal area, or sampling locations may be identified as being contaminated in a reconnaissance survey. These comparisons can be made in both space and time. In pre-dredging studies, a sampling design can be developed to assess the contamination of samples representative of the project area to be dredged (Chapter 5). Such a design should include subsampling cores taken to the project depth.

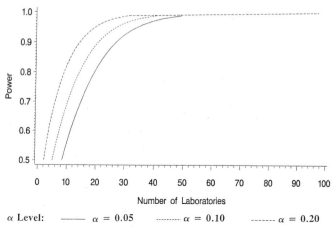

Figure 17-3 Number of laboratories as a function of Type I error and power

When dealing with a given sampling area (*i.e.*, river, lake, estuary), the appropriate sampling design is of importance since the goal might be to describe existing conditions for the entire area by collecting a discrete number of samples. The choice of the sampling scheme is also important for the intended data manipulation. Fewer samples are needed if the objective of the study is just to describe the average conditions over the entire area. On the other hand, if the sampling is to be used to draw a map of distribution (*i.e.*, highlight point sources, trends of distribution, location and area of contamination), the choice of the sampling net (regular, random, or fixed grid) may dictate which type of mapping system has to be used (Baudo 1990). In addition, if temporal variability is expected, sampling should be repeated as many times as possible to reduce this source of uncertainty. Sampling frequency is particularly critical since the timing of the successive samples can only be used to evaluate the change on the selected time scale.

17.3.2.1 *Estimation and measurement of magnitude of uncertainty*

The overall uncertainty of sampling depends on several factors, including sample 1) type, 2) volume, 3) equipment, 4) handling, 5) number, and 6) replicates.

Sample type. Sediment is a complex mixture of solid, aqueous, and gaseous phases, in addition to biotic compartments. Hence, study objectives must clearly indicate which type of medium is to be sampled. For example, different methods may be needed to subsample pore water in sediment or to sample benthic organisms in sediment. On the other hand, the objective of the study may be to sample the whole "active" layer (*e.g.*, to calculate a diversity index) or sample the vertical microstructure of some sediment characteristic (*e.g.*, the vertical profiles for redox or oxygen). Once the type of sample required has been identified, the choice of the sampling gear must be made accordingly. It is often difficult to determine the appropriate depth of sediment to sample (*e.g.*, How deep must a core be? How deep is the bioturbation? Where is the boundary between the oxic and the anoxic layers?). A common mistake is to sample at the maximum depth in the sediment which will potentially provide an unrealistic estimate of exposure (*i.e.*, sampling below the biologically active zone). Finally, the performance of the selected sampling gear is not constant (depending on the kind of substrate and the operating conditions, including skill of operators). In most cases, when sampling is made without actually seeing what is being collected, the uncertainty can be assessed only after the sample is recovered (*e.g.*, via visual observation of core length, texture, or color) or processed in the laboratory. As a consequence, the degree of uncertainty is different when sampling is done blind or is done visually by checking the performance of the sampling gear (Tables 17-1 and 17-2).

Sample volume. Some variables may exhibit pseudo-continuous distribution in space (*e.g.*, grain size where particles are sorted according to the hydraulics of the system), whereas other variables often have a pronounced patchiness (*e.g.*, benthic invertebrate distributions). Hence, the sample volume should account for the known or estimated local micro-spatial heterogeneity (both horizontal and vertical) at the scale of the sampling tool. Larger samplers will provide an "averaged" sample with an increased uncertainty of measuring smaller-scale heterogeneity. If variables with substantially different distributions have to be measured, repeated sampling with different samplers should be considered. Subsampling of sediments is often required; therefore, samples should be thoroughly homogenized before splits are made (USEPA 1994a; ASTM 1995a). The amount of sediment required for analyses can range from a few micrograms (*e.g.*, carbon-hydrogen-nitrogen [CHN] analysis) to several kilograms (*e.g.*, radioactive isotopes, laboratory toxicity tests). Hence, the minimum amount of sample needed to perform all analyses must be estimated before choosing the sampling equipment and may limit the number of measurements performed on a sample. For example, if a coring tube is used to sample sediment, the diameter and the thickness of sections need to provide enough material for all of the planned analyses must be calculated. The same sample is typically used for more than one analysis. Hence, a compromise would be to use the original sections for measurements requiring small aliquots and to use pooled sections for measurements requiring larger aliquots. It should be noted that this procedure potentially limits subsequent statistical analysis of data (*e.g.*, correlations, principal component analysis, cluster analysis) which can be done only with paired data. A special case where the sample volume is particularly critical is evaluating porewater composition of sediment. In this case, the water content of the sediment must be estimated in advance to be sure enough water can be extracted from each sample.

In summary, to increase confidence and lower uncertainty, the sample volume must be large enough to provide a representative sample for each measurement endpoint of interest, enough material to perform all analyses plus further measurements that may be needed (*i.e.*, TIE; Chapter 16). Often these two requirements are difficult to achieve due to lack of knowledge and the need to minimize the sampling effort. However, it should be kept in mind that collection of additional sample material will result in a different sample.

Table 17-1 Degree of uncertainty in blind sampling

Sample	Knowledge	Systematic errors	Random errors
Type	L	L	L
Volume	L	L	M
Equipment	L	H	H
(nonconventional)	H	H	H
Handling	L	L	M
Number	H	H	H
Repetitions	H	L	H

L= low, M =medium, H = high

Table 17-2 Degree of uncertainty in visual sampling

Sample	Knowledge	Systematic errors	Random errors
Type	L	L	L
Volume	L	L	L
Equipment	L	L	L
(nonconventional)	H	H	H
Handling	L	L	M
Number	H	L	H
Repetitions	H	L	L

L= low, M =medium, H = high

Sampling equipment (dredges, grabs, corers). All types of sampling equipment vary in performance, and each device has specific advantages and disadvantages (*i.e.*, size, weight, triggering mechanisms; ASTM 1995b). Few comparisons have been reported dealing with sampling efficiency between types of equipment (Baudo 1990); however, different types of dredges, grabs, and corers provide unique types of samples. Thus, choice of equipment for both whole-sediment and porewater sampling depends on the study objectives, measurement endpoint of interest, characteristics of the study area, sediment type and compactness, and the presence of interfering flora or fauna (*e.g.*, roots, shells) (Baudo 1990; Adams 1991; Mudroch and MacKnight 1991; ASTM 1995b). Moreover, allowance should be made for the different performance of the same sampler depending on the environment in which it is used (*e.g.*, soft bottom or sand). The equipment can alter or contaminate the sample (*e.g.*, metal or plastic in contact with sample, cleaning of the sampler) or the equipment may produce artifacts due structural limitations (*e.g.*, washing out of finer material, gas or temperature changes). Mudroch and MacKnight (1991) and ASTM (1995b) provide additional details regarding sampling equipment.

In addition to dredges, grabs, and corers, a number of nonconventional sampling tools have been applied for specific purposes, including peepers (dialysis chambers for collection of pore water), sedimentation traps, and artificial substrates (for benthic colonization studies). Although these and other nonconventional tools may provide useful information, lack of standardization may lead to a high uncertainty with their use. To summarize, any sampling equipment may introduce a marked uncertainty since it is largely unknown whether a representative sample has been collected.

Sample handling. Techniques of sample conservation and manipulation should be carefully examined using specific equipment to prevent not only the contamination of the samples but also possible alterations (Mudroch and Bourbonniere 1991; ASTM 1995b). Alteration of the sediment usually remains undetected unless specific studies and designs are used (*e.g.*, repeated measures at different times or after each sampling step). Hence, identification of new handling artifacts may make results of previous studies questionable. In order to minimize uncertainty, a consensus method should be established and followed; however, periodic revisions may be required.

Number of samples. Switzer (1975) pointed out a common statistical problem: the estimation of the required number of sampling points can be determined statistically only after the data have been gathered, or if some estimate of crucial sediment characteristics are obtained in preliminary studies. A number of approaches can be used to estimate the minimum number of samples needed to estimate an average value for the measurement of interest (Baudo 1990). Information needed to determine the number of samples includes heterogeneity in physical and chemical data (Kratochvil and Taylor 1981; Sokal and Rohlf 1981; Hakansson and Jansson 1983). The required number of samples also depends on the distribution of the data (normal, Poisson, negative binomial distributions). A detailed description of the statistical properties of these distributions can be found elsewhere (Bliss and Fisher 1953; Sokal and Rohlf 1981).

For the definitive sampling of an area, additional considerations including directionality and point sources may dictate the sampling plan. The sampling strategies in these cases can be classified in three main types (Hakanson and Jansson 1981; Baudo 1990): 1) a deterministic system, with a sampling design based on previous information and varying density; 2) a stochastic system, when the sampling stations are randomly selected; and 3) a regular grid system, with the sampling stations randomly or deterministically selected. Advantages and disadvantages of each method are discussed in Baudo (1990).

Repetitive sampling. More often than not, only one sample per station is collected. Repetitive sampling allows for an estimate of the local spatial or temporal heterogeneity (USEPA 1994a, 1994b; ASTM 1995a–1995e). An accurate estimation of the sampling variability (within the station and among stations) is needed to avoid false positive or

288   ECOLOGICAL RISK ASSESSMENT OF CONTAMINATED SEDIMENTS

false negative decisions and assumes an even greater importance when assessing spatial or temporal variability. The obvious disadvantages of repeated sampling are the increased cost and time although the increased costs of collecting two or more samples from the same station can be quite low, especially if a multi-sampler can be used. However, it will be much more expensive to perform all of the analyses on each sample. Alternatively, a representative measure could be made on all sample pairs to evaluate local heterogeneity (*i.e.*, CHN analysis which may be related to distributions of metals and organic contaminants).

An extrapolation to biological variables may be more difficult. Samples for biological measures are often pooled to provide an average of the local variability. For example, if the purpose of the study is to conduct a reconnaissance survey to identify contaminated areas for further investigation, the experimental design might include collection of just one composited sample from each area to allow for sampling a larger area. The lack of replication at an area usually precludes statistical comparisons (*e.g.*, ANOVA), but these surveys can be used to identify contaminated areas for further study or these data can be evaluated using statistical regressions (USEPA 1994a, 1994b; ASTM 1995a). In other instances, the purpose of the study might be to conduct a quantitative sediment survey to determine statistically significant differences between control (or reference) sediments and test sediments from one or more areas. The number of replicates per site should be based on the need for sensitivity or power. In a quantitative survey, field replicates (separate samples from different grabs collected at the same area) would need to be taken at each site. Separate subsamples from the field replicates might be used to determine within-sample variability or for comparisons of test procedures (*e.g.*, comparative sensitivity among test organisms), but these subsamples cannot be considered true field replicates for statistical comparisons among areas.

17.3.2.2 Accounting for and reducing uncertainty

MacKnight (1991) identified the following factors as most important to identifying sampling options: 1) purpose of sampling, 2) study objectives, 3) historical data and other available information, 4) bottom dynamics at the sampling area, 5) size of the sampling area, and 6) available funds versus estimated (real) cost of the project. Factors 1 and 2 are obviously critical and must be agreed upon in advance between managers and scientists involved in the project. This could easily be the most important source of uncertainty, since "there is no one formula for design of a sediment sampling pattern which would be applicable to all sediment sampling programs" (MacKnight 1991). Since inadequate strategies and unclear goals of sediment sampling are among the most important sources of uncertainty, the early involvement of managers in the sampling plan should be sought.

In addition, to reducing the sampling bias, the proposed project should be peer reviewed by scientists with expertise in each one of the fields of study covered by the project. This peer review should evaluate the adequacy of the sample media, sample volume, sampling equipment, sample handling, sample number, and repetitions. Information related to Factors 3 (historic data) and 4 (dynamics of bottom sediment) listed above are often not available to assist in the selection of sampling areas (number and location) with the required degree of confidence. This information is needed to assure an unbiased assessment of Factor 5 (size of the sampling area) and Factor 6 (costs). In any case, the choice of sampling plan should support a statistical evaluation of the sampling variability (both spatial and temporal). Hence, a pilot study is usually needed to establish local spatial and temporal heterogeneity before the definitive sampling is performed. Whenever feasible, *in situ* tools should be used in the pilot study to estimate distributions in relevant physical, chemical, and biological measurements (*e.g.*, sediment compactness, echo-sounding, pH, oxygen, redox, biotic communities). The extrapolation to field conditions from data gathered on samples transferred to the laboratory is always subject to uncertainty. For this reason, there is a need for developing *in situ* techniques for measuring chemistry, toxicity testing, and benthic community. Alternatively, uncertainty associated with sampling could be substantially lowered by visually checking the sampling sites (*e.g.*, by using divers, submersible cameras, manned vehicles; Tables 17-1 and 17-2).

17.3.2.3 *Interpretation of uncertainty in relation to decision-making*
Assuming all samples are used for analysis, the number of samples determine the cost of the study. Hence, there is a need to limit the number as much as possible and still retain confidence that the final results will be both sound and defensible. Too few samples will result in large variability and may result in the need for additional sampling, whereas too many samples will result in a waste of resources. On the other hand, the uncertainty associated with measurement endpoints outlined in Chapter 18 should be weighed relative to the cost-benefit analysis for the potential remedial options. An overestimate of the actual contamination, both in terms of concentration and the distribution in the study area, may lead to an inflated cost for remediation and for disposal. An underestimate could result in a wrong choice, for example, no remedial action, improperly managed disposal, or limited intervention.

17.3.2.4 *Summary*
Uncertainty in sampling is typically unaccounted for in most risk assessments. This uncertainty can result from either poor knowledge of the real performance of the samplers or from an inadequately designed sampling plan. In addition, both systematic and random errors can occur and usually remain unknown or undetected. Furthermore, the uncertainty associated with sampling is much higher in cases when the actual sampling is done under "blind" conditions (*i.e.*, sampling from a boat; Table 17-1) compared to when the operators can see the sampling medium (shallow areas sampled by hand or with visual aids; Table 17-2). The uncertainty in describing temporal variability is usually greater since it compounds the uncertainty in spatial heterogeneity with uncertainty in repeated sampling. Tables 17-1 and 17-2 provide an indication of overall uncertainty in sampling, and assuming that standardized procedures are followed, the reliability of the selected sampling method is considered, and allowances are made for systematic or random errors. Sampling uncertainty has components which can be reduced by 1) planning the sampling according to the existing information, 2) using appropriate collection and handling methods, 3) conducting pilot studies, and 4) measuring the different sources of variability in the definitive study.

17.3.3 Error and uncertainty in models applied in sediment ecological risk assessments

Exposure models are used to predict concentrations of contaminants in the physical environment (sediments, pore water, and overlying water) and concentrations in organisms as a function of space and time (Chapter 15). Two types of system-level models are typically applied to evaluate contaminated sediments: 1) contaminant transport and fate models and 2) food chain models. Sections 18.5 and 18.6 (Chapter 18) discuss uncertainty associated with specific models used to evaluate bioavailability of contaminants associated with sediments. System-level models should be able to predict contaminant concentrations as affected by natural or anthropogenic events. The goal of modeling should be to develop a predictive model (*i.e.*, a model that is based on parameters which can be measured accurately in the laboratory or by means of simple field tests). Computations should then be based on these parameters with, ideally, no calibration or fine-tuning. In this way, one can have confidence in the predictions and can also use the model to evaluate different environmental conditions and different systems, again with little or no additional calibrating.

17.3.3.1 Model calibration

When a model is calibrated, the calibration is only valid for the data used in the calibration. In order to illustrate this point, consider predicting contaminant concentrations in a lake over 25 years. During this time, a few large storms can occur and these storms may be responsible for most of the sediment and contaminant transport. If the model is calibrated to data taken in an "average" year (*i.e.*, when the large storms did not occur), then parameters and results will be incorrect because they did not include extreme events. If, on the other hand, the model is calibrated to data taken in a very stormy time, the parameters and extrapolated results will also be incorrect. A predictive model needs to be based on the concept that the future is statistically the same as the past. What is known is only that events of a certain magnitude have a certain probability of occurring. Predictive models should then be able to predict the most probable result and the probabilities of the results of different sequence of events.

As an example of the effect of different input parameters on the variability of model output, consider the prediction of PCB half-life in Lake Ontario made by three independent groups (Limnotech, Manhattan College, and University of Toronto). All of these models are based on the concept of a well-mixed sediment layer (Table 17-3; Ziegler and Connolly 1995). The effect of this layer on sediment fluxes is dependent on the thickness of the layer (typically a poorly defined characteristic). Because of this, estimates of half-life differ by almost one order of magnitude, from 3 years to 25 years.

17.3.3.2 Errors and uncertainties in contaminant transport-and-fate process models

In order to quantitatively understand and predict the environmental effects of contaminated sediments, especially as influenced by natural large episodic events or by remedial actions, a knowledge of the transport and fate of the sediments and the contaminants

Table 17-3   Results of three different predictions of half-life of PCBs in Lake Ontario

Investigators	Assumed thickness of layer	PCB half-life
Limnotech	15 cm	25 yrs
Manhattan College	8 cm	15 yrs
University of Toronto	0.5 cm	3 yrs

Source: Ziegler and Connolly (1995).

associated with these sediments is necessary (Chapter 15). Some of the more significant processes that need to be understood and quantified include 1) the resuspension, erosion, and sediment bed dynamics; 2) sorption of contaminants to particles and colloids; 3) flocculation, settling speeds, and deposition rates of particles and flocs; 4) hydrodynamics, including currents and wave action; 5) air-water exchange of contaminants; 6) biochemical reactions and degradation of contaminants; and 7) the inputs of contaminants from the surrounding land, atmospheric, and point discharges.

The process most relevant to estimates of contaminant fluxes at the sediment-water interface are processes 1 to 4 listed above. These sediment–water exchange processes are important because these factors control phase distributions, bioavailability, and contaminant concentrations in the sediments and overlying water (Chapter 15). The flux of contaminants to surface waters from the surrounding land resulting from nonpoint discharges is also not well quantified. Point discharges are generally better known and controlled.

17.3.3.3 *Relevance to decision-making*

In some form or another, models are always used in organizing and interpreting data and, therefore, in decision-making. These models may be simple conceptual models or they may be complex models involving many physical, chemical, and biological processes described by large numbers of differential equations. The solutions to these models may be simple estimates or large arrays of numbers. Models should help in making decisions (*e.g.*, selection of remedial options and understanding the effects of these remedial actions). Because of errors in models, more complex models are not necessarily more accurate or more helpful than simple models. However, simple models are often based on larger numbers of assumptions and may thus be inaccurate. Although complex models may address these assumptions more appropriately, they require more input information, which makes them less useful. In some cases, complex models may have more assumptions (*i.e.*, input parameters) which, if not verified, lead to higher uncertainty. In all modeling, the potential errors of the model should be understood and quantified. This information needs to be transmitted to the risk manager both at the problem formulation stage and at the risk management stage (Figure 1-1). Because of natural variability, models should give the most probable outcome of a sequence of events.

17.4 Conclusions and recommendations

- Guidelines have been developed for conducting ERAs to promote consistency in design, analysis, and interpretation of data. These guidelines also allow for the establishment of quality standards and consistent terminology for assessments. Consistency in the use of guidelines can help inform all stakeholders as to the relative degree of confidence and scientific knowledge under which a decision was made. Additionally, the best way to reduce initial uncertainties in the risk assessment is to involve all interested parties through stakeholder input and expert opinion surveys. This involves asking all interested parties (including the risk managers, the scientific community, and the public) to define the problem in a form of a concise narrative. Once the problem has been identified, appropriate assessment and measurement endpoints can be selected and applied.

- Sources of uncertainty unique to characterizing exposure or effects in SERAs include 1) sample location, collection, and handling errors; 2) spatial and temporal heterogeneity of the stressor; 3) references for comparison of stressor levels; 4) substrate type and interactions with stressor; 5) nonequilibrium of chemical stressor between sediment and water; 6) effects of noncontaminant stressors in toxicity tests; and 7) laboratory-to-field extrapolations.

- Risk characterization requires consideration of the relative importance of the errors associated with making both false positive and false negative determinations. Risk assessment usually focuses on reducing the risk of false positive results associated with statistical analysis (by establishing a low α level). However, from an environmental protection perspective, emphasis should also be placed on reducing the risk of making false negative decisions (*i.e.*, falsely concluding that an area is not contaminated when it actually is).

- Sediment sampling has uncertainty components which can be reduced by planning the sampling program according to the existing information, using appropriate collection and handling methods, conducting pilot studies, and measuring the different sources of variability in the definitive study. Additionally, uncertainty associated with sampling of sediment can be substantially lowered by visually checking the sampling sites.

- Exposure models are used to predict concentrations of contaminants in the physical environment and concentrations in organisms as a function of space and time. The models should be predictable of contaminant concentrations as affected by natural or anthropogenic events. The goal of modeling should be to develop a predictive model (*i.e.*, a model which is based on parameters which can be measured accurately in the laboratory or by means of simple field tests). Computations should then be based on these parameters with, ideally, no calibration or fine-tuning. In this way, one can have confidence in the predictions and can also use the model to evaluate different environmental conditions and different systems, again with little or no additional calibration.

- The outcome of all risk management actions should either be the acceptance or the reduction of the risk. Risk reduction involves many potential actions, which range from technical through socioeconomic to the political. In undertaking risk management, it is necessary to decide which risks must be managed and in what priority.
- Strategic actions for addressing uncertainty in SERAs are listed in Section 18.8.

17.5 References

Adams DD. 1991. Sampling sediment pore water. In: Mudroch A, MacKnight SC, editors. Handbook of techniques for aquatic sediments sampling. Ann Arbor MI: CRC Pr. p 171–202.

[ASTM] American Society for Testing and Materials. 1995a. Standard test methods for measuring the toxicity of sediment-associated contaminants with freshwater invertebrates. In: Volume 11.05, Annual book of ASTM standards. Philadelphia: ASTM. E1706-95b. p 1204–1285.

[ASTM] American Society for Testing and Materials. 1995b. Standard guide collection, storage, characterization, and manipulation of sediments for toxicological testing. In: Volume 11.05, Annual book of ASTM standards. Philadelphia: ASTM. E1391-94. p 835–855.

[ASTM] American Society for Testing and Materials. 1995c. Standard guide for conducting 10-day static sediment toxicity tests with marine and estuarine amphipods. In: Volume 11.05, Annual book of ASTM standards. Philadelphia: ASTM. E 1367-92. p 767–792.

[ASTM] American Society for Testing and Materials. 1995d. Standard guide for designing biological tests with sediments. In: Volume 11.05, Annual book of ASTM standards. Philadelphia: ASTM. E1525-94a. p 972–989.

[ASTM] American Society for Testing and Materials. 1995e. Standard guide for determination of bioaccumulation of sediment-associated contaminants by benthic invertebrates. In: Volume 11.05, Annual book of ASTM standards. Philadelphia: ASTM. E1688-95. p 1140–1189.

Ankley GT, Lodge K, Call DJ, Balcer MD, Brooke LT, Cook PM, Kreis RG, Carlson AR, Johnson RD, Niemi GJ, Hoke RA, West CW, Giesy JP, Jones PD, Fuying ZC. 1992. Integrated assessment of contaminated sediments in the Lower River and Green Bay, Wisconsin. *Ecotoxicol Environ Safety* 23:46–63.

Baker JL, Barefoot AC, Beasley LE, Burns LA, Caulkins PP, Clark JE, Feulner RL, Giesy JP, Graney RL, Griggs RH, Jacoby HM, Laskowski DA, Maciorowski AF, Mihaich EM, Nelson Jr HP, Parrish PR, Siefert RE, Solomon KR, van der Schalie WH, editors. 1994. Aquatic dialogue group: pesticide risk assessment and mitigation. Pensacola FL: SETAC Press.

Baudo R. 1990. Sediment sampling, mapping, and data analysis. In: Baudo R, Giesy JP, Muntau H, editors. Sediments: chemistry and toxicity of in-place pollutants. Chelsea MI: Lewis. p 15–60.

Bliss CI, Fisher RA. 1953. Fitting the negative binomial distribution to biological data. *Biometrics* 9:176–200.

Burton Jr GA, Ingersoll CG. 1994. Evaluating the toxicity of sediments. The ARCS assessment guidance document. Chicago: U.S. Environmental Protection Agency (USEPA). EPA/905-B94/002.

Burton GA, Norberg-King TJ, Ingersoll CG, Ankley GT, Winger PV, Kubitz J, Lazorchak JM, Smith ME, Greer IE, Dwyer FJ, Call DJ, Day KE, Kennedy P, Stinson M. 1996. Interlaboratory study of precision: *Hyalella azteca* and *Chironomus tentans* freshwater sediment toxicity assays. *Environ Toxicol Chem*: In press.

Cardwell RD, Parkhurst BR, Warren-Hicks W, Volosin JS. 1993. Aquatic ecological risk. *Water Environ Technol* 5:47–51.

[CWQG] Canadian Water Quality Guidelines. 1987. Task force on water quality guidelines of the Canadian council of resource and environment ministers. Ottawa ON.

Environment Canada. 1996a. Biological test method: Test for growth and survival in sediment using the freshwater amphipod *Hyalella azteca*. Ottawa ON: Environment Canada. Technical report number pending: In press.

Environment Canada. 1996b. Biological test method: Test for growth and survival in sediment using larvae of freshwater midges (*Chironomus tentans* or *Chironomus riparius*). Ottawa ON: Environment Canada. Technical report number pending: In press.

Francinques Jr NR, Palermo MR, Lee CR, Peddicord RK. 1985. Management strategy for disposal of dredged material: Contaminant testing and controls. Vicksburg MS: U.S. Army Engineer Waterways Experiment Station. Miscellaneous Paper D-85-1.

Grigalunas TA, Opaluch JJ. 1989. Economic considerations of managing contaminated marine sediments. In: Contaminated marine sediments--assessment and remediation. Washington DC: National Research Council, National Academy Pr. p 291–310.

Hakanson L, Jansson M. 1983. Principles of lake sedimentology. Berlin: Springer-Verlag. 316 p.

Harwell MA, Cooper W, Flaak R. 1992. Prioritizing ecological and human welfare risks from environmental stresses. *Environ Management* 16:451–464.

Health Council of the Netherlands. 1993. Ecotoxicological risk assessment and policy-making in the Netherlands: dealing with uncertainties. *Network* 6(3)/7(1):8–11.

[IJC] International Joint Commission. 1988. Options for the remediation of contaminated sediments in the Great Lakes: Sediment Subcommittee and its Remedial Options Work Group to the Great Lakes Water Quality Board Report to the International Joint Commission. Windsor ON.

Jessen RJ. 1978. Statistical survey techniques. New York: Wiley.

Kemble NE, Brumbaugh WG, Brunson EL, Dwyer FJ, Ingersoll CG, Monda DP, Woodward DF. 1994. Toxicity of metal-contaminated sediments from the upper Clark Fork River, MT to aquatic invertebrates in laboratory exposures. *Environ Toxicol Chem* 13:1985–1997.

Klaine SJ, Cobb GP, Dickerson RL, Dixon KR, Kendall RJ, Smith EE, Solomon KR. 1996. An ecological risk assessment for the use of the biocide, dibromonitrilopropionamide (DBNPA) in industrial cooling systems. *Environ Toxicol Chem* 15:21–30.

Kratochvil B, Taylor JK. 1981. Sampling for chemical analysis. *Anal Chem* 53: 924A–938A.

Lynam WJ, Glazer AE, Ong JH, Coons SF. 1987. An overview of sediment quality in the United States. Washington DC: U.S. Environmental Protection Agency (USEPA). EPA-905/9-88-002.

MacKnight SD. 1991. Selection of bottom sediment sampling stations. In: Mudroch A, MacKnight SC, editors. Handbook of techniques for aquatic sediments sampling. Boca Raton FL: CRC Pr. p 17–28.

McBean EA, Rovers FA. 1992. Estimation of the probability of exceedance of a contaminant concentration. *Ground Water Monitoring Rev* 12:115–119.

Mudroch A, MacKnight SD. 1991. Bottom sediment sampling. In: Mudroch A, MacKnight SC, editors. Handbook of techniques for aquatic sediments sampling. Boca Raton FL: CRC Pr. p 29–95.

Mudroch A, Bourbonniere RA. 1991. Sediment sample handling and processing. In: Mudroch A, MacKnight SC, editors. Handbook of techniques for aquatic sediments sampling. Boca Raton FL: CRC Pr. p 131–169.

Parkhurst BR, Warren-Hicks W, Noel LE. 1992. Performance characteristics of effluent toxicity tests: summarization and evaluation of data. *Environ Toxicol Chem* 11:771–791.

Parkhurst BR, Warren-Hicks W, Etchison T, Butcher JB, Cardwell RD, Voloson J. 1995. Methodology for aquatic ecological risk assessment. Alexandria VA: Report prepared for the Water Environment Research Foundation. RP91-AER-1 1995.

Russell M, Gruber M. 1987. Risk assessment in environmental policy-making. *Science* 236:286–290.

Solomon KR, Baker DB, Richards P, Dixon KR, Klaine SJ, La Point TW, Kendall RJ, Giddings JM, Giesy JP, Hall Jr LW, Weisskopf CP, Williams M. 1996. Ecological risk assessment of atrazine in North American surface waters. *Environ Toxicol Chem* 15:31–76.

Sokal RR, Rohlf FJ. 1981. Biometry. New York: Freeman. 859 p.

Steel RG, Torrie JH. 1980. Principles and procedures of statistics. New York: McGraw Hill.

Stephan CE, Mount DI, Hansen DJ, Gentile JH, Chapman GA, Brungs WA. 1985. Guidelines for deriving numerical national water quality criteria for the protection of aquatic organisms and their uses. Springfield VA: National Technical Information Service (NTIS). PB85-227049.

Switzer P. 1975. Statistical considerations in network design. *Water Resour Res* 15: 1512–1516.

[USEPA] U.S. Environmental Protection Agency. 1989. Assessing human health risks from chemically contaminated fish and shellfish: a guidance manual. Washington DC: USEPA. EPA 503/8-89-002.

[USEPA] U.S. Environmental Protection Agency. 1992a. Framework for ecological risk assessment. Washington DC: USEPA. EPA/630/R-92/001.

[USEPA] U.S. Environmental Protection Agency. 1992b. Sediment classification methods compendium. Sediment Oversight Technical Committee. Washington DC: USEPA. EPA 813-R-92-006.

[USEPA] U.S. Environmental Protection Agency. 1994a. Methods for measuring the toxicity and bioaccumulation of sediment-associated contaminants with freshwater invertebrates. Duluth MN: USEPA. EPA/600/R-94/024.

[USEPA] U.S. Environmental Protection Agency. 1994b. Methods for measuring the toxicity of sediment-associated contaminants with estuarine and marine amphipods. Duluth MN: USEPA. EPA/600/R-94/025N.

[USEPA and USACE] U.S. Environmental Protection Agency and U.S. Army Corps of Engineers. 1991. Evaluation of dredge material proposed for ocean disposal. Washington DC. EPA-503/8-91/001.

[USEPA and USACE] U.S. Environmental Protection Agency and U.S. Army Corps of Engineers. 1994. Evaluation of dredged material proposed for discharge in inland and near coastal waters (draft). Washington DC. EPA-823-B-94-002.

Warren-Hicks W, Parkhurst BR. 1992. Performance characteristics of effluent toxicity tests: Variability and its implications for regulatory policy. *Environ Toxicol Chem* 11:793–804.

Ziegler CK, Connolly JP. 1995. The impact of sediment transport processes on the fate of hydrophobic organic chemicals in surface water systems. Proceedings of WEF Toxics Substances in Water Environment Conference. p 1-13 – 1-24.

SESSION 6
CRITICAL ISSUES IN
METHODOLOGICAL UNCERTAINTY

Chapter 18

Workgroup summary report on uncertainty evaluation of measurement endpoints used in sediment ecological risk assessment

Christopher G. Ingersoll, Gerald T. Ankley, Renato Baudo, G. Allen Burton,
Wilbert Lick, Samuel N. Luoma, Donald D. MacDonald, Trefor B. Reynoldson,
Keith R. Solomon, Richard C. Swartz, William J. Warren-Hicks

18.1 Introduction

Earlier chapters have recommended procedures for conducting SERAs related to product safety assessments (Chapters 3 and 4), navigational dredging decisions (Chapter 5), cleanup decisions (Chapters 6 and 7), and general issues related to evaluation of uncertainty associated with conducting SERAs. This chapter describes the uncertainty associated with specific measurement endpoints commonly used in SERAs and discusses approaches for addressing these sources of uncertainty. These stepwise procedures provide explicit guidance on the four major elements of ERAs: problem formulation, exposure assessment, effect assessment, and risk characterization (Chapter 1). In addition, several types of measurement endpoints were identified at the workshop which have been used to conduct SERAs including 1) toxicity tests (both the fraction tested [Section 18.2] and the endpoints selected [Section 18.3]), 2) benthic invertebrate assessments (Section 18.4), 3) bioaccumulation assessments (Section 18.5), 4) sediment chemistry (Section 18.6), and 5) SQGs (Section 18.7).

Although each type of measurement endpoint may contribute to risk-based decision-making, several important questions remain regarding uncertainty. For example, "Which toxicity tests provide the most sensitive and realistic measurements for evaluating effects of contaminants on benthic communities? Which measurement endpoints for benthic invertebrate assessments provide the best linkage to sediment contamination? Which SQGs should be used to evaluate the potential biological significance of sediment-associated chemicals at contaminated areas?" To answer these and other similar questions, information is needed on the uncertainty that is associated with each of the measurement endpoints that are commonly used in SERAs. It would be ideal to quantify the absolute uncertainty associated with each measurement; however, such an assessment is currently not possible. Nonetheless, it is possible to evaluate the relative uncertainty associated with some of the measurement endpoints commonly used in SERAs.

To facilitate this evaluation, we established a series of criteria to support consistent assessments of the uncertainty associated with each measurement endpoint, including 1) precision, 2) ecological relevance, 3) causality, 4) sensitivity, 5) interferences, 6) standardization, 7) discrimination, 8) bioavailability, and 9) field validation. Each of these criteria were defined in the context of the type of measurement endpoint that was considered. For example, the definitions of precision are different for the evaluations of sediment chemistry data in Section 18.6 compared to the definition for precision in Section 18.7 dealing with SQGs. In this way, the relevance of the evaluation criteria to each assessment was assured. Although no attempt was made in this evaluation to recommend specific measurements to apply when conducting risk assessments with sediment, it is antici- pated that the information provided will allow the reader to identify the measurement endpoints which are most relevant for their specific applications. Uncertainty in mea- surement endpoints associated with lack of knowledge is indicated with an asterisk in Tables 18-1 to 18-3 and Tables 18-5 to 18-7 to differentiate from systematic uncertainty which can be rectified (methodologically) or quantified (sampling decisions and design).

The fact that no recommendations were made with regard to a single preferred endpoint reflects a lack of scientific consensus about whether any single approach provides the best answer for risk assessment. There was a consensus that each endpoint had strengths and weaknesses, many of them inherent, and that the best approach at present is probably to use multiple endpoints. Implementation of robust multiple endpoint studies remains in the experimental phase. The success of multidisciplinary tools such as the sediment qual- ity triad will expand upon these applications. The discussions of each endpoint in the following sections were developed by specialists in that field participating at the Work- shop, with inputs from colleagues in other specialties. Perhaps it is a sign of the relative immaturity of the field of SERA that specialists tended to be more optimistic about the uncertainties associated with the endpoints in their area of specialty and sometimes more critical of alternative endpoints. Rankings in Tables 18-1 to 18-3 and Tables 18-5 to 18-7 reflect the preferences of the specialists in that field. Comparisons of rankings are valid within endpoints (e.g., within Table 18-1) but not necessarily among endpoints (e.g., be- tween Table 18-1 and 18-5). As this field matures, it is expected that interrelationships among specialists will grow in sophistication. For example, the results of future commu- nity assessments may identify more sensitive species for laboratory toxicity testing, the bioaccumulation approaches may be used to help understand the complexities of natu- ral exposures at a site, and the results of the laboratory toxicity tests could be used to help explain changes in populations or communities. The paucity of knowledge about such interrelationships is an example of uncertainties associated with inadequate knowledge that are broader than those measured by the approaches discussed in Chapter 17. The knowledge uncertainties are difficult to quantify, but they are the reason for the caution that many scientists express about overly broad applications of SERAs.

18.2 Evaluation of uncertainty associated with phases used to conduct laboratory toxicity tests with sediments

Various methods have been developed to evaluate sediment toxicity. These procedures range in complexity from short-term lethality tests which measure effects of individual contaminants on single species to long-term tests which determine the effects of chemi-

Table 18-1   Uncertainty associated with sediment phases used in laboratory toxicity tests

Evaluation criteria	Whole sediment: benthos	Whole sediment: pelagic	Organic extracts	Suspended solids	Elutriates	Pore water
Precision	1	1	1	3	1	1
Ecological relevance	1	2	3	2	3	2
Causality: link	3	3	3	3	3	3
Causality: source	1	2	3*	3	3	2
Sensitivity	1	2	3	3	3	2
Interference	2*	2	3	3	3	2*
Standardization	1	2	3	3	1	2
Discrimination	1*	1*	1*	1*	1*	1*
Bioavailability	1*	1*	3	1*	3	1*
Field validation	1*	2*	3	3*	3	3*

1 = low uncertainty (good)
3 = high (bad)
* = lack of knowledge

cal mixtures on the growth or reproduction of test organisms or structure and function of communities. The test organisms might include bacteria, algae, macrophytes, fishes, and benthic, epibenthic and pelagic invertebrates (Lamberson and Swartz 1988; Burton 1992). Discussions of uncertainty relative to laboratory toxicity tests were divided into two operational categories: 1) uncertainties related to the phase tested (in this section) and 2) uncertainties related to the selection of endpoints measured in toxicity tests (Section 18.3). A diverse array of exposure phases have been used in sediment toxicity tests (Ankley, Schubauer-Berigan, and Dierkes 1991; Burton 1992; Ingersoll 1995). The present discussion of uncertainty focuses on six principal phases typically evaluated in toxicity tests: 1) whole sediment using benthic invertebrates, 2) whole sediment using pelagic organisms, 3) organic extracts of whole sediment, 4) suspended solids, 5) elutriates, and 6) pore water isolated from whole sediment. Within the category of whole sediment tests, we chose to differentiate between tests conducted with benthic versus pelagic organisms because of uncertainty with respect to route of exposure with pelagic organisms.

Whole sediment toxicity tests were developed to evaluate the effects of in-place sediments (USEPA 1994a, 1994b; ASTM 1995a–1995e; Environment Canada 1996a, 1996b). Toxicity tests with porewater samples isolated from sediment were developed for evaluating the potential *in situ* effects of contaminated sediment on aquatic organisms (Ankley,

Phipps, *et al.* 1991; Ankley, Schubauer-Berigan, and Dierkes 1991). For many benthic invertebrates, the toxicity and bioaccumulation of sediment-associated contaminants such as metals and nonionic organic contaminants have been correlated with concentrations of these chemicals in pore water (Di Toro *et al.* 1991). Toxicity tests with organic extracts were developed to evaluate effects of the maximum concentrations of organic contaminants associated with a sediment (Chapman and Fink 1984). Tests with elutriate samples and suspended solids measure the potential release of contaminants from sediment to the water column during disposal of dredged material or during sediment resuspension events (Shuba *et al.* 1978; ASTM 1995d).

Each of these phases is evaluated in relation to the following major sources of uncertainty: precision, ecological relevance, causality, sensitivity, interference, standardization, discrimination, bioavailability, and field validation (Table 18-1). The uncertainty associated with each phase is a function of inherent limitations (*e.g.*, testing of whole sediments has greater ecological significance than organic extracts) and the stage of development of the response as a toxicological endpoint (*e.g.*, whole sediment tests are much better developed than porewater tests). The available information does not support a quantitative evaluation of the uncertainty associated with each of the sediment phases considered in this assessment. For this reason, an evaluation of the relative uncertainty of various sediment phases was conducted to provide risk assessors with guidance for the application of appropriate toxicity tests.

18.2.1 Precision
Precision was evaluated in terms of the replicability of the tests commonly performed on various sediment phases. With respect to laboratory precision (*i.e.*, precision not related to sample collection, handling, and storage), tests with most of the fractions listed in Table 18-1 have a low degree of uncertainty. Round-robin (ring) tests conducted with whole sediments indicate relatively low intra- and interlaboratory variability (Mearns *et al.* 1986; USEPA 1994a, 1994b; ASTM 1995a; Burton *et al.* 1996). The one exception would be suspended-solids tests, where the workgroup noted the methods used for generating these types of exposure are quite variable (ASTM 1995d).

18.2.2 Ecological relevance
With respect to the evaluation of sediment toxicity tests conducted on various sediment phases, ecological relevance was evaluated in terms of its linkage to the receptors which are to be protected. Whole sediment tests using resident species were considered to provide the most realistic information for assessing organism responses (Table 18-1). Because organic extracts may alter the bioavailability of sediment-associated contaminants, toxicity tests conducted using this phase were considered to have a relatively lower level of relevance. Similarly, elutriate and suspended-solids tests are conducted using a phase which may artificially alter the availability of contaminants. Although water column species have been demonstrated to be sensitive indicators of whole sediment toxicity (Burton *et al.* 1992), benthic species in direct contact with sediment provide a more direct assessment of sediment contamination (Ankley, Schubauer-Berigan, and Dierkes 1991).

The most ecologically relevant test systems from the standpoint of uncertainty were whole sediment tests with benthic organisms, followed by whole sediment tests with pelagic species, suspended sediment tests, and porewater tests (Table 18-1). All of these phases represent meaningful and interpretable routes of exposure with respect to aquatic organisms, albeit for slightly different reasons. For example, pore water seems to be a reasonable surrogate test fraction for whole sediments (Giesy and Hoke 1989; Ankley, Phipps, *et al.* 1991; Ankley, Schubauer-Berigan, and Dierkes 1991; Carr and Chapman 1995), whereas whole sediment tests with pelagic species seem reasonable for assessing the potential toxicity of contaminants released from sediments into the overlying water (Burton *et al.* 1992; ASTM 1995a). The least ecologically relevant test systems were organic extracts and elutriates. A fraction of the organic extracts are comprised of contaminants that are not (or ever would be) bioavailable and represent an arbitrary situation not expected to exist in the environment. There is inherent variability to studies that have used elutriates as predictive of potential toxicity of sediments *in situ*. In terms of interrelationships, it was recognized that most toxicity tests are closed, simplified systems. As such they may have difficulty simulating the effects of contaminants within complex food webs, feedback loops, and other interactions that characterize natural ecosystems, no matter what sediment phase is employed (Luoma and Ho 1993).

18.2.3 Causality

Determination of causality (*i.e.*, correctly identifying stressors) is of fundamental importance in SERAs. For this reason, uncertainty in the phases evaluated in toxicity tests was evaluated in terms of their ability to determine or predict the factors causing adverse effects in contaminated sediments. With respect to causality, it is highly uncertain exactly which contaminants may be causing observed toxicity without the use of sediment chemistry data, irrespective of the system tested. Because toxicity, in and of itself, consists only of a generic biological response, no test with the phases listed in Table 18-1 can indicate specific contaminants of concern. For this interpretation, it is necessary to link the toxicity test to appropriate mixture toxicity models, spiked-sediment tests, or TIE procedures designed specifically for defining specific compounds or classes of compounds responsible for toxicity (Chapter 16).

On another level, toxicity tests can be related to causality in a more generic sense. For example, if an effect is observed with a measure of benthic community structure, appropriate toxicity tests can be used to indicate, with a good deal of certainty, causality in terms of the field response resulting from toxicity as opposed to other environmental stressors. Also, the spatial distribution of toxicity can indicate the location of sources of contaminants which cause adverse responses. In these instances, whole sediment tests with benthic species have a relatively high degree of certainty, while whole sediment tests with pelagic species and porewater tests have lower certainty. The test systems with the greatest uncertainty in this category included organic extracts, elutriate, and suspended-solid tests. In the case of the organic extract and elutriate tests, the exposure regimes do not represent a realistic measure of *in situ* conditions of sediment. The workgroup was unaware of any evidence that has been collected to evaluate this issue for suspended-solid tests.

18.2.4 Sensitivity

Sensitivity is another important characteristic of measurement endpoints because there is a need to reliably identify sediments that have the potential to affect sensitive species in aquatic ecosystems. Sensitivity should minimize uncertainty associated with both false positives (nontoxic sample incorrectly classified as toxic) and false negatives (toxic sample incorrectly classified as nontoxic). Whole sediment tests with benthic organisms typically exhibit an appropriate level of sensitivity. This is not to say that all toxicity tests conducted with benthic organisms are the most sensitive, but that there was a high degree of certainty that the response measured was predictive and accurate of effects of contaminants in sediment (*i.e.*, minimize both false positives and false negatives). However, these conclusions are based primarily on results of toxicity tests which monitor only short-term effects (*i.e.*, 10-d survival). Additional emphasis should be placed on developing and evaluating chronic toxicity test methods (*i.e.*, longer exposures measuring effects on growth and reproduction; Kemble *et al.* 1994; Benoit *et al.* 1996; Sibley *et al.* 1996). There was not a consensus in the group that some of the subtle, but important, effects which might occur in nature were measured with a high degree of sensitivity in the whole sediment bioassays, even though they were often the best choice of sediment phase. An intermediate level of uncertainty was assigned to the sensitivity of whole sediment tests with pelagic species and tests conducted with porewater samples. The highest relative uncertainty is associated with test systems consisting of organic extracts, elutriates, and suspended sediments. Relative sensitivity is highly dependent on the species selected for testing in addition to the phase tested. It is recommended that additional studies be conducted comparing the sensitivity of single species tested in a variety of sediment phases to the response of populations and communities in the field.

18.2.5 Interferences

Interferences are considered to be related to biotic or abiotic factors which could influence toxicity test results beyond the direct effects of specific contaminants. That is, interferences are considered to be nontreatment factors (*i.e.*, noncontaminant effects). All of the test systems have moderate to high uncertainty with regard to interferences. Factors such as particle size, OC content, salinity, BOD, and the presence (or absence) of nutrients can potentially affect the results of whole sediment tests with both benthic and pelagic species (DeWitt *et al.* 1988, 1989; Ankley, Benoit, *et al.* 1994; Suedel and Rodgers 1994). In the case of organic extracts, sample integrity is so disrupted that observed toxicity could be solely artifactual (*e.g.*, resulting from the presence of solvent). An important factor which can cause uncertainty with regard to tests with suspended sediments is the differentiation of toxicity caused by contaminants from that caused by physical effects of suspended solids (Chapman *et al.* 1987). A major source of uncertainty in the interpretation of aqueous phase (elutriate, pore water) sediment tests is pH drift which can change by one or more units (usually upward) over the course of tests with these fractions. Such changes can alter the toxicity of common sediment contaminants such as metals, hydrogen sulfide, and ammonia (Ankley and Thomas 1992; Ankley and Schubauer-Berigan 1995). If aqueous phases are removed from partially anoxic sediments, rapid precipitation of ferrous iron in the oxidized conditions of the bioassay will remove metals and bias toxicity downward (Luoma 1995).

18.2.6 Standardization

In the context of toxicity tests, standardization was evaluated in terms of the level of peer review and the publication of standard methods. There is little uncertainty related to the availability of standard methods for preparing and testing sediments with benthic species or with elutriates (ASTM 1995a–1995e; USEPA 1994a, 1994b; Environment Canada 1996a, 1996b). A moderate amount of uncertainty exists with regard to standard methods for testing pelagic species in whole sediment tests (ASTM 1995a) or for isolating and testing pore water (ASTM 1995b). However, several recently published studies may aid in the standardization of porewater preparation for toxicity testing (Schults et al. 1992; Ankley and Schubauer-Berigan 1995; Carr and Chapman 1995). ASTM has recently described general guidance for conducting tests with suspended solids (ASTM 1995d). To our knowledge, there are no standardized (or widely accepted) methods for testing organic extracts prepared from sediments.

18.2.7 Discrimination

This criterion was intended to evaluate the ability of a sediment phase used in a toxicity test to discriminate between contaminated sediments by providing a graded response. All six types of test fractions have the potential for low uncertainty with respect to discriminating among samples (i.e., the ability to distinguish among samples along a stressor gradient). However, it should be noted that the ability to discriminate among samples is more a function of the biological endpoint chosen than of the sample fraction tested (Chapter 16).

18.2.8 Bioavailability

Information from a variety of sources indicates that the characteristics of the sediments and the associated pore water can influence the bioavailability and toxicity of sediment-associated contaminants (Pavlou 1987; Di Toro et al. 1990; Swartz et al. 1990). For this reason, toxicity tests that explicitly considered test phases that can be used to evaluate the influence of these factors have a lower level of uncertainty than those which did not address bioavailability (Di Toro et al. 1991). Toxicity tests cannot be used alone to evaluate the bioavailability of contaminants. Synoptic generation of appropriate chemistry data is needed in combination with the toxicity test (Chapter 14). If this is done, the question of uncertainty is then related to the relevance of the test from a standpoint of realistic exposure (i.e., ecological relevance). Thus, tests with the highest degree of uncertainty are those with a questionable formulation from an environmental standpoint (the solvent extract and elutriates). The other four test phases — whole sediment tests with benthic and pelagic species, suspended solids, and porewater tests — all have low uncertainty with respect to assessing the biological availability of contaminants in different scenarios.

18.2.9 Field validation
Field validation is an essential element of toxicity testing because it provides a basis for assessing reliability in sediment quality assessments. In this context, toxicity tests were considered to have low uncertainty if they were shown to be predictive of responses of benthic communities in the field. None of the phases listed in Table 18-1 has received enough attention from the standpoint of field validation; further research is required in all instances. However, some data do exist to suggest that certain whole-sediment tests can be predictive of population- and community-level responses in the field (Swartz *et al.* 1994; Canfield *et al.* 1994, 1996; Chapter 8). Therefore, the general consensus was that there was a reasonable amount of certainty with respect to this extrapolation. Similarly, good correspondence was observed between the results of whole sediment toxicity tests with pelagic cladocerans and measures of benthic community structure in several Great Lakes tributaries (Indiana Harbor, Buffalo River, Saginaw River; Burton and Ingersoll 1994). Others in the workgroup suggested it may be more difficult to simulate the complexities of nature in simple whole-sediment tests and pointed out examples where toxicity tests did not predict outcomes in nature (Schindler 1987; Bryan and Gibbs 1991; Luoma and Carter 1991). The degree of uncertainty associated with laboratory-to-field extrapolation of biological tests with organic extracts of sediments, suspended sediments, elutriates, and pore water have not been thoroughly examined.

18.2.10 Summary
Whole sediment tests were considered to provide the most realistic phase for assessing organism response. Because organic extracts may alter the bioavailability of sediment-associated contaminants, toxicity tests conducted using this phase were considered to have a relatively lower level of relevance. Similarly, elutriate and suspended-solids tests are conducted using a phase which may artificially alter the availability of contaminants. In order to establish cause-and-effect relationships, it is necessary to link the toxicity test to appropriate mixture toxicity models, spiked-sediment tests, or TIE procedures designed specifically for identifying specific compounds or classes of compounds responsible for toxicity (Chapter 16).

18.3 Evaluation of the uncertainty associated with endpoints measured in laboratory toxicity tests with sediments
A diverse array of response endpoints have been measured in sediment toxicity tests (Lamberson and Swartz 1988; Burton *et al.* 1992; Lamberson *et al.* 1992; Burton and Ingersoll 1994). Uncertainty associated with toxicity endpoints was evaluated rather than assessing the types of species that could be tested in laboratory exposures. Others have addressed the advantages and disadvantages of specific tests or specific organisms used in sediment studies (*e.g.*, Swartz 1989; Burton 1992; Ankley, Collyard, *et al.* 1994; USEPA 1994a, 1994b; Ingersoll 1995). The emphasis on endpoints is not intended to imply that

any particular endpoint is preferable over another with respect to ecological relevance or overall sensitivity. For example, a "chronic" test with sublethal endpoints with a relatively insensitive species (*e.g.*, growth or reproduction with polychaetes) might be far less ecologically relevant or sensitive compared to an acute test with survival as the primary endpoint with a sensitive species (*e.g.*, survival of amphipods).

The present discussion of uncertainty focuses on seven principal classes of response endpoints that are often measured in toxicity tests, including survival, growth, reproduction, behavior, life tables, development, and biomarkers. Each of the response endpoints is evaluated in relation to the major sources of uncertainty: precision, ecological relevance, causality, sensitivity, interference, standardization, discrimination, bioavailability, and field validation (Table 18-2). The uncertainties associated with each of the endpoints are a function of inherent limitations (*e.g.*, reproduction has greater ecological significance than biomarkers) and the stage of development of the response as a toxicological endpoint (*e.g.*, acute lethality tests are much better developed than chronic reproductive tests). Uncertainty also varies among specific toxicity tests that use the same endpoint (*e.g.*, interference of low salinity on amphipod survival is substantial for *Rhepoxynius abronius* but insignificant for *Eohaustorius estuarius*; ASTM 1995c). The following comparison of uncertainty is based on the best developed examples of toxicity tests for each of the response endpoints.

18.3.1 Precision
Precision was evaluated relative to the replicability of endpoint responses. The lowest level of uncertainty was assigned to survival, growth, behavior, and development. With respect to the behavior and development endpoints, our evaluation of uncertainty was based on the following tests: emergence and reburial of estuarine and marine amphipods (ASTM 1995b), 48-h development of echinoderm and mollusc larvae (USEPA 1991), and emergence of midges (ASTM 1995a; Benoit *et al.* 1996; Sibley *et al.* 1996). Other behavioral and developmental endpoints may be much more variable than the measures evaluated in this exercise. The precision of measurements of survival, growth, behavior, and development is sufficient to allow for adequate statistical power with relatively few replicates. For example, comparisons (n=365) of mean survival of the amphipod *Rhepoxynius abronius* in treatment and control means of sediments with five replicates showed that differences of 25% or more were always statistically significant (Mearns *et al.* 1986). High uncertainty in precision was assigned to life table measurements and biomarkers because of the inherent variability in these endpoints (Benson and Di Giulio 1992; DeWitt *et al.* 1992). See USEPA (1994a, 1994b), ASTM (1995a), and Burton *et al.* (1996) for additional discussion of statistical power associated with sediment toxicity tests.

18.3.2 Ecological relevance
Ecological relevance was evaluated in terms of the linkage between response endpoints and the ecological resources to be protected. Survival is a highly relevant endpoint for measuring both acute and chronic effects. For example, survival of amphipods exposed to field-collected sediments in 10- to 28-d toxicity tests has been correlated with abun-

Table 18-2   Uncertainty associated with endpoints measured in laboratory toxicity tests with sediment

Evaluation criteria	Survival	Growth	Repro-duction	Behavior	Life tables	Develop-ment	Biomarkers
Precision	1	1*	2*	1*	3*	1	3*
Ecological relevance	1	2*	1*	2*	1*	2*	3*
Causality: link	3	3	3	3	3	3	2
Causality: source	1	2*	2*	2*	3*	1	2
Sensitivity	1	2	1	2	2*	2	1*
Interference	1*	2*	3*	2*	3*	2*	3
Standardization	1	2	2	1	3	2*	3
Discrimination	2	1	1	2	2*	1	2*
Bioavailability	1	1	1	1	1	1	1
Field validation	1	2*	2*	1	3*	2	3

1 = low uncertainty (good)
3 = high (bad)
* = lack of knowledge

dance of amphipods, species richness, and other measures of community structure in the field (Ferraro *et al.* 1991; Schlekat *et al.* 1994; Swartz *et al.* 1994). Also, growth of the amphipod *Hyalella azteca* or the midge *Chironomus tentans* have been correlated to measures of community structure (Wentsel *et al.* 1977; Giesy *et al.* 1988; Burton and Ingersoll 1994; Canfield *et al.* 1994, 1996). In contrast, longer term studies (several generations) show how amphipod populations can suffer eventual extinction at toxicant levels below those that affect survival (Sundelin 1983), and several authors have discussed the challenges of predicting the more subtle effects of toxicants in nature with short-term tests of survival (Schindler 1987; Luoma 1995). Reproduction and population dynamics (life tables) are relevant measures of chronic effects. However, substantial uncertainly is associated with many existing biomarkers relative to ecological responses (Benson and Di Giulio 1992) because biomarkers often reflect exposure to contaminants rather than adverse effects. More study is needed to determine if changes in population dynamics and biomarker-type responses precede more relevant ecological responses as warning signals (Schindler 1987; Widdows and Donkin 1989; Luoma 1995).

18.3.3 Causality

Uncertainty associated with causal relations in toxicity endpoints was evaluated in two ways. First, causality was assessed with respect to the correct linkage to the specific stress factors which caused the toxicological response. Second, causality was evaluated with respect to the location of the source of the stress factors. Determination of causal linkages (*i.e.*, correctly identifying stressors) can be of fundamental importance in SERAs. However, toxicological responses are usually generic indicators of stress and cannot, by themselves, discriminate specific factors which caused the response. This limitation is the principal justification for including all three elements of the sediment quality triad (*i.e.*, biology, chemistry, toxicity) in comprehensive SERAs (Long and Chapman 1985; Chapman *et al.* 1992). Except for biomarkers, all of the response endpoints were assigned a high degree of uncertainty with respect to linkages (Table 18-2). Some biomarkers have less uncertainty because they can be used to identify specific classes of contaminants (*e.g.*, metallothionein and certain metals; cytochrome P-450 induction and certain organic compounds; Brown *et al.* 1985; Benson and Di Giulio 1992; Huggett, Kimerle, *et al.* 1992; Huggett, Van Veld, *et al.* 1992). The different measurement endpoints can be used to identify source location as a reflection of spatial distribution of toxicity. This use of the endpoints listed in Table 18-2 to identify sources responsible for effects is best developed for survival, growth, and development endpoints (Chapman *et al.* 1991; Swartz *et al.* 1994; Canfield *et al.* 1994). Higher uncertainty is associated with use of life table measurements primarily because of a lack of application of these analyses to evaluate contaminated sediment.

18.3.4 Sensitivity
Sensitivity is an important characteristic of risk assessment measurement endpoints because there is a need to reliably identify sediments with high, moderate, and low probabilities of causing adverse biological effects. Therefore, endpoints were considered to have a relatively low level of uncertainty in this respect if they reliably determine effects at relatively low contaminant concentrations (*i.e.*, minimize false negatives, although allowing a higher probability of false positives). All of the endpoints have low to moderate uncertainty with respect to sensitivity. However, the sensitivity of biomarkers may not be ecologically relevant (Benson and Di Giulio 1992). Furthermore, measurement of chronic effects does not always equate with increased sensitivity. For example, measurements of reproduction in the oligochaete *Lumbriculus variegatus* is typically less sensitive than measurements of acute 10-d lethality with the amphipod *Hyalella azteca* (Phipps *et al.* 1995). Sensitivity tends to be species and contaminant-specific. Use of multiple test species is, therefore, recommended for SERAs. It should also be noted that in at least some of the cases where effects of moderate contamination have been thoroughly studied, ecosystem changes may occur at contaminant levels below those which had effects in laboratory toxicity or bioaccumulation tests (Schindler 1987; Bryan and Gibbs 1991). So the uncertainties inherent in the toxicity testing approach may introduce insensitivities which will not be quantifiable until responses of organisms and communities to sediment contamination in nature are better understood or until better chronic testing methods are developed.

18.3.5 Interferences
Interferences are biotic or abiotic factors which could influence response endpoints beyond the direct effects of specific contaminants. That is, interferences are considered to be not-treatment factors. Many natural sediment features (grain size, salinity, tempera-

ture, hardness) can affect responses if they exceed the tolerance limits of the test species (DeWitt *et al.* 1988, 1989; Ankley, Benoit, *et al.* 1994; Kohn *et al.* 1994).

The potential for interferences often increases with the duration of the exposure. Thus, uncertainty with respect to interference is lowest for survival in acute exposures (*i.e.*, 10-d tests). Interference may be greater in studies of chronic effects on reproduction and population dynamics. The presence of indigenous organisms can confound chronic endpoints (*i.e.*, growth; Reynoldson *et al.* 1994). Some biomarker responses (*e.g.*, stress proteins) may be associated with sediment features or other noncontaminant factors (Benson and Di Giulio 1992). Ultraviolet light causes the photoactivation and increased toxicity of some PAHs (Ankley, Collyard, *et al.* 1994). Thus, the absence of ultraviolet (UV) light in the fluorescent lighting typically used in laboratories is an interference that may result in an underestimation of potential toxicity under natural conditions. Interferences can be minimized through the selection of appropriate methods and adherence to rigorous quality assurance plans. Of particular importance is the selection of a test species that would tolerate the test sediments in the absence of contaminants.

18.3.6 Standardization

Uncertainty in the degree of standardization of response endpoints was evaluated on the basis of appropriate peer review. That is, the degree of uncertainty was considered to be lower for those approaches that have been published in the peer-reviewed literature and in standard methods. Standard methods have been developed for virtually all of the endpoints listed in Table 18-2 except for life tables and biomarkers. However, ASTM recently convened a symposium dedicated to biomarkers and established a subcommittee to develop standard guidance. Standard methods for acute toxicity tests have been developed by a variety of organizations (ASTM 1995a–1995e; USEPA 1994a, 1994b; Environment Canada 1992, 1996a, 1996b) and standard sediment methods are being developed for the chronic endpoints (DeWitt *et al.* 1992; ASTM 1995a; Benoit *et al.* 1996; Sibley *et al.* 1996).

18.3.7 Discrimination

Discrimination was evaluated in terms of the potential for response endpoints to define the degree of toxicity associated with contaminated sediments. Chronic endpoint responses have the potential to be more discriminatory than survival. Survival often approximates a binary (all or none) response and may not distinguish between marginally contaminated samples. However, growth, reproduction, and population dynamics may be able to discriminate among nonlethal sediments (Kemble *et al.* 1994). Biomarkers and behavior have greater uncertainty because of limited applications with contaminated sediments. Discrimination of the relative contamination and toxicity of sediments that cause complete mortality can be accomplished by testing a series of dilution mixtures of the toxic sediment with clean (negative control) sediment (Swartz *et al.* 1989). However, dilution of contaminated sediments with a clean sediment may result in artifacts with respect to concentration-response relationships (Nelson *et al.* 1993).

18.3.8 Bioavailability

Both chemical and toxicological data are needed to establish bioavailability of sediment contaminants. In sediment spiking experiments, there is low uncertainty about bioavailability when a biological response results from sediment exposure if 1) the contaminant and response data are correlated, 2) an appropriate experimental design is used, and 3) quality assurance objectives are met. The bioavailability of contaminants in field-collected sediments is less certain because of the difficulty of establishing causal relations between response endpoints and specific chemicals (Chapter 16). Toxicity tests that incorporate TIE procedures can be used to determine the bioavailability of specific chemicals (Chapter 16).

18.3.9 Field validation

Field validation of endpoints is essential to reduce laboratory to field extrapolation error. Uncertainty in field validation is lowest for survival tests. For example, field-collected sediments which kill amphipods in laboratory tests have been shown to be associated with areas where amphipods are rare or absent (Ferraro *et al.* 1991; Swartz *et al.* 1994). Chronic endpoints have also been field validated to a lesser degree (DeWitt *et al.* 1992; Canfield *et al.* 1994; Chapter 8). The life table assessment is a promising endpoint, but there has been little opportunity to date for field validation. The field validation of biomarkers in relation to adverse biological effects has been equivocal. As methods for evaluating effects on communities become more sophisticated, rigorous and quantitative comparisons with toxicity endpoints will become possible. Such comparisons should reduce the uncertainties which lead to the caution many ecologists feel about comparisons with simple measures of communities (Diaz 1992).

18.3.10 Summary

The uncertainty associated with survival is less than that of the other endpoints used in sediment toxicity tests (Table 18-2). This is because mortality is an extreme response with obvious biological consequences. Also, a substantial literature concerning survival in sediment toxicity tests has been created during the last two decades. Biomarkers have significant sources of uncertainty as sediment toxicological endpoints, especially with respect to ecological relevance and interferences by nontreatment factors. The continued development of more sensitive and ecologically relevant endpoints (*e.g.*, chronic effects on growth and reproduction, life-cycle tables) has the potential to produce superior measurement endpoints for use in SERAs.

Toxicity tests, in and of themselves, are not useful for identifying contaminants responsible for observed responses. Even linkage of test results to the list of chemicals generated during an exposure assessment might prove to be of limited use for defining potential causes of toxicity for a number of reasons: 1) chemicals responsible for toxicity may not have been measured, 2) the bioavailability of chemicals in either pore water or in whole sediment can be quite uncertain, and 3) correlative techniques (*i.e.*, comparison of responses to chemical concentrations) are often unable to deal with multiple contributions from complex mixtures.

Toxicity identification evaluation methods provide a useful approach for assessing toxicity contributions in sediment phases where unmeasured contaminants may be responsible for toxicity or where there are questions as to appropriate bioavailability or mixture toxicity models (Chapter 16). Toxicity identification evaluation methods consist of toxicity-based fractionation schemes which are capable of identifying toxicity due either to single compounds or broad classes of contaminants with similar properties. Overall, TIE methods offer a logical solution for bridging the gap between effects and exposure assessments, and would result in a more accurate and comprehensive characterization of risk. Sediment TIEs have typically been conducted using pore water as the test phase; however, methods are being developed for testing whole sediments (Ankley and Schubauer-Berigan 1995).

18.4 Evaluation of uncertainty associated with benthic invertebrate assessments of sediments

Under the category of benthic invertebrate assessments, an array of measurement techniques are included. We have not considered the use of the tools at a fine level of detail (*e.g.*, type of sampling device, statistical method), as this is beyond the scope of this review. A recent volume edited by Rosenberg and Resh (1993) describes the current state of benthic biomonitoring. Rather, we have evaluated uncertainty at different scales of organization and have made the assumption that users will use the optimal or most appropriate techniques available and be familiar with the current state of knowledge. In those cases where new techniques are in development, they have been specifically mentioned.

Benthic assessment methods were classified at different organizational scales, from the individual to the community level (Table 18-3). The types of endpoints included at these different organizational scales are 1) individual (*e.g.*, morphological changes, biomarkers), 2) population (*e.g.*, indicator or keystone species abundance, population size structure and life history modifications), 3) community structure (*e.g.*, indices, metrics, multivariate approaches), and 4) community function (*e.g.*, functional groups, energy transfer, size spectra). Although community function was considered, there is little information on the use and application of community function in sediment assessment. Therefore, the degree of uncertainty associated with their use is high because of lack of knowledge (Chapter 9).

The primary purpose of benthic invertebrate measurement endpoints (*e.g.*, indicators or measurement tools) is to identify departure of the endpoint from either an expected or predicted condition, given normal variability in both time and space (Chapter 10). Furthermore, measurement endpoints should relate such a departure to a directional stressor. There are an array of uncertainties associated with such a judgment and the decision on whether the measured state has departed from the expected value requires criteria or guidelines for making the judgment. We have identified three major sources of uncertainty which are associated with 1) lack of knowledge of the system about which measurements are being made (*i.e.*, what is a normal or expected state for this system), 2) a

Table 18-3 Uncertainty associated with benthic community assessments

Evaluation criteria	Individual	Population	Structure	Function
Precision	1	1	2	3*
Ecological relevance	3	2	1	3*
Causality: contamination	2*	2*	2*	3*
Causality: source	2*	3*	3*	3*
Sensitivity	1*	1	2	3*
Interference	2*	3*	3*	3*
Standardization	3*	1	1	3*
Discrimination	2	1	1	3*
Bioavailability	2*	NA	NA	3*
Field validation	3*	1	1	3*

1 = low uncertainty (good)
3 = high (bad)
* = lack of knowledge
NA= not applicable

systematic error in the method being used, and 3) the sampling scale selected and inherent to the variable being measured.

The rest of this section discusses relative degree of uncertainty associated with the various measurement endpoints (Table 18-3) and attempts to distinguish uncertainty associated with lack of knowledge (indicated with an asterisk) from systematic or stochastic uncertainty which can be rectified (methodologically) or quantified (sampling decisions and design). We have categorized each type of measurement endpoint into various potential sources of uncertainty, each are described separately and recommendations are made on how to reduce uncertainty in each category.

18.4.1 Precision

Sampling precision refers to the ability to define and reduce variability associated with the measurement. The ability to quantify and reduce this source of uncertainty is methodological, cost dependent, and primarily related to the scale of sample in relation to the "population" being estimated. In reality, the precision decreases as the scale of organization increases, thus measurement of communities is less precise than measurement of an individual. However, the uncertainty of measurements at the community level can be quantified and reduced by appropriate design and effort (Elliott 1977; Green 1979). It is recommended that appropriate studies be conducted to identify cost-effective benthic endpoints in relation to study objectives and available resources to reduce or quantify the uncertainty associated with problems of precision.

A study to develop predictive models of invertebrate community structure for the Fraser River basin in British Columbia involved the sampling of 250 reference sites throughout the 240,000-km² basin. To establish preliminary levels of precision, calibration studies with 10 replicate samples were used to establish expected variation based on replicates taken from separate riffles. This suggested that five replicate samples were sufficient to produce coefficients of variation of 20% to 70 % for the major families (Figure 18-1). More importantly, further replication did not improve precision. A subsequent study at 22 sites using multivariate analysis of three replicates from each site showed that within-site variation (based on replicates) was much less than between site replication. This is illustrated by 21 of 22 site replicates being clustered adjacent to each other (Reynoldson TB, unpublished data). As a consequence of this study, it was decided that, for identification of communities at a regional scale, single samples from a riffle were sufficient. This resulted in a five-fold reduction in cost for sampling and processing invertebrate samples without affecting the certainty of a site being correctly classified.

18.4.2 Ecological relevance

In this context, ecological relevance refers to the relation of the measured endpoint to the benthic ecosystem. Accordingly, direct measures of the populations of organisms present have a higher certainty of being related to ecosystem than measurements at a finer organizational scale (Schindler 1987). The concept of using biological indicators of pollution began in Germany at the end of the last century (Kolkwitz and Marsson 1908), and by the early 1980s, more than 50 different methods had been identified (De Pauw and Vanhooren 1983). Three separate assessment approaches using macroinvertebrates have been distinguished (Metcalfe 1989; Johnson et al. 1993), based on either taxonomic or pollution tolerance data: 1) saprobic, 2) diversity, and 3) biotic index approaches.

Saprobic approaches are based on the dependence of organisms on organic food sources, and saprobic values have been published for more than 2000 European species (Sladacek 1973). This approach also incorporates organisms other than macroinvertebrates. Diversity indices were first applied to pollution assessment by Wilhm and Dorris (1968) and were first seen as a method for incorporating ecological principles into assessments and avoiding the value judgments inherent in score systems. However, their use has been frequently criticized. Proponents of diversity indices argue that the indices measure real ecological properties and processes and are based on ecological theory, e.g., diversity/stability hypothesis (Goodman 1975) and competitive interaction (Hurlbert 1971). The most widely used index is the Shannon-Wiener index (Shannon 1948). These indices reach their maximum value when all species are evenly distributed, and high diversity is considered desirable condition. However, such an even distribution of species is not normal in most communities, and therefore the underpinnings have to be questioned. The actual meaning of diversity and how it should be measured is also questionable and has been the subject of much debate (Washington 1984). Furthermore, different types of

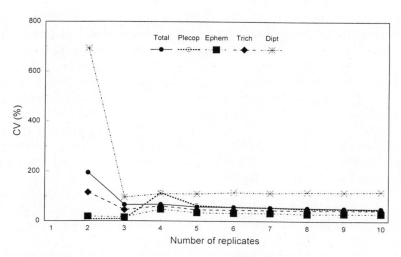

Figure 18-1 Variation based on the number of replicates taken from separate riffles

pollution may have different impacts in different habitats, and major changes in species composition may occur with no change in diversity. However, in a survey of methods being used in assessment (Resh and Jackson 1993), 40% of studies had included diversity as part of a biological assessment.

The biotic index approach uses a single index value to summarize information and to assess pollution effects on aquatic communities and has appeal in conducting SERAs. The majority of biotic indices are based on indicator organisms rather than community composition. However, diversity indices, which are also considered by some as a form of biotic index, use attributes of community structure such as numbers of taxa and the nature of their distribution. The indicator organism concept assigns scores to species based on their sensitivity or tolerance to various stressors, usually organic pollution. This approach tends to be specific to one or two types of pollution (usually organic) and consequently such systems are not universally applicable. For all indices, the calculations require a quantitative or qualitative estimate of the individuals or taxa present, counts (or estimates of the biomass) of specific groups of organisms (*e.g.*, Ephemeroptera, Plecoptera, Trichoptera, Chironomidae), and lists of the responses of different taxa to pollution or separation of invertebrate groups based on function (*e.g.*, feeding strategy). Their use is either based on a comparison of a test site to a theoretical or empirically derived score (often based on low-order stream communities) or based on an upstream-downstream comparison.

The biotic index approach has been defined (Tolkamp 1985) as one which combines diversity on the basis of certain taxonomic groups with the pollution indication of individual species (or higher taxa) into a single index or score. As such, it can be considered as a combination of saprobic (scores) and diversity concepts. An array of such biotic indices or biotic score systems have been developed; however, the majority evolved from the Trent Biotic Index (Woodiwiss 1964). From the proliferation of indices from 1960 to 1980, five current lines can be identified as being used in benthic assessment. The Belgian Biotic Index (Europe) and the Indice Biologique (France) are both true biotic index approaches. The metrics approach, widely used in the United States, has developed from several lines but is primarily adapted from Karr's (1991) Index of Biotic Integrity and, as such, is also a biotic index. The other main line derives from the RIVPACS approach (Wright *et al.* 1984), used in the United Kingdom. The RIVPACS approach was a major divergence from the biotic index approach as it incorporated a predictive capability and relied on the assessment of community structure rather than the use of indicator organisms. This forms the basis for methods being used in the National River Health Programme in Australia (Commonwealth Environmental Protection Agency [CEPA] 1994) and the BEAST models being developed in Canada (Reynoldson *et al.* 1995) for the Great Lakes and Fraser River. Community structure measurements have a low degree of uncertainty in terms of their ecological relevance, and changes at the community level are highly relevant. However, these approaches have seldom been applied to habitats with soft-bottom substrates where contaminated sediments are of concern. Specific morphological endpoints may have relatively high uncertainty with regard to their ecological relevance (Johnson *et al.* 1993). Even the loss of an entire population of a species may or may not have ecological relevance at the ecosystem scale depending on the functional attributes of the particular species.

18.4.3 Causality

Two types of causality have been identified: 1) the relation of measured effects to specific contaminants and 2) the linkage with specific sources (*i.e.*, spatial or geographic location). Relating effects to cause in field assessments is inherently empirical and correlative, requiring associated chemical and laboratory effects-based data. Each level of organization listed in Table 18-3 can provide information on spatial origin of effect with an appropriately designed sampling grid and effects mapping methodologies. Lack of knowledge at present on the specificity of response at the individual level (Graney and Giesy 1987; Giesy 1988) precludes linkage to specific contaminants, perhaps with the exception of metallothionein or acetylcholinesterase inhibition measurements (Table 18-3). However, further research in this area may provide useful measures that could reduce the uncertainty of cause-and-effect linkages (Farris *et al.* 1989; Day and Scott 1990).

At the other extreme, community measurements have more potential to identify effects (reduce false negatives) but provide limited information on specific contaminants or stressor causing the response. The endpoints that include evaluation of multiple species are more useful in mapping source origin compared to population studies. Furthermore, research on community response in dose-response experiments might potentially lead to the use of community response in diagnosing specific contaminants (categories) in multiple source situations (Chapter 9). It is recommended that further research is needed to identify contaminant-specific biomarkers and species sensitivity to contaminant categories.

18.4.4 Sensitivity

This category relates to the likelihood that the measured endpoint is responding to a contaminant and is analogous to an analytical detection level for chemical analysis. There is a need to minimize uncertainty associated with both false negatives (toxic sample incorrectly classified as not toxic) and false positives (nontoxic sample incorrectly classified as toxic). Measurements of individuals and biomarkers are inherently more sensitive than measurements at the community level (Kingett 1985; Giesy 1988). In the context of sediment assessments, the relative usefulness of a method can be determined by using criteria adapted from Giesy and Hoke (1989) and Calow (1993):

- Relevance. The ecological significance of the measurement at scales greater than that being tested. In all testing, we are concerned about overall biological structure and function. We cannot test all system attributes. Therefore, the measure being tested is simply a surrogate for the system as a whole. The ease with which we can relate the test to the system is a measure of relevance (Chapter 12). In many cases, this cannot be done quantitatively but is a subjective evaluation.

- Robustness. In order to provide consistency in interpretation and decision-making, it is important that the tests provide consistent results irrespective of when and where they are conducted. This is clearly more difficult in the case of field tests where environmental variation can be large. However, standardized methods can reduce a significant portion of the variation, and some measurement variables are inherently more robust in either time or space.

- Methodology. Both ease of use and detectability of the measurement endpoint (or availability of test organisms in laboratory tests) are important criteria. If the method is too complex, expensive, or uses an inappropriate duration, it is of limited value for routine use.

- Sensitivity. The variable chosen for measurement should be sensitive to toxicants and in an appropriate range, sufficient to detect a response but also to discriminate between measurement locations.

- Appropriateness/Application. The method being selected should be appropriate to the system for which the study is being designed; many useful methods are only applicable to specific situations or environments. Some methods have much broader application than other techniques.

Available methods cover the entire range of biological organization from the simplest (*e.g.* the cell) to most complex (*e.g.*, ecosystem). As Calow states (1993) "...moving down the hierarchy [ecosystem to molecule], systems become more easy to control, reaction times reduce and criteria for robustness, methodology and sensitivity improve, and generality should also increase. [However,] since ecological principles largely involve individuals and groups of individuals relevance [and application] increase in the opposite direction."

In selecting methods, the individual has to balance these two dilemmas, and inevitably compromise decisions have to be made. The effect has occurred by the time a change is measured at the community level. In that regard, there is a low uncertainty associated with the sensitivity of the individual response. Concomitantly, there is more uncertainty of the absolute sensitivity of the response at the level of the individual because of less knowledge of the test; however, this approach is inherently more sensitive and has an early warning component not present in community measurements.

18.4.5 Interferences

Interferences as a source of uncertainty refers to the confounding effects of biotic and abiotic factors in the interpretation of responses distinguishable from normal variation and contaminants associated with sediment. Until recently, the use of biological methods was often confounded by the temporal and spatial variability inherent in biological systems. However, over the past 10 years, methods developed in the United Kingdom (Wright *et al.* 1984; Armitage *et al.* 1987; Moss *et al.* 1987; Ormerod and Edwards 1987) and elsewhere (Corkum and Currie 1987; Johnson and Wiederholm 1989) have demonstrated the ability to predict the community structure of benthic invertebrates in clean (or "uncontaminated") sites using simple habitat and water quality descriptors. This approach has allowed the development of site-specific biological objectives to be set for ecosystems selected by habitat characteristics and also provides an appropriate reference for determining when degradation at a site is due to anthropogenic contamination. The acceptance by regulatory agencies of biological water and sediment quality objectives are being considered in Canada (Reynoldson and Zarull 1993; Reynoldson *et al.* 1995), the United States (Barbour *et al.* 1995), and the United Kingdom (the RIVPACS method; Wright *et al.* 1984), and recent initiatives in Australia (CEPA 1994). Because of the smaller scale of factors accounting for responses of individuals, the uncertainty associated with interference is lower. However, few studies have been conducted to examine the role of interference on these approaches. There is limited information available on normal levels for many biomarkers. Population- and community-level measurements have continually been confounded by the influence of abiotic factors and an inability to distinguish between directional (*e.g.*, response to a trend or gradient) and nondirectional (*e.g.*, seasonal or annual) variability.

Approaches using reference areas have allowed the development of predictive models of community structure that minimize nondirectional variability (Wright *et al.* 1984; Johnson and Wiederholm 1989; Reynoldson *et al.* 1995). This is the primary reason for the use of multivariate rather than metric community structure assessment methods. Recent developments in the analysis of biological data using multivariate statistical techniques have shown promising results in interpreting changes in community structure based on environmental parameters. As a first step toward the identification of the best achievable community for a specific type of habitat, there is a need to define reference communities based on chemical, physical, geological and geographical features in areas free from contamination. The greatest effort in this direction has been made in the study of riffle habitats in streams and rivers of the United Kingdom (Wright *et al.* 1984;

Armitage *et al.* 1987; Moss *et al.* 1987). Similar studies have been conducted in North America (Corkum and Currie 1987), continental Europe (Johnson and Wiederholm 1989), and South Africa (King 1981). An extensive collection of data at nonpolluted sites in the United Kingdom has resulted in the identification of a number of natural communities (Wright *et al.* 1984). Environmental variables such as latitude, substrate type, temperature, and depth were used to correctly predict the benthic communities at 75% of 268 sites, and at more than 90% of the sites, the observed community was similar to either the predicted or the next most similar predicted community. At lower levels of community detail, even greater accuracy of prediction was observed. Other studies have shown a similar predictive capability (see Table 18-4), and the accuracy of prediction in these studies ranges from 68% to 90% in habitats varying from large lakes to small streams and on three continents. The expected community assemblages at a location can be defined from a predictive model using a limited number of physical variables. Such predicted communities or key species in the communities can be used to establish site-specific guidelines, which may be compared with the actual species composition. Thus, determinations on whether or not the guideline is being met can be made. It is recommended that a standard set of environmental attributes be established to foster future development of models for predicting variability of community structure in reference areas and that these approaches be further applied to habitats with soft-bottom sediments where contaminated sediments are of concern.

18.4.6 Standardization

The possible degree of standardization is high for population and community measurements; therefore, the degree of uncertainty resulting from standardization can be minimized. There are many reviews of sampling devices, methods, and handling procedures, and consistent methods are an absolute requirement (Elliott 1977; Plafkin *et al.* 1989; Norris and Georges 1993). Many studies have illustrated that apparent responses resulting from methodological changes (*e.g.*, mesh size, grab) can outweigh effects resulting from contaminants. The uncertainty resulting from a lack of standard methods with use of biomarkers and with individual tests is high because of the relative newness of these methods. It is recommended that studies be undertaken to establish methods appropriate to the issue of concern and that methods not be changed once established. Where possible, common methods should be used. If a methodological change is essential, then appropriate calibration is necessary.

18.4.7 Discrimination

Discrimination refers to the ability of a measure to identify a response range. This is related to the continuous nature of the measurement endpoint and the variables associated with that measure. The large number of species involved in community measures reduces the uncertainty that an endpoint will not be unable to discriminate between contaminant levels. While it could be argued that the loss of one or two species may be missed at the community level of analysis, the loss of single species is not necessarily of significance unless they affect the overall community structure, at which point such effects will be

Table 18-4   Ability to predict benthic invertebrate community structure

Habitat	Number sites	Number community groups	Predictor variables	Prediction success rate	Reference
Lakes	68	7	11	90.0	Johnson and Wiederholm 1989
Lakes	95	6	9	90.0	Reynoldson *et al.* 1995
Rivers	35	5	6	75.4	Reynoldson and Zarull 1993
Rivers	45	6	5	68.9	Ormerod and Edwards 1987
Rivers	54	5	9	79.6	King 1981
Rivers	286	16	28	76.1	Wright *et al.* 1984
Streams	79	5	26	70.9	Corkum and Currie 1987

observed by comparison of changes at the community level. New approaches to community-level analysis can identify normal variation in unimpacted communities and detection of change at a pre-selected probability level (Reynoldson *et al.* 1995). Furthermore, assessment methods being used in the United Kingdom and Australia allow the prediction of the number of taxa at different taxonomic levels (*e.g.*, family, species) and therefore the absence of single species can be established, again at predetermined probability levels (CEPA 1994; Wright 1995). Conversely, a single individual or biomarker has higher uncertainty of being a useful discriminator. Decisions require comparison of the measurement with some reference or control value. There is inevitably a level of uncertainty associated with such decision-making. Community- and population-level criteria are being established (Karr 1990; Davis and Simon 1995; Reynoldson *et al.* 1995; Yoder and Rankin 1995). However, we are unsure as to whether analysis has been conducted on the degree to which error has been established in the use of such criteria. It is recommended that an analysis of the error associated with benthic criteria be undertaken.

18.4.8 Bioavailability
Community-level assessment methods (*e.g.*, diversity indices, biotic indices) can be affected by interference from noncontaminant effects; however, newer approaches being used to help address this issue. The use of reference sites to establish expected normal variation is an effective method of addressing this issue, and approaches to establishing reference conditions are addressed (Hughes 1995; Omernick 1995; Reynoldson *et al.* 1995; Wright 1995). Because of the lack of knowledge on the specificity of response in biomarkers, we are more uncertain of the meaning of a negative result in biomarker tests.

18.4.9 Field validation
The responses in populations or communities are field responses. We are uncertain as to the degree of validation required for biomarkers; however, some relation to other field measurements is necessary. Field validation is understood to imply the likelihood of the

measure being used is an appropriate indicator of risk associated with sediment contamination. The uncertainty associated with the accuracy of benthic community assessments is primarily either systematic or results from a lack of knowledge (*i.e.*, background variability). Insufficient information exists on the response of endpoints at the level of the individual or the population to establish that a change in a measured endpoint is a specific response to a contaminant. Insufficient information exists to establish sensitivity of species or individuals to sediment-associated contaminants. However, community-level assessments are more likely to include sediment effects as the number of species examined increases the probability of observing a sediment-associated response. It is recommended that the relations should be examined between responses measured using biomarkers and responses measured at the population and community level.

18.4.10 Summary

The use of benthic invertebrates in the assessment and management of contaminated sediments has been recommended as part of the sediment quality triad approach, which incorporates aspects of sediment chemistry, toxicity testing, and analysis of the community structure of the benthos (Long and Chapman 1985; Canfield *et al.* 1996). The IJC (1987, 1988) has suggested a sediment management strategy that incorporates both assessment and remediation. The United States has embarked on a program to examine various approaches to the assessment and remediation of contaminated sediments (ARCS program; Fox and Tuchman 1996), and Canada is also addressing these issues via the Great Lakes Action Plan. However, there are still major objections to the use of the structure of the benthic community to set sediment criteria or objectives. The main criticisms are their lack of universality (*i.e.*, they are, of necessity, site-specific) and the inability of researchers to establish quantitative objectives for their application (*i.e.*, what should the community "look" like?). Universal guidelines may not be possible due to the very nature and complexity of sediment–contaminant–biota relationships and the diversification of biological and geological components over a large regional area.

Recent developments in the analysis of biological data using multivariate statistical techniques have shown promising results in interpreting changes in community structure based on environmental parameters. As a first step toward the identification of the best achievable community for a specific type of habitat, there is a need to define reference communities based on chemical, physical, geological, and geographical features in areas free from contamination. Selecting only a few physical variables, the expected community assemblages at a location has been defined from predictive models. Such predicted communities or key species in the communities can be used to establish site-specific guidelines, which may be compared with the actual species composition. Thus, determinations could be made on whether or not the guideline is being met. While this reference community approach has been applied to riffle habitats at a variety of locations, additional studies are needed to further evaluate this approach in habitats with soft-bottom sediments where contaminated sediments are of concern. To date, this approach has only been used on community assemblages with species as the classifying variables. There is no reason why the same approach could not be applicable to either structural or functional variables such as reproduction, growth, or biomarker tests that can be included in a multivariate design to classify both predictable communities and responses in tests.

18.5 Evaluation of uncertainty associated with bioaccumulation assessments of sediments

The principal use of bioaccumulation is to estimate the exposure or dose which organisms encounter in a sediment. Bioaccumulation is not an appropriate assessment approach for contaminants that are metabolized or otherwise not accumulated in the tissues of the organisms being evaluated. Four general approaches for bioaccumulation are used in risk assessment: 1) a laboratory approach exposing organisms to sediment under controlled conditions, 2) a field approach collecting species from a study area, 3) assessment of food web transfer, and 4) models to predict bioaccumulation processes (Chapter 15). Uncertainties associated with the approaches are summarized in Table 18-5.

Laboratory approach

Individuals of a single species are exposed under controlled laboratory conditions to sediments collected from the study area being assessed (USEPA 1994a; ASTM 1995e). After an established period of exposure, the tissues of the organisms are analyzed for the contaminant of concern. Bioaccumulation has occurred if the final concentration in tissues exceeds concentrations that were present before the exposure was started. This requires, of course, that individuals representative of initial conditions are also analyzed.

Field approach

Tissue concentrations of contaminants are determined by collecting one or more species exposed to sediments at the study area being assessed. These concentrations are compared to some reference areas for interpretation. Two methods have been used to determine bioaccumulation in the field: 1) organisms resident at the area are collected *in situ* for analysis or 2) organisms are transplanted from another location (presumably with a history of little contaminant exposure) to the area of concern, then re-collected, and tissues are analyzed after an established period of exposure. The first approach takes advantage of organisms which have integrated exposure to the sediments in the area over their lifetimes. However, if sediments in the area are highly contaminated or if environmental conditions in the area are otherwise unsuitable, no appropriate organisms may be available (Cain *et al.* 1992). For example, hydropsychid caddisflies have been successfully used to assess changes in metal exposures over 400 km of a river receiving input of mining wastes (the Clark Fork River in Montana). Although relatively tolerant to metal contamination, these organisms do not survive in reaches of the river adjacent to a mining-smelting complex. Thus, downstream assessments of exposure in less contaminated areas were possible, but no data could be obtained in the reach of the stream most affected by the contamination (Cain *et al.* 1992).

Comparisons among several species are employed in some bioaccumulation studies (although different species should never be combined in a single sample, unless a lack of

Table 18-5   Uncertainty associated with bioaccumulation assessments

Evaluation criteria	Food web	Models	Laboratory	Field
Precision	3	3	1	2
Ecological relevance: protection of ecology	3	3	3	3
Ecological relevance: protection of human health	1	1	1	1
Causality: source identification	3	1	1	3
Causality: sensitivity (detection limit)	3*	3*	1	2
Interferences	NA	NA	2	2*
Standardization	2	2	1	1
Discrimination	1	1	1	1
Bioavailability	1	1	1	1
Field validation	2*	2*	2*	1

1 = low uncertainty (good)
3 = high (bad)
* = lack of knowledge
NA = not applicable

difference is proven). NOAA's Mussel Watch employs mussels in marine ecosystems, oysters in brackish waters, and freshwater clams in the least saline or freshwater habitats. Data from different species have been useful in some descriptions of exposures (Campbell *et al.* 1988; Cain *et al.* 1992), but comparing tissue residues among species can be complex if the objectives of the exposure assessment are quantitative.

Transplantation of organisms assures that organisms will be available for analysis (unless the exposure results in mortality or avoidance). However, the exposure is typically limited in duration (less than a generation time), and uncertainties about the kinetics of bioaccumulation may affect the accuracy of the results. The behavior of the transplanted organism or whether the organism was exposed to contaminants via the routes which are important for resident species is a source of uncertainty in this approach. For example, mussels (*Mytilus edulis* or *Mytilus californianus*) are often exposed in "bagged bivalve" deployments, sometimes where mussels do not live naturally. Inadequate availability of food or other stresses in the deployment area could, at least theoretically, result in different bioaccumulation than occurs in resident species. Careful comparisons of bioaccumulation in resident and transplanted species are rare (Cain and Luoma 1985).

Models
Models which describe bioaccumulation are relatively well developed for both organic and inorganic contaminants (Thomann 1989; Chapter 15). Toxicokinetic models have a

long history, as do simpler models of bioaccumulation processes. Site-specific models predict bioaccumulation on the basis of laboratory-determined characterization of biological processes in the species of interest and field-determined chemical measurements at the area of concern. Predictive modeling is relatively promising in this field because bioaccumulation is, arguably, less complex than some of the other endpoints. Some uncertainties remain unresolved in most models and consensus does not exist about the appropriate model to apply for some (if not all) contaminants (Chapter 15).

Models of two types are considered in bioaccumulation. Equilibrium models are common in risk assessment and in sediment criteria. Geochemically robust models of equilibrium partitioning are available for application for both organic and inorganic contaminants. These can be relatively simple to apply. The models assume contaminant concentrations among all compartments of the environment are controlled by thermodynamics and at least approach equilibrium conditions. If thermodynamic equilibrium exists, and if one route of uptake is known or can be predicted, overall bioaccumulation is inferred. Recent applications use an extension of the equilibrium models, termed *kinetic* or *pathway models* (ASTM 1995e). These models incorporate recognized geochemical principles but recognize the uncertainties in the assumptions of equilibrium. Kinetic models assume that routes of bioaccumulation are additive and must be determined independently. Kinetic models and equilibrium models may yield similar results if contaminant distributions and concentrations in an environment are at equilibrium (although not always) but can yield very different results where environmental compartments are not at equilibrium (*e.g.*, if biological processes control concentrations, speciation, or phase partitioning of contaminants). Determining whether equilibrium conditions exist in the area is difficult.

18.5.1 Precision
Precision is the stochastic, and perhaps systematic, uncertainty that occurs in repeated determinations of bioaccumulated tissue concentrations (*stochastic* is used in a purely statistical sense, in that much of the uncertainty has its origins in biological processes which are at least partly understood). Variability is a common problem in bioaccumulation studies and can lead to imprecise estimates of exposure (Lobel 1987), although methods for determining bioaccumulation have identified procedures for avoiding extreme sources of uncertainty (Phillips 1980; Campbell *et al.* 1988; Cain et al, 1992; Crawford and Luoma 1992; Phillips and Rainbow 1993; USEPA 1994a; USEPA and USACE 1994; ASTM 1995e).

Laboratory bioaccumulation tests are potentially the most precise of bioaccumulation approaches (Table 18-5). However, their precision is directly dependent upon biological factors such as the selection of appropriate test organisms; use of healthy and acclimated organisms; avoidance of anaerobic conditions and other sources of ancillary stress in the test; and use of organisms of similar size, age, and life-history stage. Many organisms are capable of avoiding potentially toxic concentrations of contaminant sediments. Therefore, exposure to extremely contaminated sediments can result in a highly variable bioaccumulation response by some test organisms (ASTM 1995e). Even under the best of conditions, individuals can respond very differently to a contaminated sediment. Thus, the number of individuals exposed in the test ($n>12$ is ideal; $n<5$ may be unacceptable) is a crucial consideration, as is the number of replicates analyzed if individuals are composited for analysis. If organisms accumulate sufficient contaminants to elicit toxicity, the kinetics of bioaccumulation may be altered and organisms exposed to lower doses in some sediments may accumulate greater concentrations (Landrum *et al.* 1994).

Evaluation of organisms exposed in the field and food web analyses require the same careful attention to methods as described above for laboratory studies. The inherent variability of nature is added to the potential for imprecision which can be associated with laboratory studies. Number of individuals sampled, number of composites, life stage, size of organisms, biases from analysis of gut content, or surface contamination are examples of uncertainty associated with field approaches. Cain *et al.* (1992) and Crawford and Luoma (1992) have addressed some of these sources of variability in the use of field-collected organisms (*i.e.*, estuarine bivalves, riverine insect larvae). Large sample sizes as well as characterization of potential biases such as species-specific and element-specific responsiveness to contaminants, organism size, seasonal growth, temporal variability, and spatial variability on different scales can improve precision and interpretability of bioaccumulation data. Rigorous methods resulted in the statistical power to separate 15% to 25% differences in mean tissue concentrations with 95% confidence (Brown and Luoma 1995). When resident species are sampled, an additional source of uncertainty is the ability to obtain sufficient numbers of individuals to reduce imprecision or replicate composite analyses. This can often be overcome by careful selection of target species. The field approaches are not discriminatory alone to determine if sediments are the source of the exposure. Models and laboratory exposures are needed to better identify routes of exposure.

Bioaccumulation models have been ranked as imprecise because of the large knowledge gaps which remain in identifying values for model parameters (Table 18-5). Some key parameter values differ by orders of magnitude among studies, and predictions differ accordingly. Rapid advances are overcoming this difficulty and generalized predictions are developing, but until scientific consensus about model approaches is reached, model imprecision in site-specific predictions must be considered a significant problem.

18.5.2 Ecological relevance

Ecological relevance includes both relevance to ecological change and relevance to human exposure pathways. A limitation to the bioaccumulation approach is its weak link to ecological change. Bioaccumulation does not mean an adverse effect is occurring. Organisms are capable of detoxifying, adapting to, or otherwise surviving some dose of all contaminants. Correlations between bioaccumulated contaminants and effects of contaminants are also not as well established. Widdows and Donkin (1989) showed a strong link between tissue residues of organic contaminants and physiological measurements of health in mussels. Similar linkages are not established for metals. Bioaccumulation of specific biochemical fractions of metals needs to be better understood before such linkages will be possible. In addition, the biochemical status of organisms that may affect toxicity needs to be better understood.

The relevance of bioaccumulation stems mainly from its value in characterizing exposures. Understanding dose is essential to understanding effects, whether dose is from the chemical itself or from its metabolites. The concentration of contaminants assimilated into tissues may be the most sensitive measure of the dose of contaminant an organism experiences, given the uncertainties in defining bioavailability from chemical measures alone. Exposures are also complex in nature (Luoma *et al.* 1985, 1990); the temporal and spatial complexities of exposures are an important reason why effects of contaminants are difficult to demonstrate with simplistic studies (Luoma and Carter 1991).

Collection of organisms exposed in the field, food web bioaccumulation estimates, and both empirical and site-specific models provide direct determination of contaminant concentrations in aquatic resource (food) species and provide information for pathways of human exposure (Table 18-5). Where tissue concentrations are directly determined in the food organism, there is little uncertainty about relevance. The precise human pathway exposure is predicted with less certainty if analyses of a surrogate species are used to estimate human exposures from a variety of species in an environment. Nevertheless, carefully designed studies can reduce uncertainty. Laboratory studies are not highly relevant predictors for human exposure because they are often poor predictors of steady-state field bioaccumulation. The exception is when laboratory studies are designed to provide parameter values for models.

18.5.3 Causality
Causality includes linkage to sources and pathways or linkage to the contaminant causing the effect. Field determinations reflect a combination of exposure pathways. Bioaccumulation data alone cannot provide information about whether the source of exposure was water, sediment, or pore water. Uncertainty about the route of exposure for field bioaccumulation leads to uncertainty in the problem formulation step of the risk analysis (Chapter 1). Without knowledge of the most important route of exposure, it is unclear whether chemical measurements should include water, sediment, pore water, or some fraction of each. Models which combine biological parameters and field data on water and sediment concentrations can help resolve this problem (Thomann 1989; Luoma *et al.* 1990). Bioaccumulation is the strongest of the endpoints in drawing linkages to the contaminant of concern, because it involves direct determinations of those contaminants. However, bioaccumulation alone cannot be used to evaluate effects of contaminants on aquatic organisms.

18.5.4 Sensitivity
Bioaccumulation is a readily detectable response because it measures exposure of an organism to relevant contaminants. False positive responses are not expected with bioaccumulation estimates. False negative determinations of whether bioaccumulation is occurring should be less prevalent compared to the other measurement endpoints discussed in previous sections, but systematic error in identifying exposure (false negatives)

can result from selection of inappropriate contaminants, test organisms, or sampling approaches. For example, bioaccumulation is not an appropriate measure of chemicals which are metabolized (*e.g.*, PAHs in fish; ASTM 1995e). Bioaccumulation is not appropriate for determining exposures to ammonia or some metals, which are not bioaccumulated before exerting toxic effects. Bioaccumulation of Zn by some organisms may be physiologically regulated (ASTM 1995e). Some species of trout in freshwater streams and some species of bivalves in estuarine environments show little or no change in tissue Zn concentrations in the presence of contamination, whereas other species from the same groups do respond (Moore *et al.* 1990). Mussels and some other filter-feeding bivalves are responsive to inputs of most metals but respond weakly to copper contamination (Brown and Luoma 1995). Laboratory studies and transplant field studies allow control of the species and contaminant, so the above uncertainties can be controlled. Few choices are sometimes available in selection of species to be collected from the field; this could lead to uncertainty about sensitivities. Models' results will be fraught with uncertainty about sensitivity until widely accepted parameter values are established.

18.5.5 Interferences
Interferences can add uncertainties to bioaccumulation studies. Sediment characteristics are an important source of uncertainty in laboratory bioaccumulation studies (Table 18-5) because collection, transport, and deployment can change sediment characteristics from conditions *in situ*. Natural factors, mostly biological, can add stochastic uncertainty to field determinations of bioaccumulation, a type of interference. Variability in the size of organisms within a sample can increase error by two-fold or more, if size is not accounted for. Brown and Luoma (1995) suggested approaches to reducing interferences which can affect interpretation of field bioaccumulation data.

Variability over small spatial scales can add to the uncertainty of discriminating between areas on larger scales. For example, temporal variability can also be a great source of uncertainty. In some environments, seasonal variability in metal concentrations or organic contaminant concentrations can reach four- or five-fold (Luoma *et al.* 1985). In other environments, such as many coastal marine systems, seasonal variability can be low. Year-to-year variability in metal concentrations in insect larvae collected at low flow from the Clark Fork River (in August) is more than two-fold between some adjacent years and more than four-fold between other years.

18.5.6 Standardization
Standardization of approaches is essential in use of the bioaccumulation estimations (see the studies described in the section discussing precision). Use of different methodologies is an important source of uncertainty when attempting to compare different datasets. Consistent approaches for field sampling and laboratory bioaccumulation can reduce uncertainty (Phillips 1980; Campbell *et al.* 1988; Crawford and Luoma 1992; Phillips and Rainbow 1993; USEPA 1994a; USEPA and USACE 1994; ASTM 1995e; Brown and Luoma 1995). Models are the least standardized of the approaches (Table 18-5; Chapter 15).

18.5.7 Discrimination

In the previous applications, this criterion was intended to evaluate the ability of a measurement endpoint to discriminate between contaminated sediments by providing a graded response. Bioaccumulation is inherently a continuous response (with the exception of some of the circumstances described under sensitivity). The ability of bioaccumulation to discriminate contamination gradients with low uncertainty is one of its advantages. Inherently, bioaccumulation is a highly quantitative approach for discriminating the risk of exposure from a sediment.

18.5.8 Bioavailability

Bioaccumulation directly measures bioavailability in both laboratory and field studies. Some qualitative uncertainty in bioavailability (if it is defined as contaminants assimilated into tissues) can occur in determination of whole-tissue concentrations. This is important because whole-tissue concentrations are typically analyzed in bioaccumulation studies. Undigested gut content can be analyzed as part of the tissue burden and cause systematic uncertainties (upward bias) in estimates of bioavailability if contaminant concentrations in food are high compared to tissues (and if food mass in the gut is sufficiently great). Some organisms accumulate contaminants on their surfaces (insect larvae in mine-contaminated streams) or cell walls (microalgae). Contaminants in gut content and on animal surfaces will be consumed by predators, so there is no consensus on the need to purge all undigested contamination from the gut of organisms. Even considering the above caveats, bioaccumulation is the least ambiguous approach for assessing contaminant bioavailability.

18.5.9 Field validation

Field approaches with resident species directly determine bioaccumulation (field validation is inherent). Laboratory bioaccumulation can be validated. When enriched tissue concentrations are observed in the laboratory, they are expected in the environment. However, there is uncertainty in quantitatively attempting to relate laboratory bioaccumulation results to the field. Laboratory bioaccumulation over days or weeks reflects a mixture of influx and elimination kinetics and may not reflect lifetime exposures of organisms in the field. Some studies, especially with small organisms, have successfully related bioaccumulation obtained in the laboratory with field-collected sediments to residue concentrations observed in synoptic collection of organisms from the field (Ankley et al. 1992; Brunson et al. 1996). Combined laboratory and model studies show greatest potential for reducing uncertainties in extrapolating from controlled studies to nature.

18.5.10 Summary

The principal use of bioaccumulation is to estimate the exposure or dose which organisms encounter in a sediment. Bioaccumulation is not an appropriate assessment approach for contaminants which are metabolized or, for other reasons, not accumulated in the tissues of the organisms being evaluated. Another limitation to the bioaccumulation endpoint is its weak link to ecological change. Bioaccumulation does not mean an adverse effect is occurring. The relevance of bioaccumulation stems mainly from its value in characterizing exposures and understanding the dose an organism experiences. This can be especially valuable information if used to expand understanding of bioavailability or if exposures are complex in space or time (as is often the case) at the site of interest. Bioaccumulation can be a highly variable endpoint, but if established methods are followed and sample size is adequate, variability, imprecision, and insensitivity can be controlled.

18.6 Evaluation of uncertainty associated with chemistry measurements of sediments

Sediment chemistry data represent a fundamental element of risk assessments focused on evaluation of the effects of toxic chemicals in sediments. Chemical concentrations are often reported on a dry weight basis, based on a total extraction of sediment samples. However, measurements of total chemical concentrations can be "normalized" to certain physical-chemical characteristics of sediment to aid in interpretation of analytical chemistry. For example, the concentrations of nonionic organic contaminants have been normalized to TOC concentrations in sediment (Swartz *et al.* 1990; Di Toro *et al.* 1991). In addition, AVS-normalization procedures may be used to interpret metals data (Di Toro *et al.* 1990, 1992; Ankley, Phipps, *et al.* 1991). Alternatively, metal concentrations may be normalized to those of a reference element (such as Al or Li; Schropp *et al.* 1990; Loring 1991) or to grain size (Chapman 1992). In some cases, the concentrations of contaminants in porewater and elutriate samples are also determined. Some of these normalization procedures are intended to better define the bioavailable fraction of the substance under consideration, whereas others provide a means of comparing contaminant concentrations to background levels.

The uncertainty associated with the following measures of sediment chemistry was evaluated: 1) whole sediment analysis using total extraction of sediments (Förstner 1990); 2) normalization of nonionic organic contaminants to total organic carbon concentration of sediment (Di Toro *et al.* 1991); 3) metal speciation as derived by AVS or by evaluating other partitioning phases (Di Toro *et al.* 1992); 4) concentration of contaminants in porewater samples (Adams 1991); 5) concentrations of contaminants in elutriate samples (USEPA-USACE 1991, 1994); and 6) concentrations of reference elements (*e.g.*, Al and Li; Schropp *et al.* 1990; Loring 1991) which are regional reference levels to which contaminant concentrations are compared. These methods are also used to eliminate particle size and other natural interferences (these include Al or Li normalization or sieving to remove particles > 64 μm diameter; Chapman 1992).

These categories consider a variety of sediment phases commonly used to assess contamination (Table 18-6). In particular, measurements of sediment chemistry have been used as an initial screen to identify sediments and substances of potential concern. The available information does not support a quantitative evaluation of the uncertainty associated with each of the sediment chemistry measures considered in Table 18-6. For this reason, an evaluation of the relative uncertainty of each approach was conducted to provide guidance on the selection of screening tools. It should be noted that this assessment addresses the uncertainty associated with the use of sediment chemistry alone in sediment assessments. A lower level of uncertainty would be assigned to several of the chemistry measures if they were used in combination with other measurement endpoints (*e.g.*, toxicity tests, benthic community assessments, SQGs). The workgroup also attempted to account for uncertainty associated with each analytical chemistry measurement (*i.e.*, whole sediment PCBs) and uncertainty associated with the use of the chemistry measurement alone to interpret sediment contamination (*i.e.*, OC normalization of PCBs).

18.6.1 Precision
Precision was defined in terms of the robustness of the analytical method. That is, procedures that generate similar concentrations in repeated analyses of the same samples were considered to have a lower level of uncertainty than those that generate variable results. The lowest level of uncertainty was assigned to whole sediment, TOC-normalization, SEM/AVS, elutriate, and reference element measurements because a high level of precision can be attained using existing analytical methods. Porewater chemistry and procedures intended to determine the species of contaminant present in the sample (speciation procedures) were assigned a higher level of uncertainty resulting from the lack of routine methods used in these analyses. Rigorous determinations of metals in pore waters require working with small sample volumes (which creates challenges in preconcentrating to measurable concentrations). All determinations of metals require sample preparation and analyses be conducted in ultra-clean conditions (*e.g.*, concentrations in the pore water or elutriates can be in the low-to-sub µg/l range, which creates possibilities for contamination). If sediments are anoxic, these procedures also require the collection and extraction of pore water under completely anoxic conditions (to prevent precipitation of Fe(II)). Failure to rigorously maintain these requirements results in imprecise results, in addition to inaccuracies.

18.6.2 Ecological relevance
With respect to the evaluation of chemistry measures, ecological relevance was evaluated in terms of linkages to receptors that are to be protected. Whole sediment chemistry, elutriates, and reference elements were rated low since these approaches are not based on measures of bioavailability or are not direct measures of ecological relevance. Total organic carbon, SEM/AVS, metal speciation, and porewater measures were rated with a moderate level of uncertainty since these measures are based on evaluating the bioavailable faction of a chemical in sediment. All of the chemistry measures listed in Table 18-6 are most useful when used in combination with approaches that directly measure biological effects (Sections 18.2 and 18.3).

18.6.3 Causality
Determination of causality (*i.e.*, correctly identifying stressors) can be of fundamental importance in SERAs. For this reason, uncertainty in each of the chemical measurements was evaluated in terms of their ability to determine specific linkage to a contaminant of concern or to sources of chemical contaminants. Low uncertainty was assigned to all of the measures of sediment chemistry, except those which determined chemical concentra-

Table 18-6   Uncertainty associated with sediment chemistry measurements

Evaluation criteria	Bulk sediment	TOC norm-alization	SEM metals & AVS	Metal speciation (non AVS)	Pore water	Elutriate	Reference element
Precision	1	1	1	2*	2*	1	1
Ecological relevance	3	2	2	2	2	3	3
Causality: contaminant	1	1	1	1	2	3	1
Causality: source	2*	2	2	2	2	3	1
Sensitivity	1*	1*	1*	1*	1*	1*	1*
Interference	2*	2*	2*	2*	2*	2*	2*
Standardization	1*	1*	1*	3*	2*	2*	1*
Discrimination	1	1	1	2*	1	1	1
Bioavailability	2*	1	1*	2*	2*	3	2
Field validation [a]	1	2	2*	2*	3	3	1

1 = low uncertainty (good)
3 = high (bad)
* = lack of knowledge
a = not related to field sampling

tions in sediment elutriates. The elutriate procedure provides a measure of the contaminants that are released during the suspension of sediment in water. Therefore, preparation of elutriates alters the sediment sample, increasing the uncertainty in the sediment contaminant concentration. For this reason, elutriate samples do not provide a direct measure of the chemical characteristics of sediments *in situ*. Although porewater concentrations provide more direct linkages to whole sediment chemistry, the procedures used to isolate pore water may also introduce uncertainty (USEPA 1994a; ASTM 1995b).

Identification of the sources of particular classes of chemical is central to the SERA process. In this respect, whole sediment chemistry and reference element-based procedures were considered to provide useful measures for evaluating contaminant sources, particularly for certain classes of organics (*e.g.*, PAHs) and for metals (Schropp *et al.* 1990; Environment Canada 1995). In contrast, elutriate chemistry provides limited information regarding the chemical composition of sediments *in situ* or contaminant sources (ASTM 1995d).

18.6.4 Sensitivity
Sensitivity is an important criterion for evaluating the applicability of various measurements of sediment chemistry in SERAs because there is a need to reliably identify sediments with high, moderate, and low contaminant concentrations. Most analytical

methods for determining chemical concentrations in sediments are very sensitive. However, systematic insensitivities (false negatives) and high variability may occur in interpretation of metal contamination in whole sediments if samples are collected from a range of particle sizes. Metal concentrations are surface-area dependent and in general are lower in sandy sediments (Schropp *et al.* 1990). Use of reference element normalization, sieving, or a combination of sieving and reference element normalization can greatly reduce this uncertainty (Loring 1991). TOC-normalization may also help reduce the possibility of false negatives regarding organic contamination in instances where sediments are low in TOC (Di Toro *et al.* 1991). SEM/AVS and other metal speciation procedures can reduce uncertainties about effects of natural processes on metal concentrations (Di Toro *et al.* 1992). For example, both Cd and Ag naturally accumulate to high concentrations in sulfide-rich deposits which occur in unpolluted conditions (*e.g.*, deep oceans). Similarly, elevated levels of petroleum hydrocarbons can occur near natural oil seeps, and dioxin can occur near natural combustion sources (*e.g.*, forest fires). Recognizing naturally high concentrations of sulfide, iron, or organic materials could prevent conclusions that serious contamination sources are present when natural processes are responsible for contaminant concentrations.

18.6.5 Interferences

With respect to chemical measurements, interferences are considered to be factors which impair accurate determination or interpretation of the concentrations of contaminants in sediment samples. In most cases, interferences are related to sample matrix problems and are analyte specific in the categories listed in Table 18-6. Interpretation interferences include particle size variability and anomalous high concentrations of natural sediment components that equilibrate with high concentrations of contaminants. Although rarely quantified, hydrodynamic and hydrologic processes can also interfere with interpretation of the methods listed in Table 18-6. For example, PAHs from local sources occur in lower concentrations relative to TOC in San Francisco Bay, where sedimentation rates are the highest. High inputs of PAH-poor sediments dilute concentrations of the contaminants in bedded sediments compared to areas where deposition of the PAH-poor sediments is lower (W. Pereira, USGS, unpublished data). Seasonal variations in TOC and Fe oxides also appear to drive seasonal variations in metal concentrations in surficial San Francisco Bay sediments (Luoma, 1990).

In the Clark Fork River and other mining-contaminated rivers, fluctuations in the intensity of metal desorption from sediments is biologically driven. Inputs of contaminated materials in these systems from terrigenous sources caused two- to four-fold seasonal variations in sediment metal concentrations (Brumbaugh *et al.* 1994). Interferences also add important uncertainties to the determination of metal species in oxidized sediments. Many of the standard extraction techniques are not directly selective for specific metal forms, although valuable interpretations are possible if specialized techniques are used (*e.g.*, determination of both total and methyl Hg).

18.6.6 Standardization

Standard methods have been developed for virtually all of the analytical procedures considered in this assessment (*i.e.*, by ASTM, USEPA, OECD, Environment Canada). However, there are still few methods available that can effectively speciate metals and metalloids in oxidized sediments or that can be used to measure nonpriority pollutants. As metal speciation is an important determinant of metal toxicity, more research in this area is needed to reduce uncertainty in SERAs. In addition, the standard methods for certain substances need to be refined to achieve biologically relevant detection limits.

18.6.7 Discrimination

In the previous applications, this criterion was intended to evaluate the ability of a measurement to discriminate between contaminated sediments by providing a graded response. In this evaluation, a chemical measure was considered to have a higher degree of uncertainty if it did not provide a means of distinguishing between samples which had different levels of contamination. Analytical methods are very good discriminators. However, the interpretational uncertainties described above for whole sediments add substantial uncertainties to discrimination of contamination using this method.

18.6.8 Bioavailability

For evaluating measures of sediment chemistry, procedures were considered to have a lower level of uncertainty if they incorporated factors which are considered to influence the bioavailability of sediment-associated contaminants. In general, such factors are incorporated through the use of various normalization procedures. Although bulk contaminant concentrations do not explicitly predict the bioavailable fraction, they have been shown in some cases to be generally related to biological responses (Long *et al.* 1995; Ingersoll *et al.* 1996). The relations are probably most common across sediments where contaminants occur in a similar geochemical form and particle size biases are small. The TOC- and AVS-normalization procedures often reduce the level of uncertainty about the bioavailability of nonionic organics and metals respectively (Di Toro *et al.* 1990, 1991). Iron- and TOC-normalization procedures also have been used to predict the bioavailability of metals in some estuarine and riverine habitats (Luoma and Bryan, 1978, 1982; Tessier *et al.* 1984). Exceptions to the successful use of normalization to predict bioavailability occur in the literature, resulting in some uncertainty about how broadly these approaches can be applied to dynamic natural systems (USEPA 1990a, 1996b). 1991, 1992, 1995). Elutriate preparation tends to alter bioavailability in unpredictable ways and, therefore, increases uncertainty.

18.6.9 Field validation

Field validation is an essential element of the SERA process. With respect to the evaluation of sediment chemistry measures, field validation was interpreted in terms of the accuracy of the method. That is, the uncertainty about the extent to which measurements of sediment chemistry reflect actual field concentrations of contaminants was evaluated. Whole sediment chemistry and reference element concentrations have low uncertainty with respect to accuracy because these methods have well-established QA/QC procedures, and because they are applied to samples collected in the field, uncertainty about their field relevance is low. However, a number of uncertainties are associated with the analysis of inorganics (*i.e.*, AVS or metal speciation; USEPA 1990a, 1992, 1995) and with elutriates. These uncertainties stem from alterations of the sediments which organisms are exposed to *in situ*, resulting from sample collection, storage, laboratory treatment, or other methodological procedures.

18.6.10 Summary
Depending on the objectives of the study, most of the methods examined in Table 18-6 represent useful measurements for conducting SERAs. However, the high level of uncertainty associated with elutriate chemistry makes interpretation of these data difficult. This concern is particularly associated to ecological relevance, causality, bioavailability, and field validation.

18.7 Evaluation of uncertainty associated with use of sediment quality criteria and guidelines
Numerical SQGs are useful tools for conducting ERAs and HHRAs. Specifically, SQGs for the protection of aquatic life provide a basis for screening sediment chemistry data to evaluate the potential for adversely affecting benthic organisms. Similarly, residue-based SQGs provide a means of determining if sediment-associated contaminants pose a potential risk to human health or wildlife (*i.e.*, fish, birds). Although SQGs are widely used in sediment quality assessments, the nature and extent of the uncertainty associated with these measures are often overlooked. Nonetheless, this uncertainty can be substantial and has the potential to influence the results of the risk assessment. For this reason, a comparative assessment of the relative uncertainty associated with selected SQGs and related measurement endpoints was conducted (Table 18-7). Recommendations for reducing the uncertainty associated with these tools were also identified.

Sediment quality guidelines are developed both for the protection of aquatic organisms and for the protection of human health and wildlife. A variety of different approaches have been used to derive numerical SQGs for the protection of aquatic organisms; however, all of these approaches fall into three primary categories: equilibrium partitioning, co-occurrence (*i.e.*, derived primarily from field-collected sediment samples), and models. Only the following SQGs are considered in Table 18-7 because these guidelines have been used extensively in a variety of sediment assessment applications:
- USEPA sediment quality criteria (Hansen *et al.* 1993a–1993e). To date, the EqP approach has been used to derive draft SQC for acenaphthene, phenanthrene, fluoranthene, dieldrin, and endrin. However, the approach is amenable to many nonionic organic substances.
- Effects range low and effects range median (Long and Morgan 1990; Long and MacDonald 1992; Long *et al.* 1995; Ingersoll *et al.* 1996). These informal SQGs were derived from an information system containing data from EqP models,

Table 18-7   Uncertainty associated with sediment quality guidelines

Evaluation criteria	SQC	ERL & ERM	AET	SLC	SEM/AVS	TU models	Residue-based SQG
Precision	1	2*	3	3	2*	3*	2
Ecological relevance	2*	1	2	3	1	1	2*(b)
Causality	1	3*(a)	3	3	1	1*	2
Sensitivity	2	1(c)/2	3	1	2*	2	1
Interference (d)	2	2*	2	3	2*	2	1
Standardization	1	1	2	2	1	3	1
Discrimination (e)	1	1	3	3	2	2	1
Bioavailability	1	2*(e)	2*(e)	1	1	2*	1*
Field validation	2*	2*	2*	3*	2*	2	2*

1 = low uncertainty (good)
3 = high (bad)
* = lack of knowledge
a = Uncertainty can be reduced through use of TU models, which help to identify chemicals most strongly associated with observed effects.
b = Few compounds, based on consumption effects
c = ERL
d = Interferences resulting from community responses and mixture effects
e = With normalization

laboratory spiked-sediment toxicity tests, or field studies. The threshold and probable effect levels (TELs and PELs) reported by MacDonald (1994) and MacDonald *et al.* (1996) were considered to be functionally similar to the ERLs and ERMs and, therefore, were not explicitly considered in this evaluation.

- Washington State's sediment management standards (Washington Department of Ecology 1990). These standards are based on the AETs that have been developed using the results of toxicity tests and benthic community assessments (Barrick *et al.* 1988).

- Screening-level concentrations (SLCs) (Neff *et al.* 1986; Persaud *et al.* 1992). These SQGs were derived using matching sediment chemistry and benthic community data from various geographic areas.

- SEM/AVS (Di Toro *et al.* 1990, 1992; USEPA 1995). Using this procedure, metals are not considered to contribute to any observed toxicity when their molar concentration is less than that of AVS.

- Toxic units models (Chapter 16). Toxic units models typically assume that the effects of contaminants are additive and are based on comparisons of chemical concentrations to toxicity benchmarks (*e.g.*, LC50s, SQGs; Swartz *et al.* 1995).

Toxicity may be predicted when the sum of the ratios calculated for each substance exceeds one.

- Residue-based SQGs (Cook *et al.* 1992). These SQGs are derived by determining the chemical concentrations in sediments which are predicted to result in acceptable tissue residues. Using this approach, appropriate tissue residue guidelines (such as FDA action levels for the protection of human health [USEPA 1989a] or the Niagara River fish flesh criteria for the protection of piscivorous wildlife [Newell *et al.* 1987]) are used in conjunction with biota-to-sediment accumulation factors (BSAFs) to derive the SQG.

As all of the approaches listed above have been reviewed and summarized elsewhere, they have not been described in this document. Instead, the reader is directed to Chapman (1989), Beak Consultants Ltd. (1987, 1988), USEPA (1989b, 1990a, 1996b). 1992, 1993, 1995), MacDonald *et al.* (1992), Persaud *et al.* (1992), and MacDonald (1995) for additional information on these approaches.

The available information does not support a quantitative evaluation of the uncertainty associated with each of the SQGs considered in Table 18-7. For this reason, an evaluation of the relative uncertainty of each approach was conducted to provide guidance on the selection of screening tools. The evaluation of uncertainty in Table 18-7 was conducted using the same general criteria which were used in the evaluation of the other measurement endpoints in Tables 18-1 to 18-3 and Tables 18-5 and 18-6. However, these criteria were refined to increase their applicability to SQGs.

18.7.1 Precision

Precision was evaluated in terms of the robustness of the guidelines derivation procedure. That is, procedures which generate similar guidelines using different datasets (*i.e.*, from different geographic areas) were considered to have a lower level of uncertainty than those which generate different guidelines using similar datasets. Precision provides a measure of the broad applicability of the SQGs. In terms of precision, the lowest level of uncertainty was assigned to the SQC because of the extensive toxicology database on which they were derived. High uncertainty was assigned to AETs and SLCs because of the site-specificity associated with their derivation. Although a moderate level of uncertainty was assigned to the ERLs and ERMs, advances in the development of the databases upon which they depend will hopefully reduce this uncertainty in the near future. A moderate level of uncertainty was also assigned to the SEM/AVS-based guidelines because of the micro-spatial distribution of AVS, particularly in relation to biogenic structures. It is important to note that the mixture models (*e.g.*, Swartz *et al.* 1995; Chapter 16) do not have an inherent high level of uncertainty, but instead the methods are not yet fully developed. Hence, it is not yet possible to determine the uncertainty associated mixture models as they apply to assessment of sediments.

18.7.2 Ecological relevance

With respect to the evaluation of SQGs, ecological relevance was evaluated in terms of its linkage to the receptors that are to be protected. Therefore, SQGs that are based on data collected in the field or in the laboratory on appropriate species were considered to have less uncertainty than those that are based primarily on modeling. The effects of mixtures of contaminants in sediments is critically important to evaluations of ecological relevance. Guidelines which directly consider mixtures were assigned a relatively low level of uncertainty (*e.g.*, mixture models, AVS/SEM guidelines, and the ERL/ERM guidelines derived using data from the field which included contaminant mixtures). Existing SQC do not consider mixtures and, hence, were assigned a moderate level of uncertainty. Similarly, AETs were assigned a moderate level of uncertainty because of their inherent potential for incorrectly identifying toxic samples as not toxic (*i.e.*, false negatives). The SLCs reflect the lower bound of ecologically relevant sediment concentrations (*i.e.*, background concentrations) but may not necessarily define actual effect concentrations (*i.e.*, false positives, nontoxic samples identified as toxic). Although the tissue residue guidelines with which the residue-based SQGs were derived are considered to be highly ecologically relevant, more uncertainty is associated with the models which are used to determine the BSAFs. Therefore, a moderate level of uncertainty was assigned to residue-based SQGs.

18.7.3 Causality
Determination of causality (*i.e.*, correctly identifying stressors) is of fundamental importance in SERAs. For this reason, uncertainty in the SQGs was also evaluated in terms of their ability to determine or predict the factors causing adverse effects in contaminated sediments. That is, a lower level of uncertainty was assigned to SQGs that directly identified stressors. The SQC, SEM/AVS, and mixture models were assigned low uncertainty because they are directly derived from experimental determinations of effects of specific chemicals. In contrast, ERLs and ERMs, AETs, and SLCs were assigned higher levels of uncertainty because these guidelines are derived primarily from field observations in which cause-and-effect relations were equivocal (*i.e.*, the sediments contained mixtures of contaminants and, hence, it is difficult to determine the causative agents directly).

18.7.4 Sensitivity
Sensitivity is an important characteristic of a measurement because there is a need to reliably identify sediments with high, moderate, and low probabilities of causing adverse biological effects. Therefore, SQGs were considered to have a relatively low level of uncertainty in this respect if they reliably estimated effects at relatively low contaminant concentrations (*i.e.*, minimize false negatives while allowing for a higher probability of false positives; Long *et al.* 1995; Ingersoll *et al.* 1996; MacDonald *et al.* 1996). The need for sensitivity (*e.g.*, minimize false negatives) should be balanced with ecological relevance (*e.g.*, minimize both false positives and false negatives (Section 18.7.2)). Low uncertainty with respect to sensitivity was assigned to the ERLs and SLCs because they tend to be the lowest SQGs. Most of the other SQGs were considered to have a relatively higher level of uncertainty because they are generally higher values (*e.g.*, SQC, ERMs, and SEM/AVS). The AETs were assigned a high level of uncertainty with respect to sensitivity because they only increase with the addition of new data, making them particularly prone to false negatives. In contrast, the residue-based SQGs were considered to have a lower level of

uncertainty because the tissue residue guidelines are based on the results of chronic toxicity tests on sensitive species.

18.7.5 Interferences

In the context of the SQG evaluation, interferences are considered to be related to biotic or abiotic factors that could influence the SQGs derivation beyond the direct effects of specific contaminants. That is, interferences are considered to be nontreatment factors. Benthic communities can be strongly influenced by environmental factors such as grain size, TOC, or salinity. These same factors can also influence the results of some sediment toxicity tests; however, careful experimental design can minimize these effects (DeWitt *et al.* 1988). Because the SLCs are based entirely on benthic community data, they were considered to have the highest level of uncertainty. Because the residue-based guidelines are from direct analytical determination, they are not subject to the same types of interferences. Hence, a relatively low levels of uncertainty were assigned to these measurement endpoints. The influence of sediment characteristics such as surface area, mineral interactions, and redox are not typically quantified and may add an additional amount of uncertainty to each of the SQGs listed in Table 18-7.

18.7.6 Standardization

Standard procedures have been established for deriving virtually all of the SQG values listed in Table 18-7. In this context, uncertainty in the degree of standardization of the approach was evaluated on the basis of appropriate peer review. That is, the degree of uncertainty was considered to be lower for those approaches that have been published in the peer-reviewed literature. Approaches for determination of SQC, ERLs and ERMs, and SEM/AVS have been published in the peer-reviewed literature and, hence, were assigned a low degree of uncertainty. In contrast, the mixture models (in the early stages of development with sediments), tissue residue guidelines, and AETs have not been widely peer reviewed in the literature.

18.7.7 Discrimination

In the previous sections, this criterion was intended to evaluate the ability of a measurement endpoint to discriminate between contaminated sediments by providing a graded response. That is, a measure was considered to have a higher degree of uncertainty if it did not provide a means of distinguishing between samples that had different levels of contamination. In the context of this evaluation, SQGs were considered to be discriminatory if they could be used to correctly classify toxic and nontoxic samples (Long *et al.* 1995). The SQC and the ERLs and ERMs have been demonstrated to provide accurate measures for correctly predicting toxic and nontoxic responses in the field. For this reason, a low level of uncertainty was assigned to these guidelines. Although SLCs are designed to be highly protective, they have a poor ability to discriminate the range of adverse effects that could occur. In contrast, sediment samples with contaminant concentrations that exceed the AETs have a high probability of being toxic. However, the AETs may not reliably discriminate samples with lower levels of contamination with respect to their potential for adverse biological effects (*i.e.*, false negatives). The residue-based guidelines are based on accurate analytical measurements; therefore, tissue residues provide a reliable basis for discriminating between the chemicals that could be causing adverse effects. However, these guidelines are available for only a limited number of substances. It is recommended that residue-based guidelines be derived for additional priority bioaccumulative substances. Additional studies should evaluate the ability of the SQGs listed in Table 18-7 to minimize both false negative and false positive errors.

18.7.8 Bioavailability

Information from a variety of sources indicates that the bioavailability and, hence, toxicity of sediment-associated contaminants can be influenced by the characteristics of the sediments and the associated pore water (Pavlou 1987; Swartz *et al.* 1990; Di Toro *et al.* 1991, 1992). For this reason, SQGs that explicitly considered the influence of these factors were deemed to have a lower level of uncertainty than those that did not address bioavailability (Di Toro *et al.* 1991). The factors which are considered to influence bioavailability are directly considered in the derivation of the SQC, SLCs, SEM/AVS, and residue-based guidelines. Although other guidelines (*i.e.*, ERLs, ERMs, AETs, and mixture models) are largely based on dry-weight concentrations, it is possible to refine the approaches to explicitly consider other normalization procedures (Ingersoll *et al.* 1996). Such research is in progress and should reduce the uncertainty associated with these methods in the near future.

18.7.9 Field validation

Field validation is an essential element of the SQGs development process because it provides a basis for assessing their reliability in sediment quality assessments. For this reason, a higher degree of uncertainty in the application of SQGs to SERAs was assigned to values if they had not been appropriately field validated. At a minimum, such a field validation would require an assessment of the predictability of the SQGs using a number of independent datasets (*i.e.*, not used to derive the SQGs; Long *et al.* 1995; Ingersoll *et al.* 1996; MacDonald *et al.* 1996). All SQGs suffer from a lack of comprehensive field validation. Although preliminary evaluations of several of the guidelines have been conducted (*e.g.*, SQC, ERL, ERMs, AETs, and SEM/AVS), the results apply to a limited number of geographic areas and sediment quality conditions. It is recommended that SQGs be further evaluated using the results of field studies conducted in various geographic regions. This is the most important general research need for reducing the uncertainty associated with the guidelines.

18.7.10 Summary

The results of this evaluation indicate that there is sufficient certainty associated with SQGs to recommend their use in SERAs. In particular, SQC, ERLs, ERMs, SEM/AVS, and residue-based SQGs generally have less uncertainty in their present applications than other guidelines. Although mixture models were generally considered to have somewhat higher levels of uncertainty, they address the critically important issue of the interaction of contaminants in complex mixtures. For this reason, it is strongly recommended that further development of SQGs encompassing mixture models be pursued. In spite of the uncertainty associated with the general application of AETs, their regional application has proven to be effective. Screening-level concentrations accurately define the lower bound for sediment-associated contaminants; however, they should not be used in SERAs because they cannot discriminate between environmentally relevant contaminant concentrations. Overall, TIE methods offer a logical solution for bridging the gap between effects and exposure assessments and would result in a more accurate and comprehensive characterization of risk (Chapter 16). A concern expressed with current risk assessment frameworks was that there does not appear to be any type of defined "box" (Chapter 1) for mixture models or TIEs described in Chapter 16, which could greatly advance the science of (at least retrospective) SERAs.

Further investigations are needed to reduce uncertainty in SQGs and increase their applicability in SERAs. For example, additional studies are required to identify factors which control the bioavailability of sediment-associated contaminants under various conditions. In turn, this information could be used to refine the SQGs for specific applications. In addition, residue-based SQGs should be derived for a number of substances that are known to bioaccumulate. Furthermore, all of the SQGs should be field validated to determine their predictive ability and their potential for resulting in false positive and false negative errors.

18.8 Conclusions and recommendations

18.8.1 Improving sediment ecological risk assessments to quantify and account for uncertainty

- While each measurement tool has an inherent level of uncertainty associated with its application, the uncertainty associated with the overall risk assessment process can be reduced by integrating these tools. For example, the use of sediment chemistry, sediment toxicity, and benthic community data together in a sediment quality triad can be used to establish a weight of evidence linking contaminated sediments to adverse biological effects (Chapman *et al.* 1992). Similarly, Long *et al.* (1995) used sediment chemistry and toxicity data, in conjunction with SQGs, to evaluate sediment quality and identify contaminants that were likely causing adverse biological effects. Therefore, the integration of multiple tools has the potential to substantially reduce uncertainty in SERAs and improve management decisions.

- Toxicity tests are generic indicators of stress and cannot, by themselves, discriminate specific factors which caused the response. Linkage of results of toxicity tests to a list of chemicals in the sediment sample might prove to be of limited use for defining potential causes of toxicity. Chemicals responsible for toxicity may not have been measured, the bioavailability of chemicals can be quite uncertain, and correlative techniques are often unable to deal with multiple contributions from complex mixtures.

- Although TIEs and mixture models were generally considered to have somewhat higher levels of uncertainty, both approaches address the critically important issue of the interaction of contaminants in complex mixtures. For this reason, further development of these approaches should be pursued. Overall, TIE methods offer a logical solution for bridging the gap between effects and exposure assessments and would result in a more accurate and comprehensive characterization of risk.

- The most ecologically relevant phases of sediment used in toxicity tests from the standpoint of uncertainty were whole sediment tests with benthic organisms, followed by whole sediment tests with pelagic species, suspended sediment tests, and porewater tests. In the case of the organic extract and elutriate tests, the exposure regimes do not represent a realistic measure of sediment *in situ* conditions.

- The uncertainty associated with survival as the endpoint monitored in laboratory toxicity tests is less than that for the other endpoints. The continued development of more sensitive and ecologically relevant endpoints (*e.g.*, chronic effects on growth and reproduction, life-cycle tables) has the potential to produce superior measurement endpoints for use in SERAs.

- Insufficient information exists on how to relate the response of endpoints at the level of the individual to the population or the community. However, community-level assessments are more likely to include sediment effects as the number of species examined increases the probability of observing a sediment-associated response. Therefore, relationships should be examined between responses measured using biomarkers and individuals and responses measured using populations and communities.

- Bioaccumulation cannot alone provide information about whether the source of exposure was water, sediment, or pore water and cannot alone be used to evaluate effects. Combinations of laboratory studies, field data on water and sediment concentrations, and food webs and pathway models can be used to increase our understanding of pathways of exposure and effects.

- An inherent uncertainty exists in quantitatively attempting to relate laboratory bioaccumulation results to the field. Nonetheless, studies have successfully demonstrated bioaccumulation results obtained in the laboratory with field-collected sediments can be highly predictive of residue concentrations observed in synoptic collection of organisms from the field. Combined laboratory and model studies show great potential for reducing uncertainties in extrapolating from controlled studies to nature.

18.8.2 Strategic actions for addressing uncertainty in sediment ecological risk assessments

Top five recommendations:

- Evaluate the ability of laboratory responses to predict responses in the field.
- Evaluate the predictive ability (*i.e.*, field validation) of SQGs in a variety of geographic regions, including the potential for false positive and false negative errors.
- Develop methods for evaluating chronic effects of contaminated sediments on growth, reproduction, and population dynamics.
- Adapt TIE procedures used to evaluate aqueous samples such as effluents or pore water for use in whole sediment toxicity tests.
- Develop residue-based SQGs for highly bioaccumulative substances.

Additional recommendations:

- Use a weight-of-evidence approach to reduce uncertainty associated with SERAs.
- Develop procedures for integrating uncertainty estimates for each of the three measures (*i.e.*, biology, chemistry, toxicity) in the sediment quality triad.
- Use consistent methods when possible to conduct SERAs.
- Identify biomarkers and their relation to ecologically relevant endpoints and to specific contaminants.
- Improve sampling methods to lower uncertainty in estimates of spatial and temporal variability.
- Further evaluate factors controlling the bioavailability of contaminants in sediments.
- Determine the mode of toxicity of contaminants.
- Determine the extent to which natural systems are at equilibrium or steady state to determine if models that rely on this assumption are relevant.
- Further evaluate relative sensitivity of the species commonly used in sediment toxicity assessments.
- Consider interferences which decrease the predictive ability of laboratory toxicity tests (*e.g.*, UV light and photoinduced toxicity of PAHs).
- Further elucidate the life history of sensitive species.
- Refine approaches for identifying reference conditions for benthic communities.
- Evaluate the altered sensitivity of populations exposed to stressors.
- Better define natural variability of key receptors and stressors.
- Conduct studies to determine the precision of benthic community assessment methods.
- Verify models of the interactions of contaminant mixtures in relation to *in situ* biological and toxicological effects of contaminated sediments.

- Evaluate relations between body burdens and toxic responses.
- Further evaluate effects of dietary exposures to contaminants, particularly for metals and PAHs.
- Field-validate tiered sediment quality assessment approaches to assure that they provide relevant management information.

18.9 References

Adams DD. 1991. Sampling sediment pore water. In: Mudroch A, MacKnight SC, editors. Handbook of techniques for aquatic sediments sampling. Ann Arbor MI: CRC Pr. p 171–202.

[ASTM] American Society for Testing and Materials. 1995a. Standard test methods for measuring the toxicity of sediment-associated contaminants with freshwater invertebrates. In: Volume 11.05, Annual book of ASTM standards. Philadelphia: ASTM. E1706-95a. p 1204–1285.

[ASTM] American Society for Testing and Materials. 1995b. Standard guide collection, storage, characterization, and manipulation of sediments for toxicological testing. In: Volume 11.05, Annual book of ASTM standards. Philadelphia: ASTM. E1391-94. p 835–855.

[ASTM] American Society for Testing and Materials. 1995c. Standard guide for conducting 10-day static sediment toxicity tests with marine and estuarine amphipods. In: Volume 11.05, Annual book of ASTM standards. Philadelphia: ASTM. E 1367-92. p 767–792.

[ASTM] American Society for Testing and Materials. 1995d. Standard guide for designing biological tests with sediments. In: Volume 11.05, Annual book of ASTM standards. Philadelphia: ASTM. E1525-94a. p 972–989.

[ASTM] American Society for Testing and Materials. 1995e. Standard guide for determination of bioaccumulation of sediment-associated contaminants by benthic invertebrates. In: Volume 11.05, Annual book of ASTM standards. Philadelphia: ASTM. E1688-95. p 1140–1189.

Ankley GT, Benoit DA, Balogh JC, Reynoldson TB, Day KE, Hoke RA. 1994. Evaluation of potential confounding factors in sediment toxicity tests with three freshwater benthic invertebrates. *Environ Toxicol Chem* 13: 627–635.

Ankley GT, Collyard SA, Monson PD, Kosian PA. 1994. Influence of ultraviolet light on the toxicity of sediments contaminated with polycyclic aromatic hydrocarbons. *Environ Toxicol Chem* 13: 1791–1796.

Ankley GT, Cook PM, Carlson AR, Call DJ, Swenson JA, Corcoran HF, Hoke RA. 1992. Bioaccumulation of PCBs from sediments by oligochaetes and fishes: Comparison of laboratory and field studies. *Can J Fish Aquat Sci* 49:2080–2085.

Ankley GT, Phipps GL, Leonard EN, Benoit DA, Mattson VR, Kosian PA, Cotter AM, Dierkes JR, Hansen DJ, Mahony JD. 1991. Acid-volatile sulfide as a factor mediating cadmium and nickel bioavailability in contaminated sediment. *Environ Toxicol Chem* 10:1299–1307.

Ankley GT, Schubauer-Berigan MK. 1995. Background and overview of current standard toxicity identification evaluation procedures. *J Aquat Ecosystem Health* 4:133–149.

Ankley GT, Schubauer-Berigan MK, Dierkes JR. 1991. Predicting the toxicity of bulk sediments to aquatic organisms using aqueous test fractions: pore water versus elutriate. *Environ Toxicol Chem* 10:1359–1366.

Ankley G, Thomas N. 1992. Interstitial water toxicity identification evaluation approach. Sediment classification methods compendium. Washington DC. EPA-823-R-92-006.

Armitage PD, Gunn RJM, Furse MT, Wright JF, Moss D. 1987. The use of prediction to assess macroinvertebrate response to river regulation. *Hydrobiologia* 144:25–32.

Barbour MT, Stribling JB, Karr JR. 1995. The multimetric approach for establishing biocriteria and measuring biological condition. In: Davis WS, Simon T, editors. Biological assessment and criteria: tools for water resource planning and decision-making. Boca Raton FL: Lewis. 415 p.

Barrick R, Becker S, Pastorok R, Brown L, Beller H. 1988. Sediment quality values refinement: 1988 update and evaluation of Puget Sound AET. Bellevue WA: PTI Environmental Services for the Environmental Protection Agency.

Beak Consultants Ltd. 1987. Development of sediment quality objectives: Phase I - options. Prepared for Ontario Ministry of Environment. Mississauga ON.

Beak Consultants Ltd. 1988. Development of sediment quality objectives: Phase I - guidelines development. Prepared for Ontario Ministry of Environment. Mississauga ON.

Benoit DB, Sibley PK, Juenemann J, Ankley GT. 1996. Design and evaluation of a life-cycle test with the midge *Chironomus tentans* for use in assessing chronic toxicity of contaminated sediments. *Environ Toxicol Chem*: In press.

Benson WH, Di Giulio RT. 1992. Biomarkers in hazard assessments of contaminated sediments. In: Burton Jr GA, editor. Sediment toxicity assessment. Chelsea MI: Lewis. p 213–240.

Brown CL, Luoma SN. 1995. Use of a euryhaline bivalve to assess trace metal contamination in San Francisco Bay: I. Evaluating the characteristics of *Potamocorbula amurensis* as a biosentinel species. *Mar Ecol Prog Ser*: In press.

Brown DA, Bay SM, Gossett RW. 1985. Using the natural detoxification capacities of marine organisms to assess assimilative capacity. In: Cardwell RD, Purdy R, Bahner RC, editors. Aquatic toxicology and hazard assessment; 7th Symposium, American Society for Testing and Materials (ASTM). Philadelphia: ASTM. STP 854. p 364–382.

Brumbaugh WG, Ingersoll CG, Kemble NE, May TW, Zajicek JL. 1994. Chemical characterization of sediments and pore water from the upper Clark Fork River and Milltown Reservoir, Montana. *Environ Toxicol Chem* 13:1971–1983.

Brunson EL, Canfield TJ, Dwyer FJ, Kemble NE, Ingersoll CG. 1996. An evaluation of bioaccumulation with sediments from the upper Mississippi River using field-collected oligochaetes and laboratory-exposed *Lumbriculus variegatus*. In review.

Bryan GW, Gibbs PE. 1991. Impact of low concentrations of tributytin (TBT) on marine organisms: a review. In: Newman MC, McIntosh AW, editors. Metal ecotoxicology: concepts and applications. Chelsea MI: Lewis. p 323–353.

Burton Jr GA. 1992. Sediment toxicity assessment. Chelsea MI: Lewis. 457 p.

Burton Jr GA, Ingersoll CG. 1994. Evaluating the toxicity of sediments. The ARCS assessment guidance document. Chicago. EPA/905-B94/002.

Burton GA, Nelson MK, Ingersoll CG. 1992. Freshwater benthic toxicity tests. In: Burton Jr GA, editor. Sediment toxicity assessment. Chelsea MI: Lewis. p 213–240.

Burton GA, Norberg-King TJ, Ingersoll CG, Ankley GT, Winger PV, Kubitz J, Lazorchak JM, Smith ME, Greer IE, Dwyer FJ, Call DJ, Day KE, Kennedy P, Stinson M. 1996. Interlaboratory study of precision: *Hyalella azteca* and *Chironomus tentans* freshwater sediment toxicity assays. *Environ Toxicol Chem* 15:1335–1343.

Cain DJ, Luoma SN. 1985. Copper and silver accumulation in transplanted and resident clams (*Macoma balthica*) in South San Francisco Bay. *Mar Environ Res* 15:115–135.

Cain DJ, Luoma SN, Carter JL, Fend SV. 1992. Aquatic insects as bioindicators of trace element contamination in cobble-bottom rivers and streams. *Can J Fish Aquat Sci* 49:2141–2154.

Calow P. 1993. General principles and overview. In: Calow P, editor. Volume 1, Handbook of ecotoxicology. London: Blackwell Scientific. Chapter 1, p 1–5.

Campbell PGC, Lewis AG, Chapman PM, Crowder AA, Fletcher WK, Imber B, Luoma SN, Stokes PM, Winfrey M. 1988. Biologically available metals in sediments. Ottawa: National Research Council of Canada. Publication # NRCC27694. 298 p.

Canfield TJ, Dwyer FJ, Fairchild JF, Ingersoll CG, Kemble NE, Mount DR, La Point TW, Burton GA, Swift MC. 1996. Assessing contamination of Great Lakes sediments using benthic invertebrates and the sediment quality triad. *J Great Lakes Res* 22:565–583.

Canfield TJ, Kemble NE, Brumbaugh WG, Dwyer FJ, Ingersoll CG, Fairchild JF. 1994. Use of benthic invertebrate community structure and the sediment quality triad to evaluate metal-contaminated sediment in the upper Clark Fork River, Montana. *Environ Toxicol Chem* 13:1999–2012.

Carr RS, Chapman DC. 1995. Comparison of methods for conducting marine and estuarine sediment porewater toxicity tests-Extraction, storage, and handling techniques. *Arch Environ Contam Toxicol* 28:69–77.

[CEPA] Commonwealth Environmental Protection Agency, Land and Water Resources and Dept. Environment, Sport and Territories. 1994. National River Health Monitoring Program Monitoring River Health Initiative. River Bioassessment Manual. 39 p.

Chapman P. 1992. Triad approach. In: Sediment classification methods compendium. Washington DC. EPA 823-R-92-006.

Chapman PM. 1989. Current approaches to developing sediment quality criteria. *Environ Toxicol Chem* 8:589–599.

Chapman PM, Fink R. 1984. Effects of Puget Sound sediments and their elutriates on the life cycle of *Capitella capitata*. *Bull Environ Contam Toxicol* 33:451–459.

Chapman PM, Popham JD, Griffin J, Leslie D, Michaelson J. 1987. Differentiation of physical from chemical toxicity in solid waste fish bioassays. *Water Air Soil Pollut* 33:295–308.

Chapman PM, Power EA, Burton Jr GA. 1992. Integrative assessments in aquatic ecosystems. In: Burton GA, editor. Sediment toxicity assessment. Chelsea MI: Lewis. p 313–340.

Chapman PM, Power EA, Dexter RN, Andersen HB. 1991. Evaluation of effects associated with an oil platform, using the sediment quality triad. *Environ Toxicol Chem* 10:407–424.

Cook PM, Carlson AR, Lee II H. 1992. Tissue residue approach. In: Sediment classification methods compendium. Washington DC. EPA 823-R-92-006.

Corkum LD, Currie DC. 1987. Distributional patterns of immature Simuliidae (Diptera) in northwestern North America. *Freshwater Biol* 17:201–221.

Crawford JK, Luoma SN. 1992. Guidelines for studies of contaminants in biological tissues for the National Water Quality Assessment Program. Lemoyne PA: U.S. Geological Survey (USGS). Open File Report #92-494. 69 p.

Davis WS, Simon TP. 1995. Biological assessment and criteria: tools for water resource planning and decision making. Salem MA: Lewis, CRC Pr.

Day KE, Scott IM. 1990. Use of acetylcholinesterase activity to detect sub-lethal toxicity in stream invertebrates exposed to low concentrations of organo-phosphate insecticides. *Aquat Toxicol* 18:101–13.

De Pauw N, Vanhooren G. 1983. Method for biological assessment of watercourses in Belgium. *Hydrobiologia* 100:153–68.

DeWitt TH, Ditsworth GR, Swartz RC. 1988. Effects of natural sediment features on the phoxocephalid amphipod, *Rhepoxynius abronius*: Implications for sediment toxicity bioassays. *Mar Environ Res* 25:99-124.

DeWitt TH, Swartz RC, Lamberson JO. 1989. Measuring the toxicity of estuarine sediment. *Environ Toxicol Chem* 8: 1035–1048.

DeWitt TH, Redmond MS, Sewall JE, Swartz RC. 1992. Development of a chronic sediment toxicity test for marine benthic amphipods. Annapolis MD: Chesapeake Bay Program. EPA CBP/TRS/89/93.

Diaz R J. 1992. Ecosystem assessment using estuarine and marine benthic community structure. In: Burton GA, editor. Sediment toxicity assessment. Boca Raton FL: Lewis. p 67–81.

Di Toro DM, Mahony JD, Hansen DJ, Scott KJ, Carlson AR, Ankley GT. 1992. Acid volatile sulfide predicts the acute toxicity of cadmium and nickel in sediments. *Environ Sci Tech* 26:96–101.

Di Toro DM, Mahony JH, Hansen DJ, Scott KJ, Hicks MB, Mayr SM, Redmond M. 1990. Toxicity of cadmium in sediments: the role of acid volatile sulfides. *Environ Toxicol Chem* 9:1487–1502.

Di Toro DM, Zarba CS, Hansen DJ, Berry WJ, Swartz RC, Cowan CE, Pavlou SP, Allen HE, Thomas NA, Paquin PR. 1991. Technical basis for establishing sediment quality criteria for nonionic organic chemicals using equilibrium partitioning. *Environ Toxicol Chem* 10:1541–1583.

Elliott JM. 1977. Some methods for the statistical analysis of samples of benthic invertebrates. 2nd ed. Ambleside, Cumbria UK: Freshwater Biological Association. Scientific publication No. 25:1–156.

Environment Canada. 1992. Biological test method: acute test for sediment toxicity using marine or estuarine amphipods. Ottawa ON: Environmental Protection Publications. Report EPS 1/RM/26.

Environment Canada. 1995. Canadian sediment quality guidelines for polycyclic aromatic hydrocarbons. Ottawa: Task Force on Water Quality Guidelines.

Environment Canada. 1996a. Biological test method: test for growth and survival in sediment using the freshwater amphipod *Hyalella azteca*. Ottawa: Environment Canada. Technical report number pending: In press.

Environment Canada. 1996b. Biological test method: test for growth and survival in sediment using larvae of freshwater midges (*Chironomus tentans* or *Chironomus riparius*). Ottawa: Environment Canada. Technical report number pending: In press.

Farris JL, Belanger SE, Cherry DS, Cairns Jr J. 1989. Cellulolytic activity as a novel approach to assess long term zinc stress to Corbicula. *Water Res* 23:1275–83.

Ferraro SP, Swartz RC, Cole FA, Schults DW. 1991. Temporal changes in the benthos along a pollution gradient: discriminating the effects of natural phenomena from pollution-related variability. *Estuarine Coastal Shelf Sci* 33:383–407.

Fox RG, Tuchman M. 1996. The assessment and remediation of contaminated sediments (ARCS) program. *J Great Lakes Res* 22:493–494.

Förstner U. 1990. Inorganic sediment chemistry and elemental speciation. In: Baudo R, Giesy J, Muntau H, editors. Sediments: chemistry and toxicity of in-place pollutants. Chelsea MI: Lewis.

Giesy JP. 1988. Clinical indicators of stress induced changes in aquatic organisms. *Verh Int ver Limnol* 23:1610–18.

Giesy JP, Hoke RA. 1989. Freshwater sediment toxicity bioassessment: rationale for species selection. *J Great Lakes Res* 15:539–569.

Giesy JP, Graney RL, Newsted JL, Rosiu CJ, Benda A, Kreis Jr RG, Horvath FJ. 1988. Comparison of three sediment bioassay methods using Detroit river sediments. *Environ Toxicol Chem* 7: 483–498.

Goodman D. 1975. The theory of diversity-stability relationships in ecology. *Quarterly Rev Biology* 50:237–66.

Graney RL, Giesy Jr JP. 1987. The effect of short term exposure to pentachlorophenol and osmotic stress on the free amino acid pool of the freshwater amphipod *Gammarus pseudolimnaeus* Bousfield. *Arch Environ Contam Toxicol* 16:167–76.

Green RH. 1979. Sampling design and statistical methods for environmental biologists. New York: J Wiley.

Hansen DJ, Berry WJ, Di Toro DM, Paquin P, Davanzo L, Stancil Jr FE, Kollig HP. 1993a. Proposed sediment quality criteria for the protection of benthic organisms: endrin. Washington DC: U.S. Environmental Protection Agency (USEPA) Office of Water and Office of Research and Development, Office of Science and Technology, Health and Ecological Criteria Division.

Hansen DJ, Berry WJ, Di Toro DM, Paquin P, Davanzo L, Stancil Jr FE, Kollig HP, Hoke RA. 1993b. Proposed sediment quality criteria for the protection of benthic organisms: dieldrin. Washington DC: U.S. Environmental Protection Agency (USEPA) Office of Water and Office of Research and Development, Office of Science and Technology, Health and Ecological Criteria Division.

Hansen DJ, Berry WJ, Di Toro DM, Paquin P, Davanzo L, Stancil Jr FE, Kollig HP. 1993c. Proposed sediment quality criteria for the protection of benthic organisms: phenanthrene. Washington DC: U.S. Environmental Protection Agency (USEPA) Office of Water and Office of Research and Development, Office of Science and Technology, Health and Ecological Criteria Division.

Hansen DJ, Berry WJ, Di Toro DM, Paquin P, Stancil Jr FE, Kollig HP. 1993d. Proposed sediment quality criteria for the protection of benthic organisms: acenaphthene. Washington DC: U.S. Environmental Protection Agency (USEPA) Office of Water and Office of Research and Development, Office of Science and Technology, Health and Ecological Criteria Division.

Hansen DJ, Berry WJ, Di Toro DM, Paquin P, Davanzo L, Stancil Jr FE, Kollig HP. 1993e. Proposed sediment quality criteria for the protection of benthic organisms: fluoranthene. Washington DC: U.S. Environmental Protection Agency (USEPA) Office of Water and Office of Research and Development, Office of Science and Technology, Health and Ecological Criteria Division.

Hurlbert SH. 1971. The nonconcept of species diversity: a critique and alternative parameters. *Ecology* 52:577–86.

Huggett RJ, Kimerle RA, Mehrle PM, Bergman HL. 1992. Biomarkers: biochemical, physiological, and histopathological markers of anthropogenic stress. Chelsea MI: Lewis.

Huggett RJ, Van Veld PA, Smith CL, Hargis Jr WJ, Vogelbein WK, Weeks BA. 1992. The effects of contaminated sediments in the Elizabeth River. In: Burton GA, editor. Sediment toxicity assessment. Chelsea MI: Lewis. p 403–430.

Hughes RM. 1995. Defining acceptable biological status by comparing with reference conditions. In: Davis WS, Simon TP, editors. Biological assessment and criteria: tools for water resource planning and decision making. Boca Raton FL: Lewis. 415 p.

[IJC] International Joint Commission. 1987. Guidance on the characterization of toxic substances problems in areas of concern in the Great Lakes basin. Windsor ON: Report to the Great Lakes Water Quality Board. 179 p.

[IJC] International Joint Commission. 1988. Procedures for the assessment of contaminated sediment problems in the Great Lakes. Windsor ON: Report to the Great Lakes Water Quality Board. 140 p.

Ingersoll CG. 1995. Sediment toxicity tests. In: Rand GM, editor. Fundamentals of aquatic toxicology. 2nd ed. Washington DC: Taylor and Francis. p 231–255.

Ingersoll CG, Haverland PS, Brunson EL, Canfield TJ, Dwyer FJ, Henke CE. 1996. Calculation and evaluation of sediment effect concentrations for the amphipod *Hyalella azteca* and the midge *Chironomus riparius. J Great Lakes Res* 22:602–623.

Johnson RK, Wiederholm T. 1989. Classification and ordination of profundal macroinvertebrate communities in nutrient poor, oligo-mesohumic lakes in relation to environmental data. *Freshwater Biology* 21:375–86.

Johnson RK, Wiederholm T, Rosenberg DM. 1993. Freshwater biomonitoring using individual organisms, populations and species assemblages of benthic macroinvertebrates. In: Rosenberg DM, Resh VH, editors. Freshwater biomonitoring and benthic macroinvertebrates. New York: Chapman and Hall.

Karr JR. 1990. Biological integrity and the goal of environmental legislation: Lessons for conservation biology. *Conservation Biology* 4:244–250.

Karr JR. 1991. Biological integrity: a long neglected aspect of water resource management. *Ecological Applications* 1:66-84.

Kemble NE, Brumbaugh WG, Brunson EL, Dwyer FJ, Ingersoll CG, Monda DP, Woodward DF. 1994. Toxicity of metal-contaminated sediments from the upper Clark Fork River, MT to aquatic invertebrates in laboratory exposures. *Environ Toxicol Chem* 13:1985–1997.

King JM. 1981. The distribution of invertebrate communities in a small South African river. *Hydrobiologia* 83:43–65.

Kingett PD. 1985. Genetic techniques: a potential tool for water quality assessment. In: Pridmore RD, Cooper AB, editors. Biological monitoring in freshwaters: Proceedings of a seminar; 1984 Nov 21–23; Hamilton; Part 1. Wellington NZ: National Water and Soil Conservation Authority. Water and Soil miscellaneous publication no. 82. p 179–88.

Kohn NP, Ward JQ, Niyogi DK, Ross LT, Dillon T, Moore DW. 1994. Acute toxicity of ammonia to four species of marine amphipod. *Mar Env Res* 38: 1–15.

Kolkwitz R, Marsson M. 1908. Okologie der pflanzlichen Saprobien. *Ber dtsch bot Ges* 26:505–19.

Lamberson JO, DeWitt TH, Swartz RC. 1992. Assessment of sediment toxicity to marine benthos. In: Burton GA, editor. Sediment toxicity assessment. Chelsea MI: Lewis. p 183–211.

Lamberson JO, Swartz RC. 1988. Use of bioassays in determining the toxicity of sediment to benthic organisms. In: Evans MS, editor. Toxic contaminants and ecosystem health: a Great Lakes focus. New York: J Wiley. p 257–279.

Landrum PF, Dupuis WS, Kukkonen J. 1994. Toxicokinetics and toxicity of sediment-associated pyrene and phenanthrene in Diporeia spp.: examination of equilibrium-partitioning theory and residue-based effects for assessing hazard. *Environ Toxicol Chem* 13:1769–1780.

Lobel PB. 1987. Intersite, intrasite and inherent variability of the whole soft tissue zinc concentrations of individual mussels *Mytilus edulis*: importance of kidney. *Mar Environ Res* 21:59–71.

Long ER, Chapman PM. 1985. A sediment quality triad: Measures of sediment contamination, toxicity and infaunal community composition in Puget Sound. *Mar Poll Bull* 16: 405–415.

Long ER, MacDonald DD. 1992. National Status and Trends Program approach. Sediment classification methods compendium. Washington DC. EPA 823-R-92-006.

Long ER, MacDonald DD, Smith SL, Calder FD. 1995. Incidence of adverse biological effects within ranges of chemical concentrations in marine and estuarine sediments. *Environ Management* 19:81–97.

Long ER, Morgan LG. 1990. The potential for biological effects of sediment-sorbed contaminants tested in the National Status and Trends Program. Seattle WA: National Oceanic and Atmospheric Administration (NOAA). NOAA Technical Memorandum NOS OMA 52. 175 p. + app.

Loring DH. 1991. Normalization of heavy-metal data from estuarine and coastal sediments. *ICES J Mar Sci* 48:101–115.

Luoma SN. 1990. Processes affecting metal concentrations in estuarine and coastal marine sediments. In: Furness R, Rainbow P, editors. Heavy metals in the marine environment. Boca Raton FL: CRC Pr.

Luoma SN. 1995. Prediction of metal toxicity in nature from bioassays: limitations and research needs. In: Tessier A, Turner D, editors. Metal speciation and bioavailability in aquatic systems. Sussex, England: J Wiley. p 609–659.

Luoma SN, Bryan GW. 1978. Factors controlling the availability of sediment-bound lead to the estuarine bivalve *Scrobicularia plana*. *J Mar Biol Assn UK* 58: 793–802.

Luoma SN, Bryan GW. 1982. A statistical study of environmental factors controlling concentrations of heavy metals in the burrowing bivalve Scrobicularia plana and the polychaete *Nereis diversicolor*. *Estuarine Coastal Shelf Sci* 15: 95–108.

Luoma SN, Cain DJ, Johansson C. 1985. Temporal fluctuations of silver, copper and zinc in the bivalve *Macoma balthica* at five stations in South San Francisco Bay. *Hydrobiologia* 129: 109–120.

Luoma SN, Carter JL. 1991. Effects of trace metals on aquatic benthos. In: Newman MC, McIntosh AW, editors. Metal ecotoxicology: concepts and applications. Chelsea MI: Lewis. p 261–287.

Luoma SN, Dagovitz R, Axtmann E. 1990. Temporally intensive study of trace metals in sediments and bivalves from a large river-estuarine system: Suisun Bay/Delta in San Francisco Bay. *Sci Total Environment* 97/98: 685–712.

Luoma SN, Ho KT. 1993. Appropriate uses of marine and estuarine sediment bioassays. In: Calow P, editor. Volume I, Handbook of ecotoxicology. London: Blackwell Scientific. p 193–226.

MacDonald DD. 1994. Approach to the assessment of sediment quality in Florida coastal waters: development and evaluation of sediment quality assessment guidelines. Tallahassee: Florida Department of Environmental Protection. 126 p.

MacDonald DD, Carr RS, Calder FD, Long ER, Ingersoll CG. 1996. Development and evaluation of sediment quality guidelines for Florida coastal waters. *Ecotoxicology* 5:253–278.

MacDonald DD, Smith SL, Wong MP, Mudroch P. 1992. The development of Canadian marine environmental quality guidelines. Report prepared for the Interdepartmental Working Group on Marine Environmental Quality Guidelines and the Canadian Council of Ministers of the Environment. Ottawa ON: Environment Canada. 50 p. + app.

Mearns AJ, Swartz RC, Cummins JM, Dinnel PA, Plesha P, Chapman PM. 1986. Inter-laboratory comparison of a sediment toxicity test using the marine amphipod, *Rhepoxynius abronius*. *Mar Environ Res* 18:13-37.

Metcalfe JL. 1989. Biological water quality assessment of running waters based on macroinvertebrate communities: history and present status in Europe. *Environ Pollut* 60:101–39.

Moore JN, Luoma SN, Peters D. 1990. Downstream effects of mine effluent in an intermountaine riparian system. *Can J Fish Aquat Sci* 60:45–55.

Moss D, Furse MT, Wright JF, Armitage PD. 1987. The prediction of the macroinvertebrate fauna of unpolluted running-water sites in Great Britain using environmental data. *Freshwater Biol* 17:41–52.

Neff JM, Bean DJ, Cornaby BW, Vaga RM, Gulbransen TC, Scanlon JA. 1986. Sediment quality criteria methodology validation: Calculation of screening level concentrations from field data. Prepared for USEPA Region V. Washington DC: U.S. Environmental Protection Agency (USEPA). 225 p.

Nelson MK, Landrum PF, Burton GA, Klaine SJ, Crecelius EA, Byl TD, Gossiaux DC, Tsymbal VN, Cleveland L, Ingersoll CG, Sasson-Brickson G. 1993. Toxicity of contaminated sediments in dilution series with control sediments. *Chemosphere* 27:1789–1812.

Newell AJ, Johnson DW, Allen LK. 1987. Niagara River biota contamination project: Fish flesh criteria for piscivorous wildlife. New York: Division of Fish and Wildlife, Division of Marine Resources, New York State Department of Environmental Conservation. Technical Report 87-3.

Norris RH, Georges A. 1993. Analysis and interpretation of benthic macroinvertebrate surveys. In: Rosenberg DM, Resh VH, editors. Freshwater biomonitoring and benthic macroinvertebrates. New York: Chapman and Hall.

Omernick JM. 1995. Ecoregions: a spatial framework for environmental management. In: Davis WS, Simon TP, editors. Biological assessment and criteria: tools for water resource planning and decision making. Boca Raton FL: Lewis.

Ormerod SJ, Edwards RW. 1987. The ordination and classification of macroinvertebrate assemblages in the cachment of the River Wye in relation to environmental factors. *Freshwater Biol* 17:533–46.

Pavlou SP. 1987. The use of the equilibrium partitioning approach in determining safe levels of contaminants in marine sediments. In: Dickson KL, Maki AW, Brungs WA, editors. Fate and effects of sediment-bound chemicals in aquatic systems. Proceedings of the Sixth Pellston Workshop. Toronto ON: Pergamon Pr.

Plafkin JL, Barbour MT, Porter KD, Gross SK, Hughes RM. 1989. Rapid bioassessment protocols for use in streams and rivers. Benthic macroinvertebrates and fish. Washington DC: Office of Water Regulations and Standards, U.S. Environmental Protection Agency (USEPA). EPA/444/4-89/001.

Persaud D, Jaagumagi R, Hayton A. 1992. Guidelines for the protection and management of aquatic sediment quality in Ontario. Toronto ON: Water Resources Branch, Ontario Ministry of the Environment. 26 p.

Phillips DJH. 1980. Quantitative aquatic biological indicators. London: Applied Science. 488 p.

Phillips DJH, Rainbow PS. 1993. Biomonitoring of trace aquatic contaminants. Oxford: Elsevier. 371 p.

Phipps GL, Mattson VR, Ankley GT. 1995. The relative sensitivity of three freshwater benthic macroinvertebrates to ten contaminants. *Arch Environ Contam Toxicol* 28:281–286.

Resh VH, Jackson JK. 1993. Rapid assessment approaches to biomonitoring using benthic macroinvertebrates. In: Rosenberg DM, Resh VH, editors. Freshwater biomonitoring and benthic invertebrates. New York: Chapman and Hall. p 195–233.

Reynoldson TB, Zarull MA. 1993. An approach to the development of biological sediment guidelines. In: Francis G, Kay J, Woodley, S, editors. Ecological integrity and management of ecosystems. FL: St. Lucie Pr.

Reynoldson TB, Day KE, Clarke C, Milani D. 1994. Effect of indigenous animals on chronic endpoints in freshwater sediment toxicity tests. *Environ Toxicol Chem* 13:973–977.

Reynoldson TB, Day KE, Bailey RC, Norris RH. 1995. Methods for establishing biologically based sediment guidelines for freshwater quality management using benthic assessment of sediment. *Australian J Ecology* 20:198–219.

Rosenberg DM, Resh VH. 1993. Freshwater biomonitoring and benthic macroinvertebrates. New York: Chapman and Hall. 488 p.

Shannon CE. 1948. A mathematical theory of communication. *Bell System Technical Journal* 27:379–423.

Schindler DW. 1987. Detecting ecosystem responses to anthropogenic stress. *Can J Fish Aquat Sci* 44:6–25.

Schlekat CE, McGee BL, Boward DM, Reinharz E, Velinsky DJ, Wade TL. 1994. Tidal river sediments in the Washington, D.C. Area. III. Biological effects associated with sediment contamination. *Estuaries* 17:334–344.

Schropp SJ, Lewis FG, Windom HL, Ryan JD, Calder FD, Burney LC. 1990. Interpretation of metal concentrations in estuarine sediments of Florida using aluminum as a reference element. *Estuaries* 13:227–235.

Schults DW, Ferraro SP, Smith LM, Roberts FA, Poindexter CK. 1992. A comparison of methods for collecting interstitial water for trace organic compounds and metal analyses. *Water Res* 26:989–995.

Shuba PJ, Tatem HE, Carroll JJ. 1978. Biological assessment methods to predict the impact of open-water disposal of dredged material. Washington DC: U.S. Army Corps of Engineers (USACE). Technical Report D-78-5Q.

Sibley PK, Benoit DA, Ankley GT. 1996. The relationship between growth and reproduction in the midge *Chironomus tentans* and its importance for bioassessment of contaminated sediments. *Environ Toxicol Chem*: In press.

Sladacek V. 1973. System of water quality from the biological point of view. *Arch Hydrobiol, Beih Erg Limnol* 7:1–218.

Suedel BC, Rodgers JJ. 1994. Responses of *Hyalella azteca* and *Chironomus tentans* to particle size distribution and organic matter content of formulated and natural freshwater sediments. *Environ Toxicol Chem* 13:1639–1648.

Sundelin B. 1983. Effect of cadmium on *Pontoporeia affinis* (Crustacea: Amphipoda) in laboratory soft-bottom microcosms. *Mar Biol* 74:203–214.

Swartz RC. 1989. Marine sediment toxicity tests. In: Contaminated marine sediments—assessment and remediation. Washington DC: National Research Council, National Academy Pr. p 115–129.

Swartz RC, Cole FA, Lamberson JO, Ferraro SP, Schults DW, DeBen WA, Lee II H, Ozretich RJ. 1994. Sediment toxicity, contamination, and amphipod abundance at a DDT- and dieldrin-contaminated site in San Francisco Bay. *Environ Toxicol Chem* 13:949–962.

Swartz RC, Kemp PF, Schults DW, Ditsworth GR, Ozretich RJ. 1989. Toxicity of sediment from Eagle Harbor, Washington to the infaunal amphipod, *Rhepoxynius abronius*. *Environ Toxicol Chem* 8: 215-222.

Swartz RC, Schults DW, DeWitt TH, Ditsworth GR, Lamberson JO. 1990. Toxicity of fluoranthene in sediment to marine amphipods: a test of the equilibrium partitioning approach to sediment quality criteria. *Environ Toxicol Chem* 9:1071–1080.

Swartz RC, Schults DW, Ozretich RO, Lamberson JO, Cole FA, DeWitt TH, Redmond MS, Ferraro SP. 1995. ΣPAH: A model to predict the toxicity of polynuclear aromatic hydrocarbon mixtures in field-collected sediments. *Environ Toxicol Chem* 14:1977–1987.

Tessier A, Campbell PGC, Auclair JC, Bisson M. 1984. Relationships between the partitioning of trace metals in sediments and their accumulation in the tissue of the freshwater mollusc *Elliptio complanata* in a mining area. *Can J Fish Aquat Sci* 41:1463–1471.

Thomann, RV. 1989. Bioaccumulation model of organic chemical distributions in aquatic food chains. *Environ Sci Technol* 23:699–715.

Tolkamp HH. 1985. Biological assessment of water quality in running water using macroinvertebrates: A case study for Limburg, the Netherlands. *Wat Sci Tech* 17:867–78.

[USEPA] U.S. Environmental Protection Agency. 1989a. Assessing human health risks from chemically contaminated fish and shellfish: a guidance manual. Washington DC: USEPA. EPA 503/8-89-002.

[USEPA] U.S. Environmental Protection Agency. 1989b. Evaluation of the apparent effects threshold (AET) approach for assessing sediment quality. Report of the Sediment Criteria Subcommittee. Washington DC: USEPA. EPA-SAB-EETFC-89-027

[USEPA] U.S. Environmental Protection Agency. 1990a. Evaluation of the sediment classification methods compendium. Report of the Sediment Criteria Subcommittee of the Ecological Processes and Effects Committee. Washington DC: USEPA. EPA-SAB-EPEC-90-018.

[USEPA] U.S. Environmental Protection Agency. 1990b. Evaluation of the equilibrium partitioning (EqP) approach for assessing sediment quality. Report of the Sediment Criteria Subcommittee of the Ecological Processes and Effects Committee. Washington DC: USEPA. EPA-SAB-EPEC-90-006.

[USEPA] U.S. Environmental Protection Agency. 1991. Recommended guidelines for conducting laboratory bioassays on Puget Sound sediments. Seattle WA: USEPA Region X, Office of Puget Sound.

[USEPA] U.S. Environmental Protection Agency. 1992. Sediment classification methods compendium. Sediment Oversight Technical Committee. Washington DC: USEPA. EPA 813-R-92-006.

[USEPA] U.S. Environmental Protection Agency. 1993. An SAB report: Review of sediment criteria development methodology for non-ionic organic contaminants. Report of the Sediment Criteria Subcommittee of the Ecological Processes and Effects Committee. Washington DC: USEPA. EPA-SAB-EPEC-93-002.

[USEPA] U.S. Environmental Protection Agency. 1994a. Methods for measuring the toxicity and bioaccumulation of sediment-associated contaminants with freshwater invertebrates. Duluth MN: USEPA. EPA/600/R-94/024.

[USEPA] U.S. Environmental Protection Agency. 1994b. Methods for measuring the toxicity of sediment-associated contaminants with estuarine and marine amphipods. Duluth MN: USEPA. EPA/600/R-94/025.

[USEPA] U.S. Environmental Protection Agency. 1995. An SAB report: Review of the agency's approach for developing sediment criteria for five metals. Report of the Sediment Criteria Subcommittee of the Ecological Processes and Effects Committee. Washington DC: USEPA. EPA-SAB-EPEC-95-020.

[USEPA and USACE] U.S. Environmental Protection Agency and U.S. Army Corps of Engineers. 1991. Evaluation of dredge material proposed for ocean disposal. Washington DC: USEPA. EPA-503/8-91/001.

[USEPA and USACE] U.S. Environmental Protection Agency and U.S. Army Corps of Engineers. 1994. Evaluation of dredged material proposed for discharge in inland and near coastal waters (draft). Washington DC: USEPA. EPA-823-B-94-002.

Washington Department of Ecology. 1990. Sediment management standards: Chapter 173–204 WAC. Olympia WA. 106 p.

Washington HG. 1984. Diversity, biotic and similarity indices. A review with special relevance to aquatic ecosystems. *Water Res* 18:653–94.

Wentsel R, McIntosh A, Anderson V. 1977. Sediment contamination and benthic invertebrate distribution in a metal-impacted lake. *Environ Pollut* 14:187–193.

Widdows J, Donkin P. 1989. The application of combined tissue residue chemistry and physiological measurements of mussels (*Mytilus edulis*) for the assessment of environmental pollution. *Hydrobiologia* 188/189:455–461.

Wilhm JL, Dorris TC. 1968. Biological parameters for water quality analysis. *BioScience* 18:477–81.

Woodiwiss FS. 1964. The biological system of stream classification used by the Trent River Board. *Chem Ind* 443–447.

Wright JF. 1995. Development and use of a system for predicting the macroinvertebrate fauna of flowing waters. *Aust J Ecol* 20:181–97.

Wright JF, Moss D, Armitage PD, Furse MT. 1984. A preliminary classification of running-water sites in Great Britain based on macro-invertebrate species and the prediction of community type using environmental data. *Freshwater Biol* 14:221–256.

Yoder CO, Rankin ET. 1995. Biological response signatures and the area of degradation value: new tools for interpreting multimetric data. In: Davis WS, Simon TP, editors. Biological assessment and criteria: tools for water resource planning and decision making. Salem MA: Lewis, CRC Pr. p 287–302.

SESSION 7
INTERNATIONAL PERSPECTIVES

Chapter 19

Ecological risk assessment for sediments: an Australasian perspective

D. J. Morrisey

19.1 Introduction

Ecological risk assessment as a formal process is at an early stage of development in Australia and New Zealand. Consequently, this discussion is preliminary in nature and much of it is likely to be overtaken by events. The following sections consider 1) the current state of ERA as a formal process in Australia and New Zealand, with particular reference to ERA for sediments (SERA); 2) particular problems of assessment of ecological effects in Australia and New Zealand; and 3) the appropriateness of a formal framework for SERA in situations where there is little information on local conditions. Since problems of assessment of exposure (routes of uptake and mode of contact of organisms with contaminants) to a given concentration of a contaminant are probably the same in Australasia as in other parts of the world (as discussed in other chapters), this discussion will focus on aspects where the perspective is different, notably problem formulation, assessment of effects, and risk management. Lack of information on local ecological systems and their components and the consequent problem of assessing the appropriateness of using data from studies done overseas are relevant to the first two of these. The relatively small concentrations of contaminants generally present in sediments in Australasian aquatic habitats compared with those of northern Europe and North America and the relevance of criteria or guidelines designed to maintain environmental quality in overseas situations, where existing contamination is at much greater concentrations, are additional considerations for risk management in Australasia.

19.2 Ecological risk assessment as a formal process

In Australia, the Commonwealth Environmental Protection Agency (a national body) and the Environmental Protection Authorities of the states of New South Wales and Victoria are currently developing formal frameworks for ERA. In addition, a number of commercial environmental consultants have also developed their own frameworks for ERA (*e.g.*, Calvert and Baker 1995; Wenning *et al.* 1995a). The Sydney Water Corporation, in New South Wales, now has a statutory requirement to conduct an ERA for wastewater discharges, and the state environmental protection authority is required to audit the methodology and results (Water Studies Centre 1995). In New Zealand, the Ministry of Fisheries is reviewing current practices of risk assessment and analysis in general, with the eventual aim of providing consistency within and among the various agencies providing advice on these matters to central government. The Ministry for the Environment in New Zealand is investigating potential frameworks for ERA, both from a site- or contaminant-specific perspective and as a general framework to help set priorities for environmental protection and management. As has been the case elsewhere, the initial focus in many aspects of risk assessment for aquatic environments has been on the water column rather than the sediments.

The New South Wales Environmental Protection Authority has reviewed methods for ERA developed overseas as potential frameworks for use in Australia for the development of new water quality guidelines (Water Studies Centre 1995). It is currently considering the Dutch method in preference to those developed in the United States and Canada (Warne and Davies 1995). This preference was based on perceived relative cost-effectiveness, ease of implementation, and the identification of action to be taken at the end of the process of ERA. With respect to the identification of action to be taken, the Dutch method derives environmental quality criteria (EQC) by extrapolation from ecotoxicological data, using the method of Aldenberg and Slob (1993, cited in Water Studies Centre 1995). Environmental quality criteria are defined as permitted concentrations of contaminants in the environment and are derived as part of the process of risk assessment. These EQC would then be used in the proposed Australian context as screening levels to identify situations in which further assessment is warranted. The method could also be used for site-specific assessments. The United States and Canadian methods do not lead so directly to the identification of action to be taken, this being the role of the subsequent process of risk management, which is separate from the risk assessment (Norton et al. 1995).

Sediment quality criteria and guidelines are currently used in Australia and New Zealand on a fairly ad hoc basis, with no national guidelines or standards yet in place. In this discussion, criteria are defined as scientific data and relationships used to derive guidelines or standards. Guidelines are defined as numerical or narrative statements designed to maintain environmental quality, while standards are guidelines which have statutory force to maintain environmental quality. These definitions follow those of Smith et al. (1994). In Australia, the development of sediment quality criteria is, apparently, not considered a high priority and if guidelines or standards are eventually adopted, they are likely to be derived largely from work done in other countries (Pollution Research Pty Ltd 1994). There will, therefore, be an associated need to assess the ability of these guidelines to identify correctly toxic sediments in Australian (or New Zealand) environments. This applies both to the ability reliably to classify as toxic those sediments that have a biological effect and as nontoxic those that do not. The Resource Management Act (1991) in New Zealand provides the opportunity for the Minister for the Environment to recommend that standards be established or set for contaminants and for the quality of water and soils, but this has not yet been used. To date, in both Australia and New Zealand, criteria that have been used and guidelines that have been proposed are based on those developed overseas. For example, a recent draft study on the dumping of dredged spoil prepared for the CEPA in Australia (Pollution Research Pty Ltd 1994) presented a preliminary list of screening concentrations and concentrations above which unconfined disposal at sea was not considered suitable for a range of organic and inorganic contaminants. These concentrations were based very largely on those developed for the disposal of dredged material in Florida (MacDonald Environmental Sciences Ltd 1992). The only modifications made were for those trace metals whose background concentrations in Australia were known to be larger than those used in the study in Florida.

There is currently no provision of overall objectives and policies for ERA by risk managers in appropriate environmental management agencies at regional or national level to give relevance, direction, and consistency to decisions made locally. In New Zealand, the Resource Management Act (1991) is designed to control the effects of human activities, rather than the activities themselves, which suggests that it could provide a suitable context for risk assessment and management. It specifies an overall objective of sustainable management of natural resources by "...managing the use, development, and protection of natural and physical resources in a way, or at a rate, which enables people and communities to provide for their social, economic, and cultural well-being and for their health and safety while...(a) Sustaining the potential of natural and physical resources...to meet the reasonably foreseeable needs of future generations; and (b) Safeguarding the life-supporting capacity of air, water, soil and ecosystems; and (c) Avoiding, remedying or mitigating any adverse effects on the environment." The Act does not, however, provide more specific objectives, such as basic ecological values to be protected, or explicit guidance for selecting assessment endpoints. In the case of the USEPA guidelines for ERA, Norton *et al.* (1995) suggest that, as a first step, ecological objectives and policies could be developed from existing statutes, regulations, and administrative goals and decisions, for example, the Endangered Species Act of 1973 (1988). The Ministry for the Environment in New Zealand is in the process of developing a national set of measurement endpoints to monitor environmental quality throughout the country, but the development of standards and guidelines is still in a preliminary phase (New Zealand Government 1995).

Contamination of the sediments, water, and biota of Lake Rotorua and its catchment, in the North Island of New Zealand, was assessed as an initial case study for the identification of environmental issues related to the use of wood-preservative chemicals and to provide information for the future development of national guidelines and standards for the management of sites contaminated with such chemicals (Gifford *et al.* 1993). The study assessed the risk to human health from drinking water from streams in the catchment and from eating trout from the lake. Other than comparison of concentrations of contaminants in the water and sediments with overseas guidelines for the protection of aquatic life, an assessment of environmental risk was not attempted.

The Australian and New Zealand Environment and Conservation Council and the Australian Water Resources Council have developed a strategy for the sustainable management of Australia's water resources. This includes the development of water quality guidelines for the protection of fresh and marine waters (Hart *et al.* 1993). These include physical-chemical and biological indicators, the latter being species richness and composition, net primary production and "ecosystem function" (expressed as the ratio of production to respiration). Guidelines for what constitutes unacceptable change were provided for each of these endpoints. In the case of the biological endpoints, these guidelines appear to have been intended largely to promote discussion of suitable values. In New Zealand, recommendations have recently been made for the development of water quality guidelines based on ERA (Water Studies Centre 1995). No such guidelines have yet been proposed for sediments. In terms of translating "broad regulatory mandates into concrete risk assessment policies and objectives" (Norton *et al.* 1995), then, Australia and New Zealand are at an early stage.

As elsewhere, monitoring and verification of the predictions of the risk characterization and exposure characterization are rare in Australia and New Zealand (Lincoln Smith 1991), but there is an increasing awareness of their importance, at least in principle.

Uncertainty over the appropriateness of using criteria and guidelines from other countries has led to a number of challenges to environmental management decisions involving such criteria. For example, a significant case in New Zealand in which the disposal of contaminated sediment in the marine environment was subject to a detailed assessment of the ecological risk involved was in 1990 and concerned the proposed dumping of spoil dredged from Auckland's Waitemata Harbour into the waters of the nearby Hauraki Gulf. The risk assessment was based on a modification of the protocol described in the USEPA (1990) Draft Dredged Material Testing Manual. The actual protocol used, however, deviated from that described in a number of ways (Roper 1991a). The first tier of the USEPA protocol only allows dumping of material without further testing if it has been obtained from a high-energy environment or a site remote from all sources of pollution, or if it is to be used for beach nourishment. The first tier of the proposed protocol in the assessment of the material from Waitemata Harbour, however, allowed dumping if the concentrations of contaminants in the material were less than the screening levels developed by the State of Washington for dumping of material in Puget Sound, Washington (Barrick *et al.* 1988). This was despite the fact that these levels were explicitly developed for Puget Sound, their uncritical use elsewhere was advised against (Barrick *et al.* 1988), and the fact that the USEPA protocol stated that "...at present, chemical analyses cannot be used to directly evaluate the biological effects of any contaminants or combination of contaminants in dredged material." Furthermore, the protocol used in the assessment allowed, potentially, for disposal after effects on the water column alone had been considered. In contrast, the protocol described by the USEPA states that effects on the water column and the benthos must both be considered. Another objection to the protocol related to the fact that dumping could, potentially, be permitted without assessment of bioaccumulation. Toxicity tests for water column effects were done on only one species, rather than several as recommended in the USEPA protocol. This test used larvae of the oyster *Crassostrea gigas* and was criticized on the grounds that it used an introduced rather than an indigenous species and that, since the species occurs in environments known to be contaminated, the larvae are presumably relatively hardy and, therefore, not likely to give adequate indication of the toxicity of the dredged material to other members of the fauna. For benthic toxicity tests, the species tested were the same North American species as used in the USEPA protocol. Tests using native species of invertebrates were available (see below) but were not used. Again, the appropriateness of using exotic species as surrogates for indigenous ones was questioned by objectors.

Some of these procedures were subsequently modified in response to objections raised during the appeal against the initial granting of the right to dump, though not to the full satisfaction of the objectors (Roper 1991b).

A further important aspect to this and many other risk assessments in New Zealand and Australia is the acceptability of the proposed activity to local indigenous people on cultural, traditional, or spiritual grounds. In the case of the proposed dumping of dredged spoil in the Hauraki Gulf, the traditional Maori guardians (the *tangata whenua*) of the waters concerned objected to the dumping on grounds that it was incompatible with their obligation to preserve the quality and integrity (spiritual as well as biological) of that environment. Although the balancing of costs and benefits of the proposed action to the various stakeholders is part of the process of risk management rather than risk assessment, the assessment of risk must cover the interests of all of the stakeholders.

A later study of environmental effects of the dumping of contaminated dredged spoil from Port Nelson, New Zealand, into the adjacent Tasman Bay (Roberts 1992) showed greater use of local information. Although material from the Port had been dumped at the same site for the previous 30 years, the introduction of the Resource Management Act (1991) required assessment of any impacts arising from the dumping of the spoil. In a sediment quality triad approach (Long and Chapman 1985), the assessment incorporated measurement of concentrations of contaminants in the sediments in the dredged area and at the dump site, toxicity testing of the dredged material on a native species of amphipod (*Paracorophium excavatum*), and comparison of the benthic faunas at the dump site and nearby control areas. Concentrations of heavy metals, PCBs, and PAHs in the dredged material were larger than in sediments from Tasman Bay, but there were no significant differences between concentrations at the dump site and control locations. Sediments from the Port were more toxic to *P. excavatum* than those from control sites, but there was no difference between the dump site and the controls, nor was there any evidence that dumping had caused a change in benthic communities. Concentrations of contaminants in a gastropod mollusc (*Austrofusus glans*), which was the subject of a trial fishery for human consumption, did not differ significantly between the dump site and control sites. An elutriate test indicated that risk of release of contaminants from the sediments to the water column was small. On the basis of these results, it was recommended that the environmental risks associated with continued dumping at the site were likely to be smaller than (unquantified) risks associated with changing to other marine sites or to dumping on land (assessment of alternatives to the proposed activity is a stipulation of the Resource Management Act). It is perhaps worth pointing out that, as is generally the case in Australasia and other parts of the world, there were no analyses of the power of the statistical tests of the (null) hypotheses of no difference between samples from the dump site and control sites. McBride *et al.* (1993), McDonald and Erickson (1994), and Erickson and McDonald (1995), among others, have discussed the inappropriateness of using such tests of difference in situations where the consequences of Type II errors (*i.e.*, accepting the null hypothesis of no adverse effect when, in fact, there is one) are likely to cause environmental damage. Alternative approaches are examined in the references cited.

Because of the lack of information about ecological effects, the process of combining exposure characterization and ecological response is generally qualitative rather than quantitative. Similarly, uncertainty analyses, when performed, are also qualitative. There have been exceptions to this pattern. Some commercial environmental consultants, for example, have employed probabilistic methods for assessing exposure to contaminants (Wenning *et al.* 1995b).

In the current *ad hoc* approach to risk assessment, the roles of risk assessor and risk manager in New Zealand or Australia are not rigorously separated, as in the USEPA framework. The applicant, the regulatory governmental organization, and the contractor (acting on behalf of either), where appropriate, may all take part in the problem formulation phase. The applicant, or a contractor acting on their behalf, may collect the information on exposure and response. The risk characterization may be done by the applicant and/or contractor but there may be discussion with the governmental organization. Often, the assessor may also advise the client (who may, effectively, be the risk manager) on methods of reducing impacts. Alternatively, the manager may be a local, regional, or national governmental organization, who operate in the role through, for example, the setting of conditions attached to the granting of permission to conduct the activity in question. Setting of these conditions is not done in a formal ERA framework, although conditions set may be based on previous applications or on set procedures. Involvement of the risk assessor in the process of risk management is in contrast to the strict separation of risk assessor and risk manager at the risk management stage, prescribed in the USEPA process (Norton *et al.* 1995). In the USEPA process, management roles (during problem formulation and the integrative phase between risk assessment and risk management) are influenced by societal and scientific considerations during an interpretive stage, but the risk management process itself is done by the risk manager in isolation. In the Dutch protocol, as mentioned above, derivation of EQC is a direct outcome of the assessment itself.

19.3 Particular problems of sediment ecological risk assessment in Australia and New Zealand

In Australia and New Zealand, methodological uncertainty arises from lack of information on native species, ecosystems (*e.g.*, their functioning and their inertia, stability and resilience *sensu* Underwood 1989), and existing stresses. Existing stresses might include chemical, physical (*e.g.*, changes in sedimentation in estuaries and other aquatic habitats as a result of deforestation and agricultural practices since the arrival of Europeans), and biological (*e.g.*, interactions with introduced species). Methods developed for severely contaminated parts of the world may be inappropriate for regions where there is relatively little chemical contamination (a risk management consideration, but one which also affects the choice of endpoints, *e.g.*, acute versus chronic, and hazard identification).

The use of criteria from environments that are already severely modified in the Australasian situation may be similarly inappropriate. For example, in the case of the proposed dumping of dredged spoil in the Hauraki Gulf, the objectors pointed out that the screening levels cited in the Puget Sound guidelines were larger than concentrations found in many of the most polluted environments in New Zealand. In some of these environments, however, ecological impacts were known to have occurred (as, for example, the demonstrated effects of chlordane in sediments at concentrations less than the screening levels, discussed below).

The fact that risks to local species are generally unknown leads to reliance on standards or criteria derived from overseas studies, often without demonstration of measurable effects (since these presumed effects are based on overseas species and habitats). This takes such use of criteria outside the area of ERA since actual effects are not being taken into account (Chapter 1).

There is evidence that reliance on criteria derived from studies on nonnative species (whether data are gathered overseas or in Australasia using exotic species) does not adequately assess risks to native species. Hickey (1989) tested a range of toxicants on native New Zealand species and on exotic species. He found that the native cladoceran *Ceriodaphnia dubia* was up to four times more sensitive in acute tests than *Daphnia magna*. Other differences in sensitivity have been found between native New Zealand species and those from overseas used in standard tests (Hickey and Vickers 1994). In contrast, Hickey and Roper (1992) reported that the sensitivities of two native species of amphipods (one tube-dwelling, one burrowing) to cadmium were similar to values expected on the basis of work overseas.

The development of toxicity tests for sediments using native species from Australia and New Zealand is in progress, although most work to date has focused on contaminants in the water column. Tests on sediments include sublethal, behavioral responses of marine bivalves (Roper and Hickey 1994) and effects on larval development in sanddollars (Nipper and Roper 1995). Sublethal tests on bivalves seem to be an appropriately sensitive and ecologically relevant means of assessing effects. Hickey and Martin (1995) compared the relative sensitivities of five species of freshwater, benthic invertebrates (an amphipod, a clam, an oligochaete, a tanaid, and a mayfly) to reference toxicants and to bleached kraft mill sediment. There was a large range in sensitivity among species and among toxicants. Clam reburial, amphipod survival, and oligochaete reproduction were the most sensitive endpoints for detecting effects of contaminants. A review of work in New Zealand is given by Hickey (1995). In Australia, tests for effects of sediment have been developed with burrowing amphipods and isopods and for effects of pore water with amphipods and larvae of oysters and prawns (many of these results have not yet been published). Ahsanullah *et al.* (1984) and Weimin *et al.* (1992) have examined bioaccumulation of heavy metals from sediments by Australian species of molluscs, polychaetes, and crustaceans. In general, there does not yet seem to have been much work on development of tests of sediment toxicity using organisms from tropical parts of Australia.

As part of the National Pulp Mills Research Program, funded by the Commonwealth Government of Australia, Moverley *et al.* (1995) have developed mesocosms with marine benthic meiofauna (animals with body sizes in the range 63 to 500 Tm) and examined their potential use for ecotoxicological testing. Responses of the fauna to perfusion of the sediments with different concentrations of 4-chlorophenol were detected using multivariate statistical methods. These results suggested that the mesocosms were a useful technique for testing chronic effects of contaminated sediments on meiofaunal communities.

Changes to the amount of sediment in suspension and rates of deposition in streams and estuaries have been widespread since the arrival of Europeans. Assessments of the ecological effects of these changes are an important contribution to ERA for sediments in Australasia. Quinn and Hickey (1993) compared the composition of benthic faunal assemblages in streams above and below outfalls of sewage effluent. The percentage of taxa whose abundance differed significantly between sites above and below the outfalls correlated inversely with the dilution of the effluent and directly with the concentration of suspended solids. In an equivalent study of assemblages above and below clay discharges from alluvial gold-mining, Quinn *et al.* (1992) found that abundances of invertebrates downstream of discharges as a proportion of abundances upstream were inversely correlated with the log of turbidity loadings. Taxonomic richness showed a similar relationship.

Determination of the ecological relevance of toxic effects of contaminants observed in laboratory ecotoxicological tests requires studies in the field. Surveys and comparisons of spatial and temporal variation in the distributions of contaminants and organisms provides correlative evidence of impacts. Demonstration of cause and effect under realistic conditions requires manipulative field experiments. Both of these approaches have been tried in Australia and New Zealand. There are numerous examples of correlative studies, such as that by Ward and Young (1982), of epibenthic faunas in seagrass beds near a lead smelter in South Australia. In New Zealand, Roper *et al.* (1989) described the distribution of benthic organisms around coastal outfalls in the towns of Gisborne and Hastings, and Roper *et al.* (1988) described apparent effects of diffuse contamination derived from urban runoff on benthic communities in the Manukau Harbour near Auckland.

In a manipulative field experiment, Pridmore *et al.* (1992) experimentally contaminated the surface sediments of a large area (300 m^2) of muddy sandflat with technical chlordane. The contaminant was applied by soaking sediment in the laboratory in a volatile solvent in which chlordane had been dissolved, drying the sediment, and spreading it on the surface of the sandflat in an area without existing chlordane contamination. This treatment resulted in a reduced abundance of two species of bivalves compared to a nearby control area. The concentration of chlordane in near-surface sediments at the experimental site during the period when numbers of bivalves showed the largest changes was 6.7 to 8.7ng.(g dry fines)$^{-1}$. This concentration was similar to those in contaminated areas of the same harbor where the experiment was done and to those reported from overseas studies of chlordane contamination.

To examine the effects of copper contamination on the fauna of sandy subtidal sediments in Botany Bay, near Sydney, Australia, Morrisey *et al.* (1995 1996) experimentally enhanced the concentration of copper in the sediment by burying blocks of building plaster impregnated with copper sulphate. This treatment caused a change in the abundance of a number of types of animals over a period of six months, relative to control treatments (unmanipulated areas and areas in which blocks of plaster without copper had been buried).

The general absence of severe historical contamination from anthropogenic sources in aquatic environments in Australia and New Zealand provides opportunities for studying the effects of diffuse forms of contamination. In New Zealand, attention has recently focused on diffuse contamination from urban runoff. An ongoing study by the National Institute of Water and Atmospheric Research Ltd in New Zealand is combining partly conceptual/partly quantitative modeling of the accumulation of contaminants derived from runoff in the sediments of urbanized estuaries, correlative field studies of the ecological impacts of these contaminants, in terms of patterns of abundance and distribution of contaminants and sediment-living animals, and laboratory-based studies of the toxicity of sediments from these estuaries to native species.

19.4 The appropriateness of a formal framework for ecological risk assessment

How appropriate is a formal framework in situations where there is little information on local conditions (type of habitat, existing concentrations of contaminants, functions of ecosystems), species, or environments?

1) The advantage in such a situation is that, in theory, a formal framework reduces the tendency for *ad hoc*, inconsistent decisions from one case to another. Such inconsistency arises from variations in experience of the risk assessors and managers and variation in the amount of information available on which to base risk assessments and decisions relating to risk management. Promotion of "consistency across EPA assessment" is one of the reasons behind the development of a framework for ERA by the USEPA (Norton *et al.* 1995). Even in a relatively well-developed framework, such as that of the USEPA, as discussed in Chapter 1, there is much variation in actual practice of ERA (relative to other forms of risk assessment) because of the early stage of development of the process and the lack of guidelines from appropriate regulatory agencies.

2) Provision of such a framework, however, may itself lead to uncritical applications of inappropriate methods by the same assessors or managers seeking easy solutions to their dilemmas. Derivation of EQC, for example, may be valid in a northern European context where there is much information on the response of organisms to contaminants or other stressors and on risks of exposure, and where assumptions used in deriving criteria or guidelines are most likely to be reasonable. In Australasia, however, derivation of criteria or guidelines based on assumptions developed elsewhere may result in inadequate characterization

and assessment of ecological risk. The use of expert knowledge of a particular system and the relevant risks might be particularly appropriate in situations where the validity of the assumptions associated with a formal framework is poorly known. Expert opinion is an important part of the problem formulation stage of ERA, but it is also relevant at other stages of the process. Guidelines being developed for ERA by the USEPA (Norton *et al.* 1995) "will not be rigid and will encourage the use of professional judgment within the construct of a logical and scientifically sound structure." In fact, the proposed framework for ERA in New South Wales stresses the importance of site-specific considerations and therefore, implicitly, of local, expert judgment (Warne and Davies 1995).

Balanced against the risk of misapplication of methods and standards developed overseas is the urgent need for some form of protocol for ERA, since environmental management decisions have to be made now, regardless of how inadequate the information underlying them may be. The use of methods of risk assessment, criteria, and guidelines for environmental quality, such as SQC, developed overseas may represent the best available syntheses of current knowledge, however incomplete.

19.5 Summary

It is unlikely that either Australia or New Zealand will be able to develop their own, independent methods of risk assessment. Which of the various models available overseas is eventually adopted, and how and to what extent resources allow these to be modified for local use, remains to be seen. In the meantime, an important contribution to future development of formal methods of risk assessment could be made by monitoring and reporting of the success or otherwise of the various *ad hoc* approaches currently being used, so that the information gained from such experience can guide later developments. Despite numerous pleas (*e.g.*, Hilborn and Walters 1981) for programs of monitoring after developments or other forms of impact have occurred, assessment of impacts and the accuracy of assessment of risk still frequently end with the awarding of a consent.

19.6 Acknowledgments

I am grateful to Peter Cochrane, Chris Hickey, Peter O'Hara, Dave Roper, Sheila Watson, and Bob Zuur for information and advice in the preparation of this paper.

19.7 References

Ahsanullah M, Mobley MC, Negilski DS. 1984. Accumulation of cadmium from contaminated water and sediment by the shrimp *Callianassa australiensis*. *Mar Biol* 82: 191–197.

Aldenberg T, Slob W. 1993. Confidence limits for hazardous concentrations based on logistically distributed NOEC toxicity data. *Ecotoxicol Environ Safety* 25:48–63.

Barrick R, Becker S, Brown L, Beller H, Pastorak R. 1988. Sediment quality values refinement: Volume 1 1988 update and evaluation of Puget Sound AET. Submitted to Tetra Tech Inc. for Puget Sound Estuary Program, Office of Puget Sound. U.S. Environmental Protection Agency (USEPA) Region 10.

Calvert S, Baker R. 1995. Ecological risk assessment - applications in Australia. Proceedings of the 2nd Annual Conference of the Australasian Society for Ecotoxicology; 1995 Jun 29–30; Sydney, Australia.

Endangered Species Act of 1973. 16 U.S.C. §1531 *et seq.* as amended by P.L. 100-478, October 7, 1988.

Erickson WP, McDonald LL. 1995. Tests for bioequivalence of control media and test media in studies of toxicity. *Environ Toxicol Chem* 14:1247–1256.

Gifford JS, Hannus IM, Judd MC, McFarlane PM, Anderson SM, Amoamo DH. 1993. Assessment of chemical contaminants in the Lake Rotorua catchment. Report prepared for Bay of Plenty Regional Council by Environmental Research Group, Wood Processing Division, New Zealand Forest Research Institute.

Hart BT, Angehrn-Bettinazzi C, Campbell IC, Jones MJ. 1993. Australian water quality guidelines: a new approach for protecting ecosystem health. *J Aquat Ecosys Health* 2:151–163.

Hickey CW. 1989. Sensitivity of four New Zealand cladoceran species and *Daphnia magna* to aquatic toxicants. *New Zealand J Mar Freshwater Res* 23:131–137.

Hickey CW. 1995. Ecotoxicity in New Zealand. *Australasian J Ecotoxicol* 1:43–50.

Hickey CW, Martin ML. 1995. Relative sensitivity of five benthic invertebrate species to reference toxicants and resin-acid contaminated sediments. *Environ Toxicol Chem* 14:1401–1409.

Hickey CW, Roper DS. 1992. Acute toxicity of cadmium to two species of infaunal marine amphipod (tube-dwelling and burrowing) from New Zealand. *Bull Environ Contam Toxicol* 49:165–170.

Hickey CW, Vickers ML. 1994. Toxicity of ammonia to nine native New Zealand freshwater invertebrate species. *Arch Environ Contam Toxicol* 26:292–298.

Hilborn R, Walter CJ. 1981. Pitfalls of environmental baseline and process studies. *EIA Review* 2:265–278.

Lincoln Smith MP. 1991. Environmental impact assessment: the roles of predicting and monitoring the extent of impacts. *Australian J Mar Freshwater Res* 42:603–614.

Long ER, Chapman PM. 1985. A sediment quality triad: measures of sediment contamination, toxicity and infaunal community composition in Puget Sound. *Mar Pollut Bull* 16:405–415.

McBride GB, Loftis JC, Adkins NC. 1993. What do significance tests really tell us about the environment? *Environ Management* 17:423–432 (Errata in 18:317).

McDonald LL, Erickson WP. 1994. Testing for bioequivalence in field studies: has a disturbed site been adequately reclaimed? In: Fletcher DJ, Manly BFJ, editors. Statistics in ecology and environmental monitoring. Dunedin NZ: Univ of Otago Pr. p 183–197.

MacDonald Environmental Sciences Ltd. 1992. Development of an integrated approach to the assessment of sediment quality. Vol. 1: In Florida. Vol. 2: Supporting document. Report prepared Florida State Department of Environmental Regulation. NTIS No. PB92-1884655.

Morrisey DJ, Underwood AJ, Howitt L. 1995. Development of sediment-quality criteria: a proposal from experimental field-studies of the effects of copper on benthic organisms. *Mar Pollut Bull* 31:372–377.

Morrisey DJ, Underwood AJ, Howitt L. 1996. Effects of copper on the faunas of marine soft-sediments: a field experimental study. *Mar Biol* 125:199:213.

Moverley JH, Ritz DA, Garland C. 1995. Development and testing of a meiobenthic mesocosm system for ecotoxicological experiments. Canberra, Australia: CSIRO. National Pulp Mills Research Program Technical Report No. 14. 117 p.

New Zealand Government. 1995. Towards a core set of national environmental indicators: a framework. Wellington: Ministry for the Environment draft report. 39 p.

Nipper MG, Roper DS. 1995 Growth of an amphipod and a bivalve in uncontaminated sediments: Implications for chronic toxicity assessments. *Mar Pollut Bull* 31:424–430.

Norton SB, Rodier DJ, Gentile JH, Troyer ME, Landy RB, van der Schalie W. 1995. The EPA's framework for ecological risk assessment. In: Hoffman DJ, Rattner BA, Burton Jr GA, Cairns Jr J, editors. Handbook of ecotoxicology. Boca Raton FL: Lewis.

Pollution Research Pty Ltd. 1994. Consultancy study on sea dumping of dredged spoil. Canberra, Australia: Commonwealth Environmental Protection Agency. 50 p.

Pridmore RD, Thrush SF, Cummings VJ, Hewitt JE. 1992. Effect of the organochlorine pesticide technical chlordane on intertidal macrofauna. *Mar Pollut Bull* 24:98–102.

Quinn JM, Davies-Colley RJ, Hickey CW, Vickers ML, Ryan PA. 1992. Effects of clay discharges on streams. 2. Benthic invertebrates. *Hydrobiologia* 248:235–247.

Quinn JM, Hickey CW. 1993. Effects of sewage waste stabilization lagoon effluent on stream invertebrates. *J Aquat Ecosys Health* 2:205–219.

Roberts R. 1992. Impact of dredging and dredgings disposal in Nelson: a consideration of chemical contaminants. Report prepared for Port Nelson Ltd by Cawthron Institute, Nelson, New Zealand.

Roper DS. 1991a. Evidence of David Stanley Roper, before the Planning Tribunal, Appeal nos. 743/90; 745/90; 747/90, in the matter of the Water and Soil Conservation Act 1967, and in the matter of 3 appeals pursuant to s25 of the Act, between the New Zealand Underwater Association Inc., Maruia Society Inc., and the Auckland City Council (Appellants) and Auckland Regional Council (Respondent) and Ports of Auckland Ltd (Applicant).

Roper DS. 1991b. Supplementary evidence of David Stanley Roper, before the Planning Tribunal, Appeal nos. 743/90; 745/90; 747/90, in the matter of the Water and Soil Conservation Act 1967, and in the matter of 3 appeals pursuant to s25 of the Act, between the New Zealand Underwater Association Inc., Maruia Society Inc., and the Auckland City Council (Appellants) and Auckland Regional Council (Respondent) and Ports of Auckland Ltd (Applicant).

Roper DS, Hickey CW. 1994. Behavioural responses of the marine bivalve *Macomona liliana* exposed to copper- and chlordane-dosed sediments. *Mar Biol* 118:673–680.

Roper DS, Smith DG, Read GB. 1989. Benthos associated with two New Zealand coastal outfalls. *New Zealand J Mar Freshwater Res* 23:295–309.

Roper DS, Thrush SF, Smith DG. 1988. The influence of runoff on intertidal mudflat benthic communities. *Mar Environ Res* 26:1–18.

Smith DG, Quinn JM, Cooper AB, Davies-Colley RJ, Hickey CW, Rutherford JC, McBride GB, Zuur BJ. 1994. The development of guidelines for standards in the Resource Management Act. Proceedings of the New Zealand Water Conference 1994; 1994 Aug 29–31; Hamilton, New Zealand. Hamilton: The Organizing Committee of the 1994 NZ Water Conference.

Underwood AJ. 1989. The analysis of stress in natural populations. *Biol J Linnean Soc* 37:51–78.

[USEPA] U.S. Environmental Protection Agency. 1990. Draft ecological evaluation of proposed discharge of dredged material into ocean waters. Washington DC: USEPA. EPA-503-8-90/002.

Ward TJ, Young PC. 1982. Effects of sediment trace metals and particle size on the community structure of epibenthic seagrass fauna near a lead smelter, South Australia. *Mar Ecol Prog Ser* 9:137–146.

Warne MStF, Davies HM. 1995. Modification of the Dutch ecological risk assessment method for use in Australia. Proceedings of the 2nd Annual Conference of the Australasian Society for Ecotoxicology; 1995 Jun 29–30; Sydney, Australia.

Water Studies Centre. 1995. A process for the development of guidelines for the protection of aquatic life in New Zealand. Report commissioned by the Ministry for the Environment, Water Studies Centre, Monash University, Melbourne, Australia. 57 p.

Weimin Y, Batley GE, Ahsanullah M. 1992. The ability of sediment extractants to measure the bioavailability of metals to three marine invertebrates. *The Science of the Total Environment* 125:67–84.

Wenning RJ, Leonte D, Sullivan B. 1995a. Recent developments in ecological risk assessment at contaminated sites and recommendations for establishing an Australasian framework Proceedings of the 2nd Annual Conference of the Australasian Society for Ecotoxicology; 1995 Jun 29–30; Sydney, Australia.

Wenning RJ, Leonte D, Sullivan B. 1995b. Use of probabilistic exposure assessment methods in ecological risk assessment. Proceedings of the 2nd Annual Conference of the Australasian Society for Ecotoxicology; 1995 June 29–30; Sydney, Australia.

SESSION 7
INTERNATIONAL PERSPECTIVES

Chapter 20
Ecological risk assessment for sediments: a European perspective

Rachel Fleming, Steve Maund, Lindsay Murray

20.1 Introduction

This section gives a brief overview of the status of SERA in Europe in terms of the three applications: product assessment, site cleanup, and dredged material disposal. In general, formal Europe-wide guidance is limited, although the need for consideration of sediment exposure and effects is now being incorporated into European legislation, and many research and development (R&D) programs are being funded by international bodies, to develop and standardize the tools required. Also, discussion of triggering and assessment frameworks is underway in several areas.

20.2 Product assessment

20.2.1 Current legislation

In recent years, regulatory authorities in Europe have started to raise concerns about the potential toxicity to benthic organisms of pesticides and other organic chemicals which accumulate in sediments. These concerns are now beginning to be implemented into legislation for pesticides, _e.g._, the new European Pesticide Directive (CEC 1991) which includes a sediment testing requirement for certain types of compound. In the general chemicals area, there is as yet no firm regulatory requirement for sediment toxicity testing, and so this brief review of current status principally reflects product assessment for pesticides, although clearly there are common themes.

20.2.2 Testing methods

Concerns over sediment toxicity have resulted in a great deal of activity in the development of testing methods over recent years and much of the European experience is captured in the comprehensive proceedings of an earlier SETAC-Europe workshop held in the Netherlands (Hill _et al._ 1994). For freshwater assessment, the most commonly used organism in Europe has been the freshwater dipteran, _Chironomus riparius_. This has most likely been due to its robustness and amenability to ecologically relevant endpoints like emergence, for which it is possible to develop ECx measures of response, rather than reliance on NOECs. The usefulness of the latter has been a matter of debate in Europe over recent years, at least partly because of their dependence on experimental design.

In June 1995, a meeting of international experts was organized by the OECD in Copenhagen, Denmark to discuss the prioritization of benthic testing methods (OECD 1995). The recommendation of the group was that _Chironomus_ spp. should be the first priority for the development of a standardized OECD sediment toxicity method (as would be used for product assessment). It therefore seems likely that this genus will become the principal sediment testing invertebrate for products in Europe, much as *Daphnia* spp. has been used as the representative organism for water column or pelagic invertebrates. The second priority recommendation for a standardized OECD guideline was estuarine and marine amphipods. In both cases, protocols that have already been developed by other international bodies will be used as a basis for development of definitive guidelines. For example, available European guidelines include acute and chronic toxicity tests with *Chironomus riparius* and *Corophium volutator* standardized for the European Commission (Fleming *et al.* 1994), an acute *Corophium volutator* protocol for evaluation of drilling muds and chemicals used in offshore operations (Paris Commission [PARCOM] 1994), and a *Chironomus riparius* protocol for measuring the toxicity of products applied to the water column of a sediment system, in order to simulate an overspray scenario (Biologische Bundesanstalt Für Land-Und Forstwirtschaft [BBA/IVA] 1994). It is intended that acute and chronic methods will be incorporated into the same generic OECD guideline and a list of appropriate species will be recommended. The use of artificial sediments, when fully developed and standardized, will also be considered.

20.2.3 Triggers

With testing methods gradually becoming established, more recently efforts have begun to focus on developing suitable procedures for triggering sediment toxicity assessment. In common with the approaches described in Chapter 4, the broad view in Europe is that prospective product testing should be based on the propensity of the chemical to adsorb and persist and be potentially toxic. Indeed this view is reflected in the European Pesticide Directive (91/414/EC), which states:

> Where environmental fate and behaviour data...report that an active substance is likely to partition to and persist in aquatic sediments, expert judgement should be used to decide whether an acute or a chronic sediment toxicity test is required. Such expert judgement should take into account whether effects on sediment dwelling invertebrates are likely....

Currently, the European Crop Protection Association (ECPA) and the American Crop Protection Association (ACPA) are together developing proposals for measures of these adsorption, persistence and toxicity triggers. Suitable measures are under discussion, and proposals include the use of soil aerobic degradation rates as an indication of persistence, OC partitioning coefficients (K_{OC}) to assess adsorption, and data from *Daphnia* acute or chronic toxicity tests to estimate toxicity (Table 20-1). The advantages of using these endpoints for triggering sediment studies are that the data are generated as a matter of course for the vast majority of pesticides and that, with suitably selected values, they are likely to provide conservative estimates of persistence, adsorption, and toxicity in sediment (Hamer *et al.* 1995). As with many triggers, more realistic data (with less uncertainty) can be used to supersede these preliminary triggers if appropriate. The values proposed for each of these measures are currently being evaluated from company data-

Table 20-1 ECPA/ACPA proposed values for triggering of sediment toxicity tests

Conditions for sediment toxicity testing	Exceptions
If preliminary risk assessments suggest potential for impacts on aquatic invertebrates, then sediment toxicity assessment is required	
1) if there is potential for partitioning to and persistence in sediments such that	
K_{oc} (OC partition coefficient) ≥ 1000 from a standard batch equilibrium study	
and	
2) DT_{50} (degradation half life) ≥ 30 days in a soil aerobic degradation laboratory study	unless it can be demonstrated that the $DT_{50} < 30$ days in the aquatic environment (by, *e.g.*, hydrolytic, photochemical, ready biodegradation, water-sediment metabolism or other appropriate studies) or the concentration of the parent compound $\leq 10\%$ in sediment after more than 14 days in a water-sediment metabolism system
and	
3) if there is potential for toxicity such that	
Daphnia 48-h $EC_{50} < 1.0$ mg/l or 21-d NOEC <0.1 mg/l	unless by means of a suitable risk assessment that takes into account of partitioning of compound between water and sediment, it can be shown that there is no potential risk to invertebrates, *e.g.*, the ratios between the acute or chronic effects and exposure concentration are >10 or >1, respectively.

bases of regulatory studies, and a publication of the findings is planned for the near future.

Triggers have also been proposed by the German BBA (Streloke and Kopp 1995). These also include estimates of adsorption and persistence from the BBA sediment metabolism study (BBA 1990) and toxicity, such that a sediment toxicity test, as described above, is required when "the active ingredients, toxic metabolites or bound residues are found in the sediment [from the sediment metabolism study] in an amount higher than 10% of the applied test substance after day 14. Furthermore, the NOEC from the chronic toxicity test with *Daphnia* should be lower than 0.1mg/l or the BCF in fish higher than 100."

20.2.4 Risk assessment
In addition to developing triggers and methods for sediment toxicity assessment, there have also been some preliminary discussions of how sediment data will be used in a risk assessment. The general principle of risk assessment always involves the comparison of an effect concentration to a PEC. Although methods for estimating concentrations in surface water are well-established, until recently few models have included PECs in sedi-

ment. One exception is TOXSWA, a new model under development in the Netherlands by the Winand Staring Centre (SC-DLO; Adriaanse 1995). Furthermore, the development of sediment PECs has also been included in the remit of the Forum for the Coordination of Pesticide Fate Models and their Uses (FOCUS), so development of a harmonized approach is expected in the near future.

20.3 Site cleanup

20.3.1 Current legislation

Until recently, philosophy in European legislation has been that maintenance of acceptable water quality will automatically lead to acceptable sediment quality. Mention of sediments in water legislation has generally been restricted to emission control directives such as the Dangerous Substances Directive (76/464/EEC). This directive provides a framework for the elimination or reduction of pollution of inland, coastal, and territorial waters by dangerous substances. The most dangerous, decided on the basis of toxicity, bioaccumulation, and persistence (List 1 or Black List) should be eliminated; the less harmful substances (List 2 or Grey List) must be reduced. List 1 substances are controlled at a European Union level by means of daughter directives which lay down standards for specific substances, and List 2 substances are controlled at a national level. Under the terms of the Directive, Member States may adopt either Limit Values based on uniform fixed emissions which are independent of the use of receiving water, or Environmental Quality Standards for the receiving water, which are dependent on the intended use. In the absence of generic sediment quality standards for receiving water, or limit values for emissions that have taken into account the potential of a substance to bind to sediment, the "stand still principle" is often adopted whereby monitoring for compliance requires that levels of dangerous substances in sediments do not increase.

At present, there is a move within Europe toward a more integrated approach to water management, focusing on protecting the aquatic ecosystem and water uses as a whole. Changes in water legislation are being brought about by the development of a new European Commission Framework Water Resource Directive, which is intended to form the basis of all future water policy. This is likely to incorporate some of the existing directives covering the protection of receiving waters. The Framework Directive is aimed at protecting ecosystems from point- and diffuse-source pollution and other anthropogenic influences, will establish links between quantity and quality of water, and will apply to both surface water and groundwater.

The water quality aspects of this Directive will be based on the recent European Commission's proposed Directive on the Ecological Quality of Water (COM(94)680), which lays out a system for global estimation and classification of the ecological quality of surface waters. The measures proposed are intended to maintain and improve the ecological quality of waters with the ultimate aim of achieving "good ecological quality." Ecological quality is defined by nine quality elements including biological, chemical, and physical measures or indices. Two quality elements are relevant to management of contaminated sediments:

1) Levels of toxic and other harmful substances in water, sediment, and biota
2) Structure and quality of the sediment and its ability to sustain the biological community of the ecosystem

For both of these quality indicators, assessment and measurement endpoints are currently being investigated by the European Commission. Should the Directive be adopted, Member States will select parameters that represent the most sensitive indicators of ecological quality of the waters concerned. This approach will provide guidance on global classification of sediment quality, and it will allow resources to be directed to those sites at which further investigation is required to determine whether remedial action is necessary, *i.e.*, site-specific SERA.

20.3.2 Testing methods

At present, there is little formal guidance within Europe for carrying out SERAs to help decision-making in terms of remediation options and habitat rehabilitation. Site-specific investigations are occasionally being used, but the chosen methodologies are fragmentary and are in need of standardization and harmonization within and between Member States. This fragmentation is often caused by different organizations being contracted to undertake the risk assessment and, on each occasion, methods being selected on the basis of in-house expertise. In addition, procedures and policies differ between Member States, with different degrees of importance being placed on the role of ERA for sediments. However, in several States, methods and guidelines are being proposed, generally in accordance with the guidance given in Chapter 7. The Netherlands and the UK are used here as two case examples.

In the Netherlands, site cleanup decisions are based largely on national SQC. National target values, limit values, and intervention values have been derived using aquatic toxicity test data and the limited spiked-sediment toxicity test data available (Dutch Ministry of Housing, Physical Planning & Environment [VROM] 1994). The former are translated into sediment data using EqP theory, and all criteria are standardized for OC and particle size distribution. Short-term decisions on remediation of highly contaminated sites are based on exceedance of the intervention value (*i.e.*, the hazard concentration [HC50], which theoretically protects 50% of all species). However, it is recognized that these guidelines, because of their generic nature, should lead to a site-specific investigation before any further decisions are made. A second-level assessment has been proposed which incorporates chemical analysis, benthic surveys (based primarily on chironomid and oligochaete abundance and diversity), and chronic bioassays (*Chironomus riparius* and *Daphnia magna*). In the risk analysis phase, emphasis is placed more on causal relationships between the contaminants identified and effects observed in the field, rather than weighting and aggregating the data. To date, only one large-scale remediation has been carried out in the Netherlands based on this type of assessment (van de Gucthe 1995).

In the UK, cleanup of contaminated sediments is less common than in the Netherlands, where the negative effects of sediment contamination in some areas of the Rhine and Meuse deltas have long been proven (van de Gucthe 1995). UK sediment cleanup operations are driven more by chemical-specific pollution incidents and are dealt with on an *ad hoc* basis by operational regulatory staff in the region affected. Assessments may include acute and chronic bioassays, surveys of fish and bird populations, and various modeling approaches, depending on the nature of the contaminant involved. However, a number of recent incidents has prompted the drafting of formal guidelines for SERA, which are expected to be completed in 1996.

Unlike the Netherlands, the UK has a wide range of sediment geochemistry and, as yet, has no national sediment quality standards nor a robust basis on which to calculate values. Therefore, the guidance will take the form of a tiered approach as outlined in Chapter 7, integrating toxicity testing and benthic community surveys with measurement of contaminant levels in sediments and biota. Effects-based guidelines, developed by other national bodies, may be used in initial tiers.

For the toxicity testing component, recommended laboratory bioassays have been adapted from toxicity test procedures developed for product assessment, primarily using *Chironomus riparius*, *Corophium volutator*, and *Arenicola marina*. Emphasis will be placed on the further development and use of chronic bioassays for which procedures are currently being standardized, and on *in situ* deployments for measurement of toxicity and bioaccumulation using species such as *Chironomus riparius*, *Gammarus pulex*, and *Mytilus edulis*. For the biological component, routinely used procedures such as the BMWP score (Biological Monitoring Working Party 1978) and the RIVPACS model (Wright *et al.* 1993) will continue to be used for assessing biological quality of freshwater systems, while techniques for estuaries and coastal waters are currently under development and are due to be tested on a large scale in 1997.

20.4 Dredged material disposal

20.4.1 Current legislation

In different European Member States, dredging operations and the disposal of dredged material, whether to sea or land, are subject to a range of statutory and other controls. Disposal of dredged material to sea falls under the remit of the London Convention (1972), which has now been ratified by more than 70 countries and controls the dumping of "noxious substances," including chemical waste, into the oceans. Other local conventions are also in place, such as the OSPAR Convention (OSPAR 1992; derived from Oslo and Paris Conventions), which is aimed at protecting the northeast Atlantic and including the North Sea from pollution by dumping by ships and aircrafts and land-based sources, and applies to all countries which border on these areas or have rivers that discharge into them. Many Member States have developed their own control procedures for dredging and disposal, to fit in with the relevant international conventions. Control is carried out by licensing authorities within each State.

Assessing the effects of dredged material disposal requires consideration of both the disposal site and the material to be dredged. Some guidance on the criteria to be considered in the licensing process is given in the international conventions. The London Convention is recently undergoing substantial revision, and an assessment procedure for dredged material disposal licensing was adopted in December 1995 (the Dredged Material Assessment Framework [London Convention 1995]). For the disposal site, consideration of the location must be taken into account, and the effects on navigation, coast protection, fishing amenity, nature conservation, and other legitimate uses of the sea considered. Disposal in sensitive areas such as live coral reefs and other areas adapted to low turbidity is not recommended. For the dredged material itself, the international conventions give general guidance on physical and chemical considerations. The Oslo Commission Guidelines for the Management of Dredged Material (OSPAR 1993) give guidance on the assessment of dredged material in terms of characteristics, composition, sampling, and analysis for both management of dredged material and subsequent monitoring programs. In these guidelines, the dredged material is exempt from testing if it has not been exposed to appreciable sources of contamination and satisfies the following criteria:

1) It is composed almost exclusively of sand, gravel, or rock.

2) It is for beach nourishment or restoration.

3) The amount is less than 10,000t/year and there is existing information on sediment quality.

Further chemical information is required for those dredged materials which do not meet these exemptions to establish whether disposal may cause undesirable effects on marine organisms or human health. Attention is paid to bioaccumulation in marine organisms, particularly in food species. If substances are present for which biological effects are not fully understood, the use of biological test procedures is recommended. These may include laboratory tests for acute toxicity, chronic toxicity, and bioaccumulation.

20.4.2 Testing methods

Some detail of procedures in the Netherlands is given in Chapter 5. This section will refer primarily to procedures adopted in the UK. In the UK, the majority of dredged material, particularly from estuaries, ports, and harbors for navigation purposes, is replaced in the sea either for disposal or beneficial use. The main UK legislation for sea disposal is the Food and Environment Protection Act (1985), and licensing under this legislation is controlled by the Ministry of Agriculture, Fisheries and Food (MAFF) in England, the Scottish Office Agriculture and Fisheries Department (SOAEFD) in Scotland, and the Department of the Environment in Northern Ireland (DoENI). Prior to dredging, an assessment is carried out for the licensing authority, which takes into account the characteristics of the material to be dredged and also the characteristics of the deposit site.

The characteristics of the dredged material that are required for licensing are quantity, sediment type, and chemical nature. For the majority of sediments, the determinands measured are those defined in the technical supplement to the Oslo Convention (OS-

CINE) guidelines. In England and Wales, samples are routinely analyzed for metals, *i.e.*, mercury, cadmium, lead, zinc, copper, nickel, chromium, and sometimes arsenic. Most sediments are also analyzed for organotin compounds. Analyses for PCBs and organochlorine pesticides, petroleum hydrocarbons, and other determinands are carried out as appropriate in particular situations where contamination is suspected, or where background information on the levels of these contaminants in UK dredged material is required. The use of sediment bioassays in the licensing process is currently being investigated, although the way in which these tests will be incorporated is yet to be determined. A number of surveys have been conducted to assess biological effects of dredged sediments using a battery of acute and chronic bioassays including *Corophium volutator* and *Arenicola marina*.

Monitoring the effects of disposal at sea is also a requirement of the international conventions and is carried out by the licensing authorities in the UK. It allows the authorities to confirm the predicted impacts of disposal and to take early action if unexpected impacts occur, and it provides a basis for future licensing policy. Monitoring investigations take place at a number of disposal sites and include these aspects:

1) Sampling and analysis (physical and chemical) of seabed sediments
2) Recording of seabed topography, using sidescan sonar and photography
3) Measurement of water currents
4) Benthic community assessment
5) Chemical analysis of fish and shellfish

In 1992, a group was set up by MAFF to develop detailed guidelines for monitoring at sewage-sludge and dredged material disposal sites (Group Co-ordinating Sea Disposal Monitoring [GCSDM]). One of the outputs of the group was a list of "Sediment Action Limits" (levels which, if exceeded, indicate that studies should be initiated to investigate the availability of the contaminant to organisms; MAFF 1994). These were derived using the EqP approach for a number of substances on UK priority hazardous substances lists. Field studies are now being conducted to assess the validity of these guidelines. Other recommendations made by the GCSDM were that chemical monitoring of disposal sites should be combined with measures of benthic community structure and sediment toxicity. The tools with which to carry out such investigations are currently under development and validation. A report on monitoring and assessment of marine benthos at UK dredged material disposal sites has recently been published by the GCSDM (SOAEFD 1996).

20.5 Conclusions
There are many areas of overlap between current procedures in Europe and those recommended in previous chapters. Some areas of common research include these:

1) Development of triggers for product assessment based on adsorption, persistence, and toxicity
2) Development of a risk assessment framework for prospective testing

3) Standardization of existing whole-sediment toxicity test methods
4) Refinement of exposure models
5) Development of artificial sediments for product testing
6) Development of integrated guidelines for site cleanup and dredged material disposal based on sediment chemistry, toxicity, and biology
7) Development of chronic and *in situ* bioassays for all SERA applications

20.6 Acknowledgments
The authors would like to thank Peter Matthiessen and Kees van de Gucthe for their help in preparing this overview.

20.7 References
Adriaanse P. 1995. Exposure assessment of pesticides in field ditches: the TOXSWA model. Society of the Chemical Industry Conference, Ecotoxicology of Organic Compounds in the Aquatic Environment; 1995 Dec.

BBA. 1990. Degradability and fate of plant protectants in the water/sediment system. Guidelines for the examination of plant protectants in the registration process, Part IV, BBA, Braunschweig 1990.

[BBA/IVA] Biologische Bundesanstalt Für Land-Und Forstwirtschaft. 1994. International toxicity ring-test on sediment dwelling *Chironomus riparius*. Protocol for the toxicity ring-test of two pesticides to the sediment-dwelling larvae of *Chironomus riparius*. BBA/IVA *ad hoc* Working Group on Sediment Toxicity Tests, BBA, Braunschweig.

[BMWP] Biological Monitoring Working Party. 1978. Final Report: Assessment and presentation of the biological quality of rivers in Great Britain, December 1978, unpublished report. Department of the Environment, Water Data Unit.

CEC. 1991. Council Directive on the placing of plant protection products on the market. *Official J CEC* 1991 Aug 19; L230.

Fleming, R, Crane M, Van de Gucthe C, Grootelaar L, Smaal A, Ciarelli S, Karbe L, Borchert J, Westendorf J, Vahl H, Holwerda D, Looise B, Guerra M, Vale C, Gaudencio MJ, van den Hurk P. 1994. Sediment toxicity tests for poorly water-soluble substances. Final Report to the European Commission. EC 3738.

Hamer MJ, Maund SJ, Hill IR. 1995. Development of trigger values for sediment toxicity assessment in pesticide registration. SETAC-Europe Annual Conference; 1995 Jun; Copenhagen. Abstract No. 0166.

Hill IR, Matthiessen P, Heimbach F. 1994. Guidance document on sediment toxicity tests and bioassays for freshwater and marine environments. SETAC-Europe proceedings from the Workshop on Sediment Toxicity Assessment; 1993; the Netherlands.

London Convention. 1972. Final Act of the Inter-governmental Conference on the Convention on the Dumping of Wastes at Sea; 1972 Nov 13; London. London: (Cmnd 5169) HMSO.

London Convention. 1995. Dredged material assessment framework. Eighteenth Commission Meeting of Contracting Parties to the London Convention; 1995 Dec; London.

[MAFF] Ministry of Agriculture, Fisheries and Food. 1994. Sixth Report of the Group Co-ordinating Sea Disposal Monitoring. Lowestoft: Directorate of Fisheries Research. Aquatic Environment Monitoring Report 43.

[OECD] Organization for Economic Cooperation and Development. 1995. Final report of the OECD Working Group meeting on aquatic toxicity testing; 29–30 June 1995. Copenhagen: OECD. TB/95132.

[OSPAR] Oslo and Paris Conventions for the Prevention of Marine Pollution. 1993. Oslo Commission Guidelines for the Management of Dredged Material. Annex 1, Fifteenth Meeting of the Oslo and Paris Commissions; 1993 Jun 14–19; Berlin.

[OSPAR] Oslo and Paris Conventions for the Prevention of Marine Pollution. 1992. Final Declaration of the Ministerial Meeting of the Oslo and Paris Commissions, Paris, 21 September 1992. London: Oslo and Paris Commission.

[PARCOM] Paris Commission. 1994. Final report of the results of the PARCOM Sediment Reworker Ring-Test Workshop. 1993 Dec; The Hague. Paris Commission Group on Air Pollution. GOP/18/4/4-E.

[SOAEFD] Scottish Office Agriculture and Fisheries Department. 1996. Monitoring and assessment of the marine benthos at UK dredged material disposal sites. Prepared by the Benthos Task Team for the Marine Pollution Monitoring Management Group, Co-ordinating Sea Disposal Monitoring. Scottish Fisheries Information Pamphlet 21, 1996.

Streloke M, Kopp H. 1995. Long-term toxicity with *Chironomus riparius*: development and validation of a new test system. BBA Heft 315, Berlin 1995.

Van de Gucthe C. 1995. Ecological risk assessment of polluted sediments. *Eur Water Pollut Control* 5:16–24.

[VROM] Dutch Ministry of Housing, Physical Planning & Environment. 1994. Environmental quality objectives in the Netherlands: A review of environmental quality objectives and their policy framework in the Netherlands. Risk Assessment and Environmental Quality Division. Directorate for Chemicals, External Safety and Radiation Protection. Ministry of Housing, Spatial Planning and the Environment.

Wright JF, Furse MT, Armitage PD. 1993. RIVPACS - a technique for evaluating the biological quality of rivers in the UK. *Eur Water Pollut Control* 3:15–25.

Glossary

The following definitions obtained from USEPA (1994) and ASTM (1995a–1995d) are applicable to many of the terms listed below.

accuracy	Combination of bias and precision for a procedure which reflects the closeness of a measured value to a true value.
assessment endpoint	An explicit expression of the environmental value to be protected.
bias	Consistent deviation of measured values from the true value caused by systematic errors in a procedure (difference between the true values from the mean value determined by using a large number of replicate determinations).
bioaccumulation	The net accumulation of a substance by an organism as a result of uptake from all environmental sources.
bioaccumulation factor	Ratio of tissue residue to source compartment (*e.g.*, sediment) contaminant concentration at steady-state.
bioconcentration	The net assimilation of a substance by an aquatic organism as a result of uptake directly from aqueous solution.
bioconcentration factor (BCF)	Ratio of tissue residue to water contaminant concentration at steady-state.
biota-to-sediment accumulation factor (BSAF)	The ratio of lipid-normalized tissue reside to organic-carbon-normalized sediment contaminant concentration at steady state, with units of g-carbon/g-lipid.
characterization of ecological effects	Portion of the analysis phase of ecological risk assessment that evaluates the ability of a stressor to cause adverse effects under a particular set of circumstances.
community	An assemblage of populations of different species within a specified location in space and time.
conceptual model	A model which describes a series of working hypotheses as to how the stressor might affect ecological components. The model also describes the ecosystem potentially at risk, the relationship between measurement and assessment endpoints, and exposure scenarios.
contaminated sediment	Sediment containing chemical substances at concentrations that pose a known or suspected threat to environmental or human health.
control sediment	A sediment which is essentially free of contaminants and is used routinely to assess the acceptability of a test.
cumulative effects	The combined effects of multiple stressors (*e.g.*, metals and pesticides) or multiple events (*e.g.*, placement of dredged material from several different projects at a site over time).
depuration	Loss of a substance from an organism as a result of any active (*e.g.*, metabolic breakdown) or passive process when the organism is placed into an uncontaminated environment.
direct effect	The stressor acts on the ecological component of interest itself, rather than through effects on other components of the ecosystem.
ecological component	Any part of an ecological system, including individuals, populations, communities, and the ecosystem itself.

ecological risk assessment (ERA)
The process that evaluates the likelihood that adverse ecological effects may occur or are occurring as a result of exposure to one or more stressors.

ecosystem
The biotic community and abiotic environment within a specified location in space and time.

elimination
General term for the loss of a substance from an organism which occurs by any active or passive means.

exposure
Co-occurrence of or contact between a stressor and an ecological component.

exposure profile
The product of a characterization of exposure in the analysis phase of ecological risk assessment. The exposure scenario summarizes the magnitude and spatial and temporal patterns of exposure for the scenarios described in the conceptual model.

exposure scenario
Set of assumptions concerning how an exposure may take place, including assumptions about the exposure setting, stressor characteristics, and activities that may lead to exposure.

indirect effect
The stressor acts on supporting components of the ecosystem, and these in turn have an effect on the ecological component of interest.

measurement endpoint
A measurable characteristic which is related to the valued characteristic chosen as the assessment endpoint. Measurement endpoints are often expressed as the statistical or arithmetic summaries of the observations which comprise the measurement.

median lethal concentration (LC50)
A statistically or graphically estimated concentration which is expected to be lethal to 50 percent of a group of organisms under specified conditions.

no observed effect concentration (NOEC)
The highest level of a stressor evaluated in a test which does not cause statistically significant differences from the controls. Same definition applies to the no observed effect level (NOEL).

population
An aggregate of individuals of a species within a specified location in time and space.

pore water
Water occupying space between sediment or soil particles.

precision
Measure of agreement among replicable analyses of a sample. Precision is measured by repeatability and reproducibility.

recovery
Partial or full return of a population or community to a condition that existed before the introduction of the stressor.

reference sediment
A whole sediment near an area of concern used to assess sediment conditions exclusive of material(s) of interest.

reference toxicity tests
A test with a high-grade reference material conducted in conjunction with sediment tests to determine possible changes in condition of the test organisms.

relative bias	difference between the mean value as determined by a method using a large number of replicate determinations and the true value of the test sample.
repeatability	Measure of the degree of agreement among replicate analyses carried out simultaneously or in rapid succession by the same operator using the same apparatus under the same conditions for the analysis of the same sample.
reproducibility	Measure of the degree of agreement among replicate analyses carried out by operators in different laboratories using different apparatus under different conditions for the analysis of the same sample.
risk characterization	A phase of ecological risk assessment which incorporates the results of the exposure and ecological effects analyses to evaluate the likelihood of adverse ecological effects associated with exposure to a stressor. The ecological significance of the adverse effects is discussed, including consideration of the types and magnitudes of the effects, their spatial and temporal patterns, and the likelihood of recovery.
sediment	Particulate material which usually lies below water.
spiked sediment	A sediment to which a material has been added for experimental purposes.
steady state	A "constant" tissue residue resulting from the balance of the flux of a compound into and out of the organism. Operationally defined by no statistically significant difference in three consecutive sampling periods.
stressor	Any physical, chemical, or biological entity which can induce an adverse response.
stressor-response profile	The product of characterization of ecological effects in the analysis phase of ecological risk assessment. The stressor-response profile summarizes the data on the effects of a stressor and the relationship of the data to the assessment endpoint.
trophic levels	A functional classification of taxa within a community which is based on feeding relationships.
true value	The known value (*i.e.*, the actual quantitative valued implied by the preparation of the sample). Since the known value does not always exist, it is often considered the value towards which the average of single results obtained by n laboratories as n approaches infinity. Consequently, such a true value is associated with a particular method.
whole sediment	Sediment and associated pore water which have had minimal manipulation. Synonymous with bulk sediment.
xenobiotic	A chemical or other stressor which does not occur naturally in the environment.

References

[ASTM] American Society for Testing and Materials. 1995a. Standard test methods for measuring the toxicity of sediment-associated contaminants with freshwater invertebrates. In: Volume 11.05, Annual Book of ASTM Standards. Philadelphia: ASTM. E1706-95b. p 1204–1285.

[ASTM] American Society for Testing and Materials. 1995b. Standard guide collection, storage, characterization, and manipulation of sediments for toxicological testing. In: Volume 11.05, Annual Book of ASTM Standards. Philadelphia: ASTM. E1391-94. p 835–855.

[ASTM] American Society for Testing and Materials. 1995c. Standard guide for designing biological tests with sediments. In: Volume 11.05, Annual Book of ASTM Standards. Philadelphia: ASTM. E1525-94a. p 972-989.

[ASTM] American Society for Testing and Materials. 1995d. Standard guide for determination of bioaccumulation of sediment-associated contaminants by benthic invertebrates. In: Volume 11.05, Annual Book of ASTM Standards. Philadelphia: ASTM. E1688-95. p 1140–1189.

[USEPA] U.S. Environmental Protection Agency. 1994. Methods for measuring the toxicity and bioaccumulation of sediment-associated contaminants with freshwater invertebrates. Duluth MN: USEPA. EPA/600/R-94/024,

Index

SETAC

A Professional Society for Environmental Scientists and Engineers and Related Disciplines Concerned with Environmental Quality

The Society of Environmental Toxicology and Chemistry (SETAC), with offices in North America and Europe, is a nonprofit, professional society that provides a forum for individuals and institutions engaged in the study of environmental problems, management and regulation of natural resources, education, research and development, and manufacturing and distribution.

Goals
- Promote research, education, and training in the environmental sciences
- Promote systematic application of all relevant scientific disciplines to the evaluation of chemical hazards
- Participate in scientific interpretation of issues concerned with hazard assessment and risk analysis
- Support development of ecologically acceptable practices and principles
- Provide a forum for communication among professionals in government, business, academia, and other segments of society involved in the use, protection, and management of our environment

Activities
- Annual meetings with study and workshop sessions, platform and poster papers, and achievement and merit awards
- Monthly scientific journal, *Environmental Toxicology and Chemistry*, SETAC newsletter, and special technical publications
- Funds for education and training through the SETAC Scholarship/Fellowship Program
- Chapter forums for the presentation of scientific data and for the interchange and study of information about local concerns
- Advice and counsel to technical and nontechnical persons through a number of standing and *ad hoc* committees

Membership
SETAC's growing membership includes more than 5,000 individuals from government, academia, business, and public-interest groups with technical backgrounds in chemistry, toxicology, biology, ecology, atmospheric sciences, health sciences, earth sciences, and engineering.

If you have training in these or related disciplines and are engaged in the study, use, or management of environmental resources, SETAC can fulfill your professional affiliation needs. Membership categories include Associate, Student, Senior Active, and Emeritus.

For more information, contact SETAC, 1010 North 12th Avenue, Pensacola, Florida; T 904 469 1500; F 904 469 9778; E setac@setac.org; http://www.setac.org.